Energy Smart Appliances

IEEE Press
445 Hoes Lane
Piscataway, NJ 08854

Energy Smart Appliances

Applications, Methodologies, and Challenges

Edited by
Antonio Moreno-Munoz
Universidad de Cordoba
Córdoba, Spain

Neomar Giacomini
Whirlpool Corporation
Benton Harbor, USA

Published by John Wiley & Sons, Inc., Hoboken, New Jersey.
Published simultaneously in Canada.

Library of Congress Cataloging-in-Publication Data Applied for:

Hardback ISBN: 9781119899426

Cover Design: Wiley
Cover Image: © andresr/Getty Images

Set in 9.5/12.5pt STIXTwoText by Straive, Chennai, India

Dedicated to our ever patient, supportive, and loving families.

Contents

About the Editors

Antonio Moreno-Munoz is a professor at the Department of Electronics and Computer Engineering, Universidad de Córdoba, Spain, where he is the chair of the Industrial Electronics and Instrumentation R&D Group. He received his PhD and MSc degrees from UNED, Spain, in 1998 and 1992, respectively. From 1981 to 1992, he was with RENFE maintenance service, the Spanish National Railways Company, where he received a scholarship for his university studies. Since 1992, he has been with the University of Cordoba, where he has been the director of its department and academic director of the Master in Distributed Renewable Energies. His research focuses on Smart Cities, Smart Grids, Power Quality, and Internet of energy. He has participated in 22 R&D projects and/or contracts and has more than 200 publications on these topics.

He is currently a member of European Technology & Innovation Platforms (ETIP) Smart Networks for Energy Transition (SNET) WG-4; WG Member of the Spanish Railways Technological Platform (PTFE); WG Member of the IEEE P3001.9 Recommended Practice for the Lighting of Industrial and Commercial Facilities; a member of the Technical Committee on Smart Grids of the IEEE Industrial Electronics Society. He has been a member of the CIGRÉ/CIRED JWG-C4.24 committee "Power Quality and EMC Issues associated with future electricity networks." He has been a member of the IEC/CENELEC TC-77/SC-77A/WG-9 committee. He has been a member of the ISO International Organization for Standardization AEN/CTN-208/SC-77-210.

He is an evaluator of R&D&I projects for the Estonian Research Council; the Fund for Scientific and Technological Research (FONCYT) of the National Agency for the Promotion of Science and Technology in Argentina; the Directorate General of Research, Development, and Innovation of the Ministry of Science, Innovation, and Universities of Spain; and academic promotion at Qatar University. He is also an evaluator for European Quality Assurance (EQA) and DNV-GL. He is the Section Board Member of the journal *Electronics* published by MDPI, an associate editor of Elsevier's *e-Prime* journal, the Section Editor in Chief of MDPI's *Smart Cities* journal, an associate editor of *Electronics* journal published by MDPI, an editor of *Intelligent Industrial Systems* journal published by Springer Nature Science, and an editor of *Frontiers in Energy Research, Sustainable Energy Systems*, and *Policies*. Also, he is a guest editor of and reviewer for numerous journals of IEEE, IET, MDPI, and Elsevier.

Neomar Giacomini is a Senior Manager for Electronics Hardware Development at Whirlpool Corporation, USA. He is an accomplished inventor, developer, and technology aficionado who has been in electronics for more than 20 years, developing hardware, firmware, sensors, and user interfaces. Neomar has filed more than 50 patents, with over 20 already granted. In his current position, he is leading a team focused on electronics hardware development for refrigeration and cooking home appliances. His work at Whirlpool Corporation has actualized into component and module deployment on a global scale and across multiple home appliance platforms. In the technology space, Neomar has faced numerous challenges while working to apply and debug electromechanical and electronic-based systems. Through this work, he has gained an understanding of the complexity and key points associated with the technologies involved on a very elevated scale.

In the Internet of Things space, Neomar has experience both delivering connected products and also as a user with nearly 120 connected devices to experience what a heavily connected home brings to the consumer. This experience enabled him to speak at events such as Sensors & IoT Virtual World Week 2020, Sensors Converge 2021, Sensors Converge 2022, and Digital Manufacturing Summit North America 2022.

From an educational perspective, Neomar holds an Executive MBA from Fundação Getúlio Vargas, Brazil, a Master's of Science in Electronics, and a Bachelor of Science in Electrical Engineering, both from the Santa Catarina State University, Brazil.

He is also an Advisory Board Member for Sensors Converge 2022/2023, holds a Certified Professional in Management® certification from the American Management Association®, and is a certified Six Sigma Black Belt by Whirlpool Corporation.

List of Contributors

Jesús Acero
Department of Electronic Engineering and Communications
Instituto de Investigación en Ingeniería de Aragón
University of Zaragoza
Zaragoza
Spain

José A. Aguado
Escuela de Ingenierías Industriales
University of Málaga
Málaga
Spain

Rolf Bienert
OpenADR Alliance
San Ramon, CA
USA

Ignacio Bravo
Department of Electronics
University of Alcalá
Madrid
Spain

José M. Burdío
Department of Electronic Engineering and Communications
Instituto de Investigación en Ingeniería de Aragón
University of Zaragoza
Zaragoza
Spain

Recep Çakmak
Samsun University
Department of Electrical-Electronics Engineering
Faculty of Engineering
Samsun
Turkey

Inmaculada Casaucao
Escuela de Ingenierías Industriales
University of Málaga
Málaga
Spain

Carlos Cruz
Department of Electronics
University of Alcalá
Madrid
Spain

Laura Daniele
TNO (Netherlands Organization for Applied Scientific Research)
The Hague
The Netherlands

Raúl García-Castro
Ontology Engineering Group
Universidad Politécnica de Madrid
Madrid
Spain

Joaquin Garrido-Zafra
Department of Electronics and Computer Engineering
Universidad de Córdoba
Córdoba
Spain

Neomar Giacomini
Senior Engineering Manager for Electronics Hardware Development at Whirlpool Corporation
Benton Harbor
Michigan
USA

Maxime Lefrançois
Mines Saint-Étienne
Univ. Clermont Auvergne
INP Clermont Auvergne
CNRS
Saint-Étienne
France

Sahana Lokesh
Deutsche Gesellschaft für Internationale Zusammenarbeit (GIZ)
New Delhi
India

Óscar Lucía
Department of Electronic Engineering and Communications
Instituto de Investigación en Ingeniería de Aragón
University of Zaragoza
Zaragoza
Spain

Indradip Mitra
Deutsche Gesellschaft für Internationale Zusammenarbeit (GIZ)
New Delhi
India

Antonio Moreno-Munoz
Department of Electronics and Computer Engineering
Universidad de Córdoba
Córdoba
Spain

and

Industrial Electronics and Instrumentation R&D Group
Córdoba
Spain

Angshu Nath
Indian Institute of Technology Bombay
Mumbai
Maharashtra
India

Esther Palomar
Department of Electronics
University of Alcalá
Madrid
Spain

María Poveda-Villalón
Ontology Engineering Group
Universidad Politécnica de Madrid
Madrid
Spain

Zakir Rather
Indian Institute of Technology Bombay
Mumbai
Maharashtra
India

Héctor Sarnago
Department of Electronic Engineering
and Communications
Instituto de Investigación en
Ingeniería de Aragón
University of Zaragoza
Zaragoza
Spain

Neyre Tekbıyık-Ersoy
Energy Systems Engineering
Faculty of Engineering
Cyprus International University
Nicosia
Turkey

Alicia Triviño
Escuela de Ingenierías Industriales
University of Málaga
Málaga
Spain

Acknowledgments

Many people have contributed to making this book a reality. We have received advice and support from our professional colleagues, our students, friends, and family, and we thank them all.

We are very grateful to Elke Morice-Atkinson, the editorial manager of this book, and Becky Cowan, our editorial assistant, as well as all the editorial staff at Wiley, for their trust, help, and kindness.

Our great gratitude goes to the authors who have accompanied us on this journey, without whom it would not have been possible.

Introduction

Antonio Moreno-Munoz[1] *and Neomar Giacomini*[2]

[1] *Department of Electronics and Computer Engineering, Universidad de Córdoba, Spain*
[2] *Senior Engineering Manager for Electronics Hardware Development at Whirlpool Corporation, USA*

Overview

Currently, household appliances are responsible for about two-thirds of the energy consumed by buildings. Energy labeling of household appliances helps improve their energy efficiency. These energy labels provide a clear and simple indication of the energy efficiency category, making it easier for consumers to save money on their household energy bills. A smart appliance is popularly recognized as having some degree of electronic processing and wireless connectivity. Sometimes called network appliances or Internet appliances, more and more appliances are incorporating smart features that make household tasks easier. It can be as simple as receiving an alert if your refrigerator door is open or remotely turning off your room's air conditioner from your smartphone if you forget before leaving home. Or as complex as having your dryer adjust the cycle time automatically with built-in sensors to help you reduce your dryer's energy use, or remotely controlling your oven from your smartphone or through a voice assistant. With the Internet of things sub-metering devices embedded in home appliances, it is even now possible to measure and record the overall energy consumption of the household and then, with the result of a machine learning model, accurately predict the behavior of individual appliances by employing this data.

Some of those appliances offer consumers novel "smart grid features" intended to include complex demand response policies. These have been properly called "Energy Smart Appliances" and can modulate or shift their electricity consumption in response to external signals such as price information, local measurements, or direct control commands. While we have seen the widespread success of smart thermostats in utility schemes, other smart energy options have seen limited adoption. Overall, we expect smart home systems to become increasingly important components of utilities' residential demand-side management, customer engagement, and decarbonization initiatives. There is therefore a need to better establish the energy and non-energy benefits of different devices and systems, and to identify best practices in utility programs. This book aims to provide utilities and appliance manufacturers with new approaches to better understand real-world performance, assess actual energy benefits, and tailor each technology to the needs of their customers.

Brief Description of the Book Chapters

The structure of this book provides a broad and comprehensive view of the Smart Energy appliance space. Smart Energy use started getting more attention nearly two decades ago with the Smart Grid initiative and evolved over the years branching into multiple areas of development that support the ecosystems under development and deployment these days.

Chapter 1 introduces Demand-side Flexibility in Smart Grids, its importance, concepts, relevance, and economics that drive the need for its implementation. Demand-side management is a key concept that forms the backbone of energy management.

A deep dive into the Smart Home follows on Chapter 2 talking about the Smart Home current and future complexity, discussing the trends, global penetration, energy management opportunities to be explored, technologies, and overall required elements to achieve the intent of Demand-Side energy management.

Chapter 3 on Household energy demand management discusses communication infrastructure, protocols, and security on Household energy demand management.

Demand-Side Management and Demand Response which are of great importance both for the electrical grids and for the consumers willing to achieve savings are discussed in Chapter 4.

Due to the need for standardization across the industry, the OpenADR Standard and Complementary Protocols are presented in Chapter 5, this chapter describes a communications data model designed to facilitate sending and receiving Demand Response signals from a utility or independent system operator to electric customers.

Chapter 6 provides a deep dive into the Smart Home and appliances with potential impact on Smart Energy applications and debates on each appliance that is capable of energy collaboration.

With the necessary background on standards, trends, and appliance opportunities already presented, Chapter 7 goes into The ETSI SAREF ontology for smart applications, an ontology focused on the smart home and smart application domains. SAREF discusses standardized interfaces and data models to ensure interoperability across the currently fragmented landscape of IoT technologies.

Chapter 8 introduces Scheduling of residential shiftable smart appliances by metaheuristic approaches, by showing how metaheuristic algorithms can be utilized in the scheduling of smart appliances operation applied to demand side management. Prominent and modern metaheuristic algorithms are simulated for a case and a comparison of the algorithms is discussed in terms of convergence performance.

Electric vehicles should not be left aside, their importance in Smart Energy Management at home is evident and for that reason Chapters 9 and 10, Distributed Operation of an Electric Vehicle Fleet in a Residential Area, and Electric Vehicle as Smart Appliance for Residential Energy Management, respectively, discuss multiple aspects related to charging methods, charging allocation and other elements related to providing energy back to the house, clearly showing how EVs are a key element for the future of energy management at home.

To close this book, a special Chapter 11 on Induction heating Appliances is presented. The main enabling technologies in the fields of power electronics, digital control, and magnetic component design are reviewed.

What Does the 2022 Energy Crisis Entail?

One of the main objectives of any advanced society and, in general, the responsibility of states is to ensure an affordable, reliable, and secure energy supply for all citizens. The use of different renewable energy sources (RES) has been presented as the most successful way to achieve this goal. They were also considered clean, compared to petroleum derivatives, because when used, they do not pollute the atmosphere, creating few or no greenhouse gases, such as:=Carbon dioxide (CO_2), Methane (CH_4), Nitrous oxide (N_2O), and other industrial gases. But there is no mention of other types of waste. RES are usually divided into dispatchable ones, such as geotherm, hydro or biomass, and non-dispatchable ones, such as solar or wind. Dispatchable generation refers to sources of electricity that can be scheduled at the request of grid operators, according to market needs. Solar or wind energies are called intermittent or variable renewable energies (VREs) because they vary depending on weather conditions. This is the reason for their non-dispatchability. The solution has been to facilitate the use of more flexible power plants when needed to balance VRE and demand. In this sense, gas-fired power plants have been supporting the massive integration of VRE into the grid because, although they run on fossil fuels, their emissions are much lower than those of existing coal and other fossil fuel plants. But this can pose a geopolitical threat: see, e.g. how the war in Ukraine has revealed Europe's deep dependence on Russian gas supplies. Therefore, the energy transition agenda must be carefully reconsidered, as the Global impact of the enormous efforts being demanded of citizens and companies in Western countries is minimal. It should not be forgotten that fossil fuels remain the largest contributor to electricity generation worldwide. Currently, coal – including peat and oil shale – accounted for approximately 37% of the global energy mix, while natural gas followed with a 24% share. In this scenario, China leads the world with more than 5300 TWh produced each year in coal-fired power generation. This represents 54%, followed by India with 12% and the United States with 9% of the world's coal-fired electricity production. In addition, construction of new coal plants is overwhelmingly taking place in Asia, with China accounting for 52% of the 176 GW of coal capacity under construction in 20 countries last year. The overall figure is little changed from the 181 GW under construction in 2020, contrary to all climate agreements.

Flexibility services can be provided not only on the supply side but also by improving power transmission and on the demand side. The current state of technologies and advances that allow for more active and dynamic consumer behavior can also provide flexibility to the power system through Demand-Side Management (DSM). DSM comprises a portfolio of measures to improve the energy system at the side of consumption. DSM ranges from improving Energy Efficiency (EE) by using better materials or upgraded technologies, over energy tariffs with incentives for certain consumption patterns, what has been termed Demand Response (DR). DR can be defined as changes in electricity use by end customers from their normal consumption patterns in response to changes in the price of electricity over time, or to incentive payments designed to induce lower electricity use at times of high wholesale market prices or when system reliability is at risk. Although the consumer DR it is less well known than the EE, it has already proven to be a resource that the grid operator can use to provide system reliability, stability, and ancillary services. EE includes both the investment in high-efficiency equipment and energy conservation (EC). EC is the decision and practice of using less energy. Both are often complementary or overlapping ways of avoiding or reducing energy consumption. Convergence between DR and EC can also be observed because the latter is also a behavioral issue. Also, there is a strong link between EC and high-efficiency equipment through information and communication technologies (ICTs), as the use of ICT to curb energy use is relevant in terms of behavioral change interventions. But unlike EE, the regulatory framework for harnessing the potential of DR is less developed in most countries. In the case of DR, more disclosure and consumer education are needed, as DR is not as well established as EE and the gains are often still rated lower than the gains from EE. It is also true that as consumer awareness of EE increases, EE can open the door to DR. In this sense, EE training tools could be redesigned to cover DR aspects as well. Synergies can be achieved by including DR aspects in energy audits.

However, the smart home appliance sector is currently at a crossroads. While it is expected to see impressive growth in these products, several challenges threaten to slow down their progression. On the one hand, consumers are reluctant to pay higher prices for first-generation smart appliances, probably because it is unclear whether these costs justify the value, perhaps because they do not yet understand the advantages of energy smart appliances. Although hardware prices have reduced, especially those related to connectivity, appliance manufacturers have encountered new technical hurdles. The most notable of these challenges is improving interoperability across different vendors due to overall standardization. Another implicit challenge is related to what value consumers receive in exchange for joining the Demand Response initiative. Consumers will invest time and money, share data and open their devices to be managed by a Home Energy management System (HEMS), but in exchange of course they want cost and process efficiency, an overall challenging equation to solve pleasing all the consumer base. This book will also address these challenges and present the different solutions adopted by the main players in this market.

What the Future may Bring for the Energy Smart Appliances

For the past decades, we have seen great improvements on home appliances in terms of efficiency and feature availability. However, although connectivity seems a common topic across many industries, the deployment of such technology on large appliances such as water heaters, heating, ventilation, and air conditioning (HVAC), and kitchen and laundry appliances is still in the initial stages of market penetration and is usually available in specialty units. Statistics on that are presented later on.

Shopping for the products listed above on any large retail store makes that evident when the consumer has few options that offer connectivity. The limitations go even further when the consumer looks for those products in terms of energy collaboration, compatibility with Smart Grid integration and such.

The fragmentation of IoT technologies which was, and still is, a limiting factor when the consumer tries to integrate multiple appliances to collaborate for energy is also an aspect discussed further on, but the future is promising! This book shows that the standards, ontologies and technologies to enable collaboration across the multiple domains of home appliances in terms of Smart Energy use are already available, so it's no longer a matter of knowledge gaps, but only time for the market to catch up to the current state of technologies.

One important factor that is helping drive attention from a consumer perspective in regards to Smart Energy is the growth of the Electric Vehicles market, and also the availability and application of RESs that are becoming more affordable. These two elements are driving consumers to question why other devices and appliances are not yet part of this energy collaboration and therefore driving companies and regulatory bodies to put more attention on that.

As aforementioned, the Smart Grid initiative started nearly two decades ago and is still not at full speed, so forecasting whether Smart Energy Homes will be at a high deployment rate anytime soon is quite elusive, but for those consumers that want to be early adopters some options are already available.

The overall global push for clean and efficient energy use will continue to drive this space and consumers' buy-in will be key to push for more product availability offering the features described in the book.

1

Demand-Side Flexibility in Smart Grids

Antonio Moreno-Munoz[1,2] *and Joaquin Garrido-Zafra*[1,2]

[1] *Department of Electronics and Computer Engineering, Universidad de Córdoba, Córdoba, Spain*
[2] *Industrial Electronics and Instrumentation R&D Group, Córdoba, Spain*

1.1 The Energy Sector

In the last decades, western economies have been making decisive efforts to promote and encourage the use of renewable energies to decarbonize the energy system. In this sense, the commitments resulting from the last 26th Conference Of the Parties (COP26) based on the Paris Agreement (UNFCCC 2015) are a clear driver of this policy. In the case of the European Union (EU), for example, the countries involved agreed upon several objectives such as increasing the ability to adapt to the adverse impacts of extreme climate events, curbing greenhouse gas emissions, and providing the necessary funding to support these measures. These actions reinforce the ambition for Europe to remain a world leader also in terms of the so-called energy transition.

This evolution is also reflected in the historical primary energy production data collected by the statistical office of the EU, Eurostat (Energy, transport and environment statistics – Publications Office of the EU 2020). Figure 1.1 depicts the relative evolution (to levels of 2010) of the primary energy production by fuel within the EU-27 during the period of 2010–2020 and a detailed distribution of primary energy sources in 2020 as a donut chart, and trends are quite evident. Production of primary energy within the EU was 573.8 million tons of oil equivalent (TOE) in 2020, 17.5% and 7.1% lower than in 2010 and 2019, respectively (see dashed line in Figure 1.1). The distribution in 2020 was as follows: Renewable energies and biofuels (40.9%), nuclear (30.6%), solid fossil fuels (14.6%), natural gas (7.2%), oil and petroleum products (3.8%), non-renewable wastes (2.4%), as well as oil shale, oil sands, and peat products (0.5%).

In general terms, the trend in primary energy production has been downward in recent years due to the descending trends of solid fossil fuels, natural gas, and oil

Energy Smart Appliances: Applications, Methodologies, and Challenges,
First Edition. Edited by Antonio Moreno-Munoz and Neomar Giacomini.

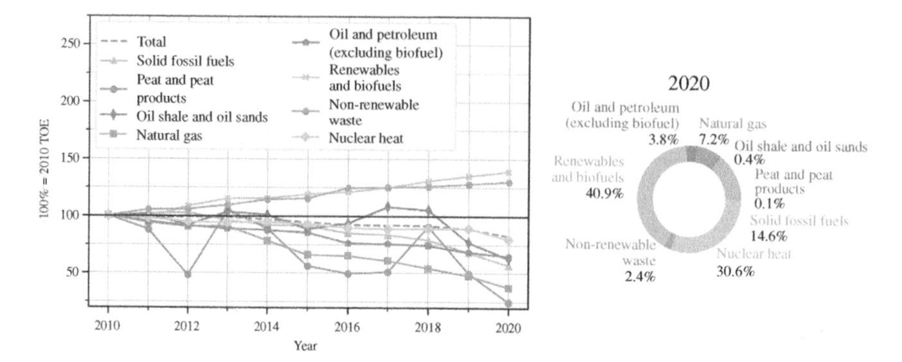

Figure 1.1 Evolution of the primary energy production by fuel in EU-27 from 2010 to 2020, and primary energy production in 2020. Source: Adapted from Energy, transport and environment statistics – Publications Office of the EU (2020).

petroleum products, with 16.5%, 21.1%, and 5.2% of reductions, respectively, in the period 2019–2020. However, this decline did not exclusively take place in the last year, the trend of these energy sources is predominantly negative since 1990 but with minor increases. The primary energy produced by solid fossil fuels, natural gas, and oil petroleum products was 43.0%, 62.4%, and 35.1% lower than in 2010.

By contrast, the highest growth was reported by renewable energies and biofuels, as well as non-renewable wastes, with 3.0% and 1.6% of variation, respectively, in 2020 (2019 as baseline). The energy production from renewable sources has increased significantly in recent decades (IEA 2021) and accounted for the highest share in primary energy production since 2015. Nuclear energy also shows a downward trend over the studied decade with a decrease of 10.7% and 20.2% in 2020 (2019 and 2010, respectively, as baseline).

Oil shale, oil sand, and peat products have had a more unstable trend over this decade. Oil shale and oil sand with maximum of 108.0% in 2017 and minimum of 63.1% in 2020 compared to 2010 levels. Concerning peat products, these peaks took place in 2012 (104.7%) and 2020 (24.8%). Although both energy sources have experienced a decrease compared to 2010 (37.0% and 75.2% of decrease, respectively), the reduction during the period 2019–2020 has been quite considerable: 18.0% and 50.2%, respectively.

1.2 The Power Grid

Power systems around the world are undergoing significant changes in response to several key drivers: The increasing availability of low-cost variable renewable energy sources (VRES), the deployment of distributed energy resources (DER),

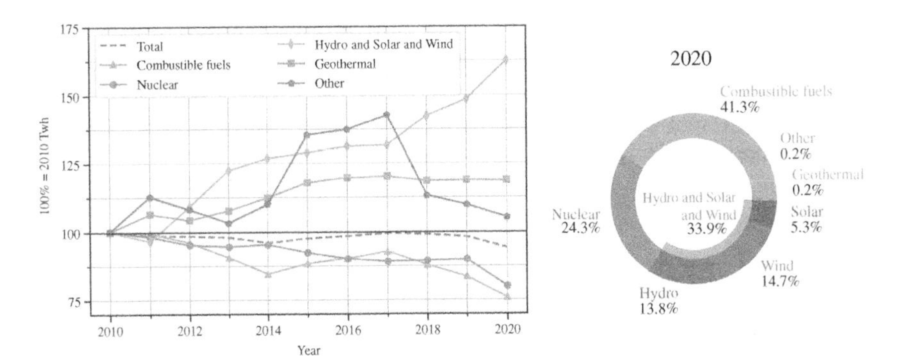

Figure 1.2 Evolution of the net electricity generation in EU-27 from 2010 to 2020, and net electricity generation in 2020. Source: Adapted from Energy, transport and environment statistics – Publications Office of the EU (2020).

advances in digitalization, and growing opportunities for electrification (International Energy Agency 2019), since electricity is playing an ever-more central role in the lives of citizens and expected to be the energy source on which people rely for almost all their everyday needs such as mobility, cooking, lighting, heating, or cooling.

In terms of electricity generation, Figure 1.2 illustrates the evolution of the net production in the decade of 2010–2020 relative to 2010 levels within the EU-27, as well as the breakdown of the different energy sources at the end of this period using a donut chart. Net electricity generation has remained about the same over the studied decade, as shown by the dashed line, but there have been two notable declines in 2014 and 2020 accounting for 95.9 and 94.1 of the 2010 levels. The EU-27 reached 2664 TWh in 2020, 4.03% lower than in 2019 and 5.91% lower than in 2010. Nearly a half of the generation (41.3%) was covered by combustible fuels such as natural gas, coal, and oil, and almost a quarter (24.3%) came from nuclear power plants. Concerning renewable energy sources, the highest share of net electricity generation in 2020 was from wind power plants (14.7%), followed by hydropower plants (13.8%) and solar power (5.3%). Furthermore, geothermal and other sources accounted for 0.23% and 0.18%, respectively.

The electricity coming from combustible fuels was at historical lows in 2020. It has followed a general downward trend since 2010, accounting for 75.8% of the electricity generated from this source in 2010, which means a 24.23% of reduction. The electricity from nuclear power plants shows a similar behavior and has never returned to 2010 levels representing 79.9% of the 2010 levels. The production of solar plants and wind turbines has reported the highest growth: From 0.82% and 4.86% of the total production in 2010 to 5.3% and 14.7%, respectively, in 2020. However, the electricity produced by hydropower plants has remained stable

(11.3–14.4%) in this period. Furthermore, in 2020 the group of hydropower, wind, and solar plants reached almost 170% of the 2010 production, and the electricity generated by wind turbines has surpassed levels like those of the hydropower plants. Geothermal and other electricity sources also illustrate a significant growth; however, they are still a minority in the total breakdown.

In summary, the relative weight of the renewable energy sources in the EU's electricity portfolio has undergone strong growth in parallel with a large decrease in the significance of combustible fuels as well as a significant decline in the amount of nuclear energy utilization. Concretely, the renewable energy sources of electricity have increased their importance by more than 14% points in the period 2010–2020. By contrast, both the electricity coming from combustible fuels and nuclear power plants registered a reduction of 10.0% and 4.3% points over the same period.

Finally, Figure 1.3 shows the evolution of the final energy consumption by sector in EU-27 over the previously considered period as well as a detailed breakdown of the different sectors in 2018. The electricity available for final consumption within the EU-27 reached 2462 TWh in 2020, practically the same as in 2019 (−3.93%), and has experienced periods of growth and decline with fluctuations between 95% and 100% throughout the period under analysis as can be seen from the figure (see dashed line). The distribution of electricity consumption among the different sectors in 2020 is depicted by the donut chart: Industry (35.9%), transport (2.2%), services (27.5%), household (29.0%), and others (5.4%). The final consumption and the consumption of the different sectors account for 94.3, 95.6, 95.2, 90.9, 97.7, and 86.0 of the 2010 levels. The highest variation in the studied decade occurred in electricity consumption of the transport sector, as well as in the "others" category. Moreover, it should be noted the generalized

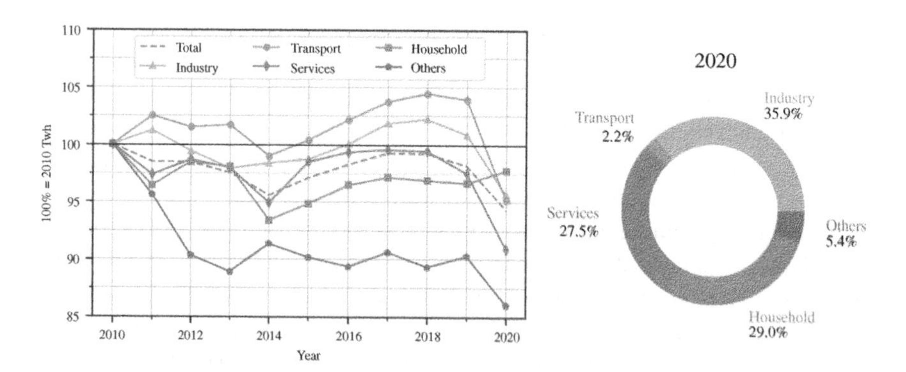

Figure 1.3 Evolution of the final energy consumption in EU-27 from 2010 to 2020, and final energy consumption in 2020. Source: Adapted from Energy, transport and environment statistics – Publications Office of the EU (2020).

fall in most sectors in 2020 agree with the global health crisis resulting from the COVID-19. The only sector that increased was the residential sector, probably due to the lockdowns. The final electricity consumption decreased moderately by 5.7% concerning the levels of 2010 as well as the electricity demanded by all sectors: Industry (−4.4%), transport (−4.8%), services (−9.1%), household (−2.2%), others (−13.9%), leading to the conclusion that the weight of this fall was mainly in the services sector and others category.

1.3 The Smart Grid

The classical power system was originally built to deliver the electrical energy generated by central power plants to the relatively nearby end-users safely and reliably. For this purpose, the voltage level is increased up to 60–750 kV at the source to be transmitted over high-voltage transmission lines and then is gradually reduced to be delivered to consumers in a two-stage distribution process: First, from substations to transformation centers at medium voltage (5–20 kV) and finally from this point to the end-users at low-voltage (120 or 230 V in America or Europe, respectively) (ENTSO-E Transmission System Grid Map n.d.; Mapa del sistema eléctrico ibérico n.d.; Carr 1996). The structure of a conventional grid can be summarized as follows: Power plants that generate the electrical power, high-voltage transmission lines that transport the power from power plants to power stations which then outputs medium- and low-voltage distribution lines that interconnect individual consumers. Notice that the energy flow is thus unidirectional from power plants to end-users. This architecture has remained practically unaltered since its conception as it has been highly effective for decades in covering the initial needs of providing electrical energy to end-users reliably and safely. However, this vision of the power system is being forced to face new conditions and more demanding requirements in both the industrial and residential sectors because of the current digital revolution as has already been mentioned at the beginning of Section 1.2. Some of them are detailed as follows (Colak 2016):

- The increased energy demand due to population growth, increase in manufacturing capability all over the globe, the trend in household appliances that traditionally used gas as a fuel and now moving electricity, and the proliferation of new technologies such as the electric vehicle.
- The need to increase the production capacity in the current power plants as well as the reduction of the transmission and distribution energy losses.
- The challenge of reducing the operational costs, while improving the management of the existing transmission and distribution infrastructures.

- The rapid growth of distributed generation (DG) due to grid-connected DER and VRES in addition to conventional power plants. These resources are mainly solar photovoltaic (PV) panels and wind turbines.
- The need to replace equipment and deploy new technologies over existing infrastructure. In most cases, the power devices employed in the transmission and distribution systems are transformers, power switches, power breakers, utility meters, and relays. These components have low reliability due to the old technology in use. Moreover, the grid capacity for collecting information and measurements during these stages is still quite limited today.

With the intend to overcome such challenges, the power grid has evolved and must continue doing so. This new paradigm of the power grid has been called Smart Grid. The European technology platform (ETP) for Smart Grids provides the following definition in its documentation for the strategic deployment of the European electricity networks of the future (European Technology Platform [ETP] Smart Grids 2010): "A Smart Grid is an electricity network that can intelligently integrate the actions of all users connected to it – generator, consumers and those that do both – to efficiently deliver sustainable, economic and secure electricity supplies." This concept has also been widely discussed in the United States (US Department of Energy: Office of Electricity 2011). Notice that now end-users take an active role and can act as energy producers and consumers, becoming what is known as prosumers (Dai et al. 2020). The Smart Grid could be seen as a digital upgrade of both transmission and distribution grids, in which the idea of a one-way flow of energy and information from energy providers to end-users turns into a complex scheme with a bidirectional flow. Moreover, aspects such as scalability, maintainability, security, and interoperability between devices are central to the Smart Grid concept (Colak et al. 2020). To this end, the Smart Grid must be undoubtedly linked to several concepts such as the information and communication technologies (ICTs) to ensure the exhaustive coordination of stakeholders, the use of renewable energy sources (RES), and the decentralization of them through the DG, the deployment of smart meters or an advanced metering infrastructure (AMI) toward the monitoring of the consumption and the creation of statistics, and the demand-side management (DSM) to achieve a better balance between generation and consumption as will be discussed later (Cecati et al. 2010).

1.4 Power Grid Flexibility

The concept of power grid flexibility has been introduced recently by academics and international organizations. Although a global definition has not been reached yet, as a rule, flexibility describes the capacity of the power grid to respond to changes in demand or supply while preserving the stability of the system. Thus, from a technical viewpoint, flexibility is essential to address the generation-demand imbalances; however, other aspects need to be considered. A more complete definition is provided by the International Energy Agency (IEA): “Flexibility is the ability of a power system to reliably and cost-effectively manage the variability and uncertainty of demand and supply across all relevant timescales, from ensuring instantaneous stability of the power system to supporting long-term security of supply” (International Energy Agency 2019). Notice how flexibility extends to other dimensions such as time, management, uncertainty, and cost (Akrami et al. 2019). These points are further detailed in the following lines:

- **Time**: Indicates how fast the system can be restored to a given state when it undergoes a deviation. Control actions are often classified into short-term, mid-term, and long-term measures.
- **Management measures** or control procedures are performed by the power grid operator to deal not only with day-to-day but also with unexpected events. These corrective actions depend directly on the time interval available to be applied.
- **Uncertainty** or absence of information about future condition. The more uncertainty in the system, the more flexibility is required for its proper operation.
- **Cost**: Although the power system scheduler should always offer flexibility, this concept implies an extra charge as the marginal cost or marginal risk is considered to serve system flexibility and, therefore, high marginal cost control actions are required to ensure low marginal risk and vice versa. Accordingly, a level of commitment must be found between the amount of flexibility and its associated cost.

1.4.1 The Need for Flexibility

All power systems have a certain degree of flexibility aiming to continuously balance the generation and consumption and ensure system stability. This flexibility is employed to maintain the foremost power grid parameters (i.e. voltage and frequency) within the safe range specified by several international standards addressed later. Although variability and uncertainty have always been considered during the power systems operation, these inherent flexibility mechanisms have demonstrated to be insufficient to perform a successful system regulation when dealing with the presence of large quantities of grid-connected VRES, as is being experienced in recent years since VRES are now cheaper to acquire for electricity generation due to the government funding and the absence of fuel costs. These VRES refers mostly to solar, wind, or hydro resources. In this regard, achieving an acceptable balance between generation and demand turns out to be a major challenge due to the intermittent and variable dynamic that characterizes these energy sources. Therefore, given these reasons, making the power planning and operation more flexible has become a global priority to achieve the power system transformation in response to these novel trends. Moreover, the current context brought by the COVID-19 pandemic has revealed that a flexible and well-functioning power system is crucial to maintaining the operation of critical infrastructures such as those in the healthcare sector (Heffron et al. 2021).

1.4.2 Sources of Flexibility

Regulators and system operators recognize that flexibility in all power systems must be addressed by ensuring the following elements (Mohandes et al. 2019; Babatunde et al. 2020; Cochran et al. 2014; Nikoobakht et al. 2019):

1.4.2.1 Flexible Generation

Flexibility is often offered by power plants with fast start-up and shut-down operation and high-power ramp capabilities. Moreover, one of the main features of these flexible sources is an efficient operation at a lower minimum level in periods with high penetration of VRES or even the ability to perform deep turndowns. In this regard, it is crucial to ensure a minimum marginal cost so that these power plants can compete in the market as a source of flexibility. Some of these conventional plants include hydro plants, gas-fired, coal-fired, and fuel-fired power plants, as well as dispatchable renewable power plants (i.e. biomass, geothermal plants, etc.). Currently, conventional power plants are the predominant source of flexibility in modern power systems. DG can also perform a fast response to power mismatches to provide local flexibility by modulating their production.

1.4.2.2 Flexible Transmission and Grid Interconnection

Transmission networks are responsible for this kind of flexibility. Among other features, transmission networks must avoid bottlenecks and have the capability to take advantage of a wide range of resources that support achieving the generation-demand needs. These resources include the use of smart network technologies that better optimize the energy transmission and the interconnection between neighboring power systems. Furthermore, grid interconnection opens the door for electricity trade which could be highly advantageous for power systems extended over multiple time zones. Consequently, their peak-load intervals take place at different times, and their renewable energy sources with a strong dependency on the time, such as the photovoltaic plants, also reach their maximum production at different times. Therefore, a coordinated strategy can contribute to smoothing out peak demand periods and making use of the energy surpluses.

1.4.2.3 Control Over VRES

Uncertainty and variability are part of the VRES's nature and often limit the amount of flexibility that can be provided or sometimes even contribute to the opposite. Therefore, greater control over the use of these resources can help alleviate the situation. A scenario with congestion of the transmission lines or when the produced power exceeds the required power system demand may be the best example to understand this issue. In such a case, flexibility can be offered via the renewable generation curtailment although this action is the least preferred choice, as it can lead to a suboptimal operation from both viewpoints: Owner's revenues or savings and loss of renewable energy in the absence of energy storage facilities.

1.4.2.4 Energy Storage Facilities

The spread of storage systems throughout the power grid is undoubtedly another source of flexibility and is especially relevant when considering a context with high penetration of generation coming from VRES. These storage infrastructures can help the power system to absorb the energy surpluses or inject the required energy to solve a momentary mismatch between supply and demand. Currently, pumped hydro energy storage accounts for the highest amount of total storage capacity worldwide. Nevertheless, other technologies such as batteries, ultracapacitors, flywheels, and compressed air are also becoming popular.

1.4.2.5 Demand-Side Management

DSM is a portfolio of measures to improve the energy system on the side of consumption and evolved during the 1970s because of economic, political, social, technological, and resource supply factors (Gellings 2017). The US Department of Energy (DoE) provides the following definition of DSM (Loughran and Kulick 2004): "DSM is the planning, implementation and monitoring activities of electric utilities that are designed to encourage consumers to modify their level and pattern of electricity usage." DSM includes both energy efficiency (EE) and demand response (DR) measures as can be depicted in Figure 1.4. These measures range from improving the EE by using less energy while providing the same or even better level of service to the consumers to the implementation of DR techniques such as the use of smart energy tariffs with incentives for certain consumption patterns or sophisticated real-time control of DER. More specifically, EE includes both the use of high-efficiency equipment and energy conservation strategies, while DR is divided into explicit and implicit measures.

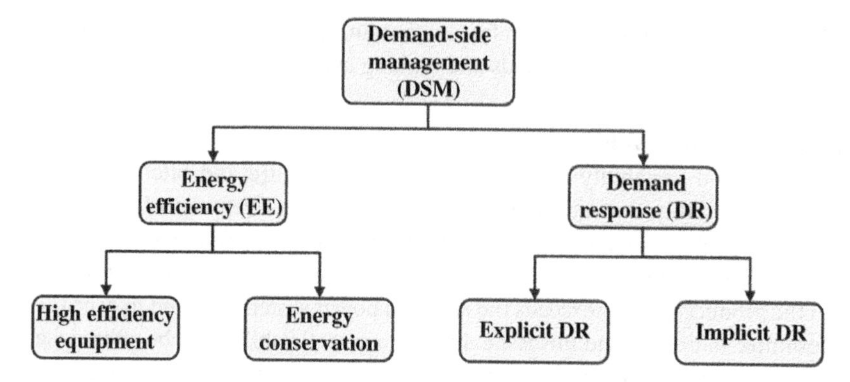

Figure 1.4 Classification of demand-side management measures. Source: A. Rezaee Jordehi (2019).

Regarding the concept of DR, the US DoE (Qdr 2006) also defines it as "Changes in electric usage by end-use customers from their normal consumption patterns in response to changes in the price of electricity over time, or to incentive payments designed to induce lower electricity use at the time of high wholesale market prices or when system reliability is jeopardized." DR has already proven to be a resource that the energy provider can offer to improve system reliability, stability, and security services. As shown in Figure 1.4, DR services are normally classified into two groups attending to the mechanism used to promote the response: Explicit and implicit DR. Explicit DR is a committed and dispatchable DR action traded on the energy market and is usually provided by an independent aggregator, virtual power plants (VPPs), or the energy provider. In this case, consumers receive an incentive to change their consumption in certain scenarios such as the grid congestion or balance problems among others. This is referred to as "incentive-driven" DR. The following programs can be found within this category:

- **Demand bidding/buyback programs (DBP)**: The utility pays an incentive to reduce electric load when notified of a DR event day. Customers submit load reduction bids for a DBP event, which can be called on a day-ahead or day-of basis. For any event, the customer may elect to submit or not submit a bid.
- **Direct load control (DLC)**: Allows the aggregator control over certain equipment, e.g. switching-off noncritical loads or modifying devices' setpoint to reduce net electrical load.
- **Emergency demand response programs (ERDP)**: Customers receive incentive payments for load reductions when needed to ensure reliability.
- **Interruptible/curtailable (I/C)**: Customers receive a discounted rate for agreeing to reduce the load on request.

- **Ancillary services market programs**: Customers receive payments from a grid operator for committing to restrict load when needed to support the operation of the electric grid (i.e. auxiliary services).
- **Capacity market programs (CMP)**: Customers offer load curtailments as system capacity to replace conventional generation or delivery resources. Customers typically receive day-of notice of events and face penalties for failure to curtail when called upon to do so. Incentives usually consist of up-front reservation payments.

Concerning implicit DR, some of the most common DR products are summarized below. Under this scheme, consumers agree to be exposed to hourly or shorter-term tariffs in which the price of the electricity varies depending on production costs. Therefore, consumers adapt their consumption (through automation or personal choice) to save on the electricity bill. Implicit DR is also known as priced-based DR.

- **Time-of-use (TOU)**: A rate with different unit prices for usage during different blocks of time, for a 24-hour day. Daily pricing blocks include an on-peak, partial-peak, and off-peak price for non-holiday weekdays, the on-peak price being the highest, and the off-peak price the lowest. These tariffs include diurnal and seasonal variations in electricity cost but are fixed several months before. It can be integrated within the operations planning stage.
- **Real-time pricing (RTP)**: A retail rate in which the price fluctuates hourly reflecting changes in the wholesale price of electricity. These are typically known to customers on a day-ahead or hour-ahead basis.
- **Critical peak pricing (CPP)**: Hybrid of the TOU and RTP. The basic rate structure is TOU. However, the normal peak price is replaced with a much higher CPP event price under specified trigger conditions (e.g. when system reliability is compromised, or supply prices are very high). It is called on the day of economic dispatch.

Finally, Figure 1.5 describes the potential impact of DR measures on customer service levels. The opportunities and potential depend on the existing building and appliances infrastructure. This figure also summarizes the load commitment timescales over which these DR schemes operate.

1.4.2.6 Other Sources of Flexibility

Other flexibility resources include ancillary services. The power grid requires ancillary services to ensure reliability and support its main function of delivering electrical energy to consumers. These services are employed by system operators as a flexibility mechanism to preserve the instantaneous and continuous balance between generation and consumption. Although most balance requirements are

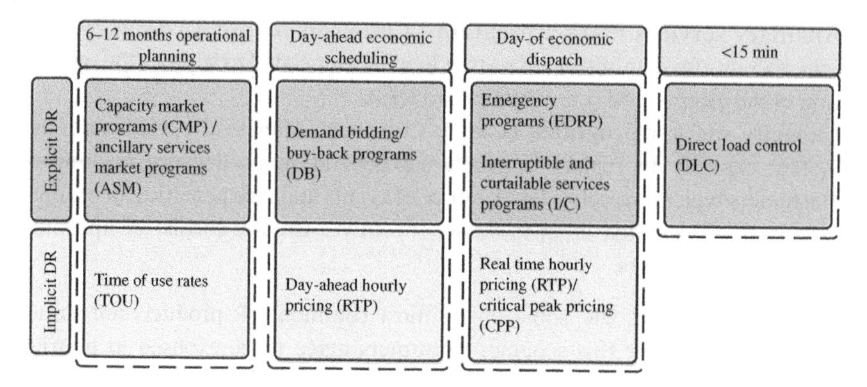

Figure 1.5 Demand response programs timescale. Source: Adapted from Qdr (2006).

being covered by regulation, spinning, and non-spinning ancillary services, new ancillary technologies such as load following, frequency response reserve, or inertia response are also proliferating.

Moreover, on a smaller scale, the utilization of electric vehicles and multi-mode operation of combined cycle units have also been revealed as another source of flexibility by providing a particular case of energy storage system or recovering exhaust heat from thermal units to drive a steam turbine and generate more electricity, respectively.

1.5 Power Quality, Reliability, and Resilience

The main goal of modern power systems is to deliver the required electrical energy to its customers as economically as possible with an acceptable level of reliability (Billinton and Allan 2003). Nowadays, the working and social habits of modern society have led end-users to expect the supply to be continuously available on demand. Although a power system with high reliability is possible, a risk-free power system is not. In this context, engineers and power system managers try to maintain reliability as high as possible within their socioeconomic constraints.

In most countries, the electricity sector is currently a deregulated and competitive environment where accurate information on system performance must be provided to ensure adequate service to customer needs. As consequence, series of indexes have been proposed under the concept of reliability. In the electric power industry, reliability reflects the ability to supply electricity in the amount demanded by users and in the time required. Specifically, reliability has to do with total electrical interruptions (outages), that is, the complete loss of voltage. These reliability indexes include measurements such as the number of interruptions and how long they last, the customers affected, and the power interrupted.

There are a wide variety of indexes to measure reliability, but the most common are SAIDI, SAIFI, and CAIDI as defined in IEEE Standard 1366 (Bollen 2003). SAIDI (System Average Interruption Duration Index) and SAIFI (System Average Interruption Frequency Index) values include sustained interruptions, which are defined as outages that last at least five minutes (although this is not uniform and may vary).

Another concept related to reliability is that of power quality (PQ), although they are two different issues. While the simplest idea of reliability is whether the power is available as it is needed, PQ can be defined as the degree to which current and voltage maintain their waveforms adjusted to a perfect sine wave with constant amplitude and frequency at a given point of the power system. An additional requirement of the current quality is that it must be in phase with the voltage waveform. Therefore, PQ is the combination of voltage and current quality (IEEE 2012). As will be detailed further in Section 1.5.1, a wide variety of electromagnetic disturbances are collected under this term and all of them can affect a critical installation to the extent that it depends on the sensitivity of each load.

Finally, the concept of power system resilience is currently attracting a lot of interest. The topic has become one of the most studied characteristics in the energy industry since Hurricane Katrina dramatically exposed the vulnerability of the power grid in Louisiana in 2005. The frequency of extreme weather events such as hurricanes, tsunamis, ice storms, and other natural disasters as well as man-made cyber and physical attacks have increased in recent years and affect an increasing number of human and environmental victims worldwide (Bhusal et al. 2020). This term comes from the Latin root "resilire," which means "the ability to spring back or rebound." Assuming that disruptive events can occur regularly, for a system, resiliency would be the ability to anticipate, compensate, adapt, and recover from a potentially damaging event (Gholami et al. 2018).

1.5.1 Power Quality Disturbances

The increasing number of electronic equipment connected to the power grid that can generate electromagnetic disturbances or be affected by them has caused the community to become interested in the classification of these disturbances as a first step to subsequently decide on the appropriate strategy to address their mitigation. In this regard, prestigious international organizations such as the International Electrotechnical Commission (IEC) and the Institute of Electrical and Electronics Engineers (IEEE) have made decisive efforts by providing relevant standards and regulations related to PQ issues from several viewpoints. Although this section is mainly focused on standards that address the classification of the principal electromagnetic phenomena causing PQ disturbances within the power system, many others intended for specifying measurement techniques or limits for these disturbances are also included to bring a more detailed insight and introduce the required background for the understanding of the remaining chapters. Accordingly, the IEEE std. 1159-2019 "Recommended practice for monitoring electric power quality" ("IEEE Recommended Practice for Monitoring Electric Power Quality" 2019) has been considered as a reference and thus the main body of this section follows its structure. This standard adopts a quantitative approach as opposed to the qualitative one assumed by other standards. Finally, electromagnetic phenomena and their characteristics are firstly classified in Table 1.1 and later discussed in the following lines.

1.5.1.1 Transients

Transients give a name to a phenomenon that is undesirable and momentary and can be classified into two categories: Impulsive and oscillatory transients depending on the waveshape of a current or voltage transient. IEEE Std. C62.41.1-2002 (IEEE 2003) deals with defining standard impulsive and oscillatory transient test waves to test electrical equipment.

- **Impulsive transients**: Impulsive transients are sudden, non-power frequency change from the nominal condition voltage, current, or both, that is unidirectional in polarity. The most common cause of impulsive transients is lightning and is often damped quickly by impedance circuit elements due to the high frequencies involved. There can be a significant difference in the transient characteristics from one location to another within the power system. Impulsive transients are often characterized by their peak value, rise, decay, or duration times.
- **Oscillatory transients**: Oscillatory transients are sudden, non-power frequency change in the steady-state condition of voltage, current, or both, that includes both positive and negative polarity values. An oscillatory transient consists of a voltage or current whose instantaneous value changes polarity rapidly and often decays within a fundamental-frequency cycle. The subclasses are high, medium, and low frequency and have been chosen to coincide with typical types of oscillatory transients within the power system. High-frequency oscillatory transients (>500 kHz) are normally provoked by switching events or can be the response of one point of the system to an impulsive transient. When the frequency of the primary frequency component of an oscillatory transient is within the range of 5–500 kHz, the category used is medium frequency. Back-to-back capacitor energization can give rise to this electromagnetic phenomenon. Finally, low-frequency oscillatory transients (<5 kHz and duration between 0.3 and 50 ms) are normally found in sub-transmission and distribution lines and can be the result of many types of events (e.g. capacitor, ferro resonance, or transformers energization).

Table 1.1 Categories and typical characteristics of power system electromagnetic phenomena.

Categories			Typical spectral content	Typical duration	Typical voltage magnitude
Transients					
	Impulsive				
		Nanoseconds	5 ns rise	<50 ns	
		Microseconds	1 μ rise	50 ns–1 ms	
		Milliseconds	0.1 ms rise	>1 ms	
	Oscillatory				
		Low frequency	<5 kHz	0.3–50 ms	0–4 pu[a)]
		Medium frequency	5–500 kHz	20 μs	0–8 pu
		High frequency	0.5–5 MHz	5 μs	0–4 pu
Short-duration RMS variations					
	Instantaneous				
		Sag		0.5–30 cycles	0.1–0.9 pu
		Swell		0.5–30 cycles	1.1–1.8 pu
	Momentary				
		Interruption		0.5 cycles – 3 s	<0.1 pu
		Sag		30 cycles – 3 s	0.1–0.9 pu
		Swell		30 cycles – 3 s	1.1–1.4 pu
	Temporary				
		Interruption		3 s–1 min	<0.1 pu
		Sag		3 s–1 min	0.1–0.9 pu
		Swell		3 s–1 min	1.1–1.4 pu
Long-duration RMS variations					
	Sustained interruptions			>1 min	0 pu
	Undervoltages			>1 min	0.8–0.9 pu
	Overvoltages			>1 min	1.1–1.2 pu
	Current overload			>1 min	
Imbalance					
	Voltage			Steady state	0.5–2%
	Current			Steady state	1–30%

Table 1.1 (Continued)

Categories	Typical spectral content	Typical duration	Typical voltage magnitude
Waveform distortion			
DC offset		Steady state	0–0.1%
Harmonics	0–9 kHz	Steady state	0–20%
Interharmonics	0–9 kHz	Steady state	0–2%
Notching		Steady state	
Noise	Broadband	Steady state	0–1%
Voltage fluctuations	<25 Hz	Intermittent	0.1–7% 0.2–2P_{st} [b)]
Power frequency variations		<10 s	±0.1 Hz

a) Per unit (pu).
b) Flicker severity index as defined in IEC 61000-4-15:2010 and IEEE Std. 1453.
Source: IEEE std 1159-2019 (2019).

1.5.1.2 Short-Duration RMS Variation

These PQ disturbances are related to or usually provoked by fault conditions, power supply of large loads with high initial currents, or intermittent loss of connection in power conductors. It consists of a variation of the RMS value of the voltage or current from the nominal during a time greater than 0.5 cycles of the power frequency but less than or equal to one minute. When the affected variable is voltage, it can be further described using a modifier indicating the magnitude of the voltage variation (e.g. sag, swell, interruption) and possibly a modifier indicating the duration of the variation (e.g. instantaneous, momentary, and temporary). These modifiers regarding the voltage are detailed as follows:

- **Instantaneous**: A type of short-duration RMS voltage variation with a duration between 0.5 and 30 cycles of the power frequency.
- **Momentary**: A type of short-duration RMS voltage variation with a duration between 0.5 cycles of the power frequency and three seconds.
- **Temporary**: A type of short-duration RMS voltage variation with a duration between three seconds and one minute.
- **Interruption**: It is considered a complete loss of voltage and refers to an RMS voltage variation in which the RMS of the voltage on one or more phases falls below 0.1 pu for a time period less than one minute. Power systems faults, equipment failures, or control malfunctions can result in an interruption.
- **Sag**: A type of short-duration RMS voltage variation where the RMS of the voltage on one or more phases is within the range 0.1–0.9 pu. Voltage sags are often

related to system faults but are also provoked by large load changes when the power consumption increases.

- **Swell**: A type of short-duration RMS voltage variation where the RMS of the voltage on one or more phase wires is above 1.1 pu from durations from 0.5 cycles to one minute. Typical magnitudes are between 1.1 and 1.2 pu. Voltage swells are much less frequent than voltage sags and can be caused by switching off a large load, switching on a large capacitor bank, or when a single line-to-ground fault occurs, resulting in a temporary voltage rise on the unfaulted phases.

1.5.1.3 Long-Duration RMS Variation

These PQ disturbances represent a variation of the voltage or current RMS value from the nominal for a time period greater than one minute. The causes are usually the load variations on the system and system switching operations rather than system faults. Similar to the short-duration RMS variation, it can be further described using a modifier indicating the magnitude of the voltage variation. In this sense, three subcategories are possible: Overvoltage, undervoltage, and sustained interruption.

- **Overvoltage**: An overvoltage involves an RMS increase in the voltage greater than 1.1 pu for a period of time exceeding one minute. Typical values in power systems are within the range 1.1–1.2 pu. Overvoltages can also be the consequence of switching off a large load or of a variation in the reactive power of the system when a large capacitor bank is connected. A wrong connection of the transformer taps can also provoke a system overvoltage.
- **Undervoltage**: A system undervoltage occurs when the RMS value of either voltage or current falls below 0.9 pu with the same time condition as system overvoltages. Concerning the causes, undervoltages are produced by the opposite events to overvoltages. Overloaded circuits can also give rise to system undervoltages.
- **Sustained interruptions**: A sustained interruption is defined as the decrease of the voltage to less than 10% of nominal for a period exceeding one minute. Voltage interruptions longer than one minute are often permanent a require manual intervention for restoration.

1.5.1.4 Imbalance

In a three-phase system, imbalance (or unbalance) is defined as the ratio of the magnitude of the negative sequence component to the magnitude of the positive sequence component, expressed as a percentage. This definition can be applied for either voltage or current. The voltage imbalance is often around 5% in normal three-phase power systems. Current imbalance can be significantly higher when single-phase loads are present in the system.

1.5.1.5 Waveform Distortion

Waveform distortion involves a steady-state deviation from the theoretical power frequency sinusoid mainly characterized by the spectral content of the deviation. There are five primary types of waveform distortion under this category:

- **DC offset**: The existence of DC voltages or currents in an AC power system can be the result of geomagnetic disturbances or due to the consequence of half-wave rectification that many devices include in their electronics. The presence of DC components in AC networks can lead to an increase in transformer saturation, among other adverse effects.
- **Harmonics**: Sinusoidal voltages or currents having frequencies that are integer multiple of the frequency at which the supply system is designed to operate (termed the fundamental frequency; usually 50 or 60 Hz) are called harmonics. Harmonics contribute to the waveform distortion in combination with the fundamental voltage or current and their presence is mainly due to the nonlinear nature of devices and loads connected to the power system. Electronic-based equipment (e.g. pulse-width modulation converters, switching power supplies, or rectifiers) is often responsible for this harmonic distortion that is currently a growing concern for many power system stakeholders. IEC 61000-4-7:2008 (IEC n.d.-c) defines a harmonics measurement technique and IEC 61000-3-2:2018 (IEC n.d.-a) establishes limits for individual harmonics.
- **Interharmonics**: These are voltages or currents in whose spectral content, there are frequency components that are not integer multiple of the fundamental frequency. Interharmonics can be found at certain frequencies or as a wideband spectrum at any point of the network. The principal origins of this waveform distortion are pulse-width modulated converters, cycloconverters, static frequency converters, induction furnaces as well as arcing devices, especially those whose control is not synchronized with the power system frequency. Power line carrier signals can also be considered interharmonics. Although the effects of interharmonics are not well known yet, they have been shown to cause flicker or low-frequency torsional oscillations in motors, among others. IEC 61000-4-7:2002 also defined the measurement technique for interharmonics.
- **Notching**: Notching is a periodic voltage disturbance characterized by a high-frequency spectral content and is caused when the current commutates from one phase to another (momentary short circuit between two phases) during the normal operation of power electronics converters. The severity of the phenomenon at any point of the power system is given by the source inductance and the isolating inductance between the converter, the magnitude of the current, and the point being monitored. This PQ disturbance can sometimes provoke frequency or timing errors on power electronics circuits that use zero crossings for synchronization purposes since the voltage notch can produce additional zero crossings (e.g. Thyristor-based converters). Notching is further described in IEEE std. 519-2014 (IEEE 519-2014 – IEEE Recommended

Practice and Requirements for Harmonic Control in Electric Power Systems n.d.). A variant termed "voltage notching ringing" can also appear when a system resonance results in a ringing response at each of the commutation notches. In such a case, the disturbance can be mitigated by using harmonic filters or capacitor banks to change the system resonance conditions.

- **Noise**: Noise consists of any undesirable disturbance with broadband spectral content (typically below 200 kHz), either voltage or current, that cannot be classified as harmonic distortion or transient. In power systems, noise can be caused by power electronics devices, control circuits, loads with solid-state rectifiers, and arcing equipment. The magnitude of the noise does not normally exceed 1% of the voltage magnitude and can usually be mitigated by using isolation transformers, line conditioners, filters, or proper grounding circuits.

1.5.1.6 Voltage Fluctuation

Voltage fluctuations involve systematic variations of the voltage envelope or a series of random voltage changes, the magnitude of which are typically within the range 0.95–1.05 pu. Voltage fluctuations can be caused by equipment that exhibits a rapid variation of the load current magnitude or reactive power. For further details, IEC 61000-3-3 (IEC n.d.-b) defines several voltage fluctuations, and IEEE Std. 1453-2015 (IEEE 2015) incorporates the IEC methodology for these measurements.

When voltage fluctuations occur in lighting systems, humans can perceive the changes in the lamp illumination intensity, and the phenomenon is referred to as flicker. However, both terms must not be confused. Voltage fluctuations cause flickers. Therefore, voltage fluctuation describes the electromagnetic disturbance, while flicker describes the impact of this on the lighting intensity.

1.5.1.7 Power Frequency Variations

Power frequency variations reflect the deviation of the power system's fundamental frequency from its specified nominal value (e.g. 50 and 60 Hz). The steady-state power system frequency is directly influenced by the rotational speed of the generators in the power system. The frequency depends on the balance between the capacity of the available generation and the consumption at any given instant. Therefore, small mismatches between these variables will cause small instantaneous frequency deviations and the magnitude and duration of these PQ disturbances will depend on the load characteristics and the dynamics of the generation system's response to load changes. Power frequency deviations are often caused by faults in the bulk power transmission system, a large block of loads being disconnected, or a large source of generation changing to islanded mode.

It should be noted that large frequency variations are rare on modern interconnected power systems; however, weak systems such as islanded microgrids are more likely to report these disturbances due to their relatively low inertia and capacity.

Concerning the limits to identify these variations, the standard EN 50160 (UNE-EN 50160:2011/A2:2020 Voltage characteristics of electricity supplied by public electricity networks n.d.) establishes the thresholds for both interconnected and islanded power systems with 50 Hz nominal frequency. The average value of the frequency measured every 10 seconds must be within the range 49.5–50.5 Hz and 47–52 Hz during 99.5% of the year and 100% of the time, respectively, for interconnected power systems. These ranges are more relaxed for islanded power systems: 49–51 Hz and 42.5–57.5 Hz for 95% of the week and 100% of the time.

1.6 Economic Implications and Issues of Poor Power Quality

While it is easy to understand how unreliability can affect all customers, the effects of poor PQ, on the other hand, are more difficult to recognize. Pure waveform deviations can create everything from a barely noticeable annoyance for the residential customer to a major disruption to the processes of industrial or commercial customers. As operations based on electronic technology become more common, high PQ requirements become more important, particularly for mission-critical facilities.

The power system reliability literature regarding the economic consequences of an energy supply outage and how they are calculated today represents a mature discipline (Larsen et al. 2019). The renowned survey (Eto 2017) found that by 2015 the power outages had cost the US economy some $59 billion, an increase of more than 68% since its previous study in 2004. Commercial and industrial businesses account for more than 97% of these costs. Particularly, according to MeriTalk (n.d.), 40% of global health organizations experience an unplanned outage in the year. The average cost is $432,000 per incident, where diagnostic imaging systems are among the main affected. To get a closer look the recently published paper (Mendes et al. 2019) presents the direct economic impact of power outages in inpatient and outpatient healthcare facilities across the United States and the District of Columbia. A combination of traditional metrics has been adopted, including the calculation of the Value of Lost Load (VoLL), achieving the appropriate granularity.

Outages due to catastrophic events are not usually included in reported index numbers. These include ice storms, hurricanes, earthquakes, and floods. Many analysts agree that the per-customer economic costs of these long, severe, and extended outages are far greater than the above and that those larger costs have not yet been well-reported or well-estimated (Silverstein et al. 2018).

On the other hand, there are fewer surveys on the cost of poor energy quality, and not entirely comprehensive. The largest survey was carried out in the European Union across 16 industry sectors (Targosz and Manson 2007) and found that the total cost of losses related to poor PQ exceeds €150 billion, where the industry represents over 90%. The survey also identified that voltage dips and interruptions were liable for around 55% of losses and mainly affected electronic equipment, which is now so widespread in the industrial and service sectors. The detailed impact of PQ disturbances facing countries is classified in Figure 1.6. It also shows that the amount of voltage sags detected is about twice as high as long interruptions.

In the industrial sector, the most important losses occur in manufacturers with continuous processes. In contrast, the average number of disturbances is lower in the service sector. These costs are probably underestimated since it is often difficult to distinguish the root cause of the electrical problems that arise in the office environment. In addition, the survey did not include data centers, which may be the most critical infrastructure in this sector. It should be noted that the highest losses occur in hospitals, which, due to their idiosyncrasy, have a slightly higher PQ cost than others within the service sector. The companies in this study invest €297.5 million annually in mitigation solutions for various PQ issues. The results of the survey are synthesized in Table 1.2.

Another well-known study (Sharma et al. 2018) based on information provided by 985 US companies, concluded that for PQ disturbances other than sags, the cost per year to the US digital economy and industrial companies is $6.7 billion. Overall, the data suggest that while the US economy (across all business sectors) is losing between $104 billion and $164 billion per year due to outages, PQ disturbances alone account for $15–24 billion per year.

A very recent paper includes the latest information on the impact of PQ issues in the United Kingdom (Vegunta et al. 2019). This study explores the various

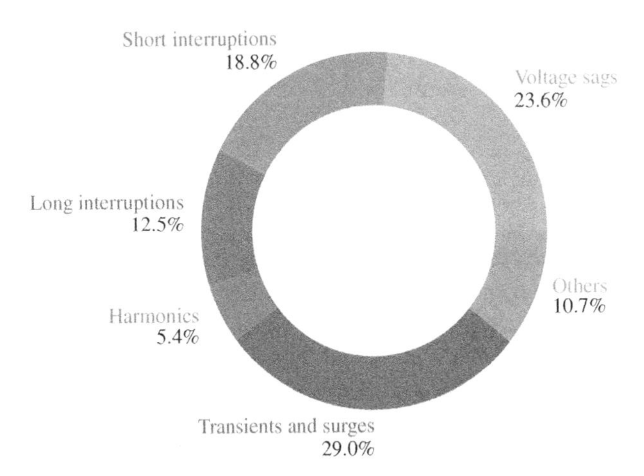

Figure 1.6 Impact of power quality disturbances. Source: Adapted from Targosz and Manson (2007).

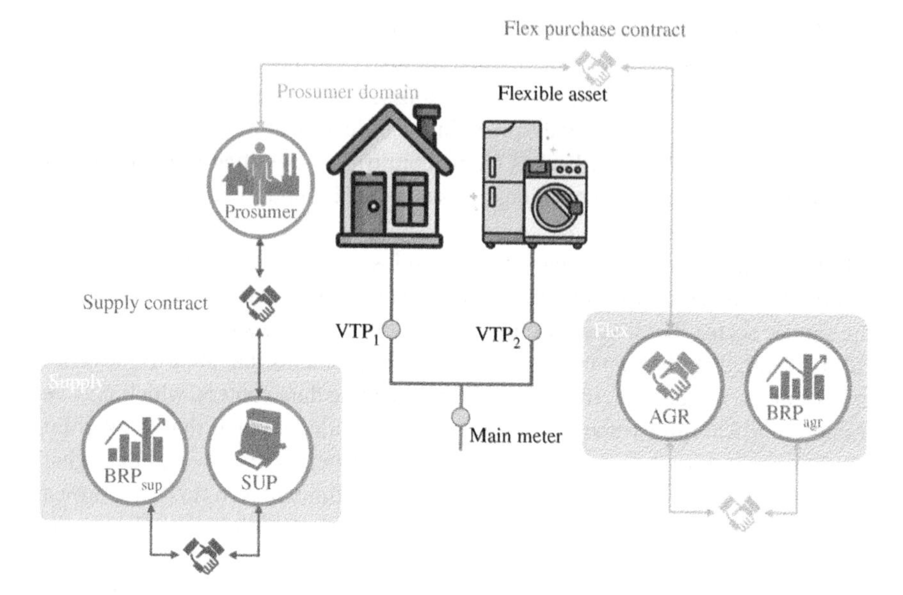

Figure 1.7 Example of location of the Virtual Transfer Points. Source: the author.

Table 1.2 Average costs by type of poor PQ event from the survey results.

PQ event	Average cost (€)
Surge or transient	12,000–18,000
Long interruption	90,000
Short interruption (service sector)	18,000–36,000
Short interruption (Industry)	7,000–14,000
Voltage sag	2,000–4,000

Source: Adapted from Targosz and Manson (2007).

challenges and gaps in British electricity distribution. No specific comprehensive and large-scale studies have been carried out in the world in recent years. The doctoral thesis (Chan and Milanovic 2010) presents an extensive summary of all the surveys carried out worldwide from the end of the twentieth century to the decade 2010 and their main conclusions. The summary of direct cost per event is presented in Table 1.3. Reported damages due to long interruptions are also included only as a reference. The results of these surveys can still be extrapolated to most current cases.

Table 1.3 Direct cost per voltage sag.

Section	Division	Activities	Financial loss	Currency
Manufacturing	General	Large industrial and commercial (United States)	7,694	US$
	Paper and paper products	Paper manufacturing (United States)	30,000	US$
	Chemical and chemical products	Chemical industry (United States)	50,000	US$
	Non-metallic mineral products	Glass industry (EU)	250,000	€
	Non-metallic mineral products	Glass plant (United States)	200,000	US$
	Basic metals	Steelworks (EU)	350,000	€
	Basic metals	Steelworks (United Kingdom)	250,000	US$
	Computer, electronic and optical products	Semiconductor (United States, Europe, and the Far East)	2,500,000	US$
	Computer, electronic, and optical products	Semiconductor (EU)	3,800,000	€
	Machinery and equipment	Equipment manufacturing (United States)	100,000	US$
	Motor vehicles, trailers	Automobile industry (United States)	75,000	US$
Information and communication	Telecommunications	Telecommunications (EU)	30,000	€
	Information service activities	US computer center (United States)	600,000	US$
	Information service activities	Computer center (EU)	750,000	€
Financial and insurance activities	Activities auxiliary to financial services and insurance activities	Credit card processing (United States)	250,000	US$

Source: Adapted from Chan and Milanovic (2010).

1.7 Internet of Things

The concept internet of things (IoTs) was originally introduced in 1985 by the co-founder of the first US cellular company, Peter T. Lewis, in a speech delivered at the Congressional Black Caucus Foundation 15th Annual Legislative Weekend in Washington, DC (Chetan Sharma Consulting n.d.). Everyday objects become smarter by including internet connectivity and ubiquitous sensors and given the progress of the fourth industrial revolution (or industry 4.0), these objects will end up connected sooner or later. Although the digital transformation of an industrial system is already a major achievement, the IoT will bring about a revolution in society. Undoubtedly, the idea of industry 4.0 involves embracing the IoT paradigm (Shrouf et al. 2014) which means that the network is extended to the real world covering all kinds of daily objects. The main areas of IoT investments include manufacturing operations, transportation, Smart Grid, smart buildings, and, increasingly, consumer IoTs, smart home automation, and retail. The unprecedented boom of IoT has been fueled by several market enablers:

- Decrease sensors and electronic components costs.
- Increase of Edge Computing. Centralized computing represents a limitation for IoT since it increases the amount of data transferred and the related costs. Edge computing allows processing exactly where and when it is required.
- Global hyperconnectivity allows processing in near real-time the information collected by the myriad of deployed IoT sensors.

The underlying aim is not just to create data, but also to extract valuable insights from the data generated by these devices. In this regard, communication technologies play a central role and should enable a dynamic and secure deployment of IoT so that this large amount of generated data can be properly managed. While challenging, the next generation of 5G and new IoT connectivity will speed up the ability to collect data and upload it to the cloud, which means massive exploitation of these technologies (Li et al. 2018). At the end of 2018, there were an estimated 22 billion connected IoT devices in use around the world and, this number will rise to 38.6 billion by 2025. Moreover, forecasts suggest that by 2030 around 30 billion of these devices will be in use worldwide, creating a web of interconnected devices (Statista 2022).

IoT paves the way not only for the transformation of products, which can be customized in quasi-real-time but also for service models. From an industrial point of view, the adoption of IoT and interconnected services should be based on the following basic design principles (Guan et al. 2017):

- **Interoperability**: All material and human resources within industry 4.0 should offer the ability to be interconnected using IoT and its applications.

- **Virtualization**: Buildings or industrial facilities should make use of digital twins representing all the information collected by sensors or systems as well as simulation models.
- **Edge-computing**: Objects located at the industrial facilities should have autonomous decision-making capabilities.
- **Real-time capabilities** through data acquisition, analysis, and decision-making by incorporating the required artificial intelligence.
- **Service-oriented** by offering an extensive portfolio of services to enable the interaction and creation of new applications and, therefore, a higher value-added process.
- **Modularity**: Systems should be as flexible and scalable as possible within the smart industry.

The implementation of these basic principles makes tailor-made manufacturing production possible, as well as automatic diagnosis, adjustment, and optimization of the process, and what is more, this is also essential to support workers in their activities and improve their working conditions (Jeschke et al. 2017).

1.8 The Relevance of Submetering

Successful approaches to demand flexibility require very different capabilities than those needed for the traditional energy supply business, such as the following:

- Adopt human-centered design to understand the consumer and involve them in flexibility plans.
- Understand in-depth and predict customer energy usage patterns.
- Manage and orchestrate massive data from large and diverse customer groups.

So, first, we are compelled to highlight the importance of detailed energy metering. Submetering, as opposed to bulk metering, involves measuring the energy consumption of individual units or appliances in a building complex. Submeters provide crucial information for a more granular measurement of energy consumption data. Unfortunately, submetering remains limited due to the costs of meters and the technical complexity of their installation, operation, and disaggregation. The study (Zhai and Salazar 2020) clearly shows that each additional level of metering correlates to increased energy savings, so the energy management system (EMS) should integrate submetering at the system and equipment level. Moreover, if the next generation of EMS included data-driven analytics and machine learning, it would unlock hidden benefits such as accurate energy control and error detection on utility bills. Ability to record actual electricity usage. Comparison of the usage of similar appliances over time. Ability to identify equipment running off-schedule, to avoid energy waste. Early access to equipment health and maintenance issues. Examples from the United States have achieved savings of up to 17%, with a payback of less than one year (Green Alliance 2020).

But not only is power consumption submetering relevant. The importance of reliable, high-quality electric power continues to grow as the ongoing digitalization of our societies expands and deepens. Electronics equipment can be both a source and a victim of poor PQ. In the US industrial and digital companies are collectively losing each year $45.7 billion due to outages, and another $6.7 billion due to PQ disturbances. Of course, these costs change with the length of the outage, but even the short ones are costly as they may imply that the systems need to be reset to come back online. What is more, DER may lead to PQ problems and operational limit violations in the power system when their penetration exceeds a particular value, called hosting capacity (HC). DR can be employed to manage Hosting Capacity, under PQ constraints. The detailed and continuous measurement of PQ indicators can contribute to a more accurate "dynamic hosting capacity" assessment (Palacios-Garcia et al. 2017).

Finally, as part of the separation of supply flexibility, submetering is necessary to isolate the controllable loads that can be used for demand response from the other loads at the Prosumer site. In addition, a Prosumer can be managed by more than one Aggregator (AGR) at the same time, if the aggregators operate a mutually exclusive set of resources. As depicted in Figure 1.7, one possible way to do this is to introduce what has been called Virtual Transfer Points (VTPs) (de Heer 2015) for controllable assets by installing accounting submeters. So that by subtracting this load from the main meter, the remaining residential load can be found. In any case, submetering will serve the following purposes (Flamm et al. 2017): To better quantify the activated flexibility as a basis for the transfer of energy, to allocate the flexibility activated by each resource and to the appropriate aggregator, and to better quantify the performance of the prosumer toward the aggregator and supplier (SUP), to better quantify the performance of the aggregator toward the customer of the flexibility: Transmission system operator (TSO), distribution system operator (DSO), or balance responsible parties (BRPs). Although requiring a submeter in the residential segment may jeopardize the business case, technological developments around the IoT are expected to largely remove investment barriers.

1.9 Energy Smart Appliances

Equipment and systems that can be used as potential demand flexibility assets can be found in all residential environments. Recently the term Grid-Interactive Efficient Buildings (GEB) has been coined (Neukomm et al. 2019). This can be seen from two points of view: As an individual building and as part of the power grid. First, these are energy-efficient buildings that incorporate both passive and active strategies. Passive design measures include building orientation, continuous insulation, and energy-efficient windows. While active strategies typically consist of heating and cooling systems, use of occupancy sensors and controls to optimize energy use, intelligent lighting, and solar photovoltaic technology. Or, in the commercial sector, an EMS. But the feature that makes it unique is its ability to interface with the local grid system, functioning as a demand flexibility resource. Building managers are always looking for the latest energy efficiency technologies to flatten their consumption curve. However, complimentary use of energy resources, such as energy storage and dynamic load management, may be the best option for this type of building. Let us focus on the role of the appliances we can find inside this building.

Popularly, Smart Appliances (SA) are recognized for having some electronic processing capability and wireless connectivity. For example, smart washing machines can independently regulate the detergent to be used depending on the weight of the load and the type of fabric. They can also automatically send alerts when the detergent runs out. However, in the energy field, within the framework of Smart Grids, the term "smart" refers to those capable of modulating their electricity demand in response to signal requests from the electrical system. Thus, incorporating different DR strategies: Would materialize into load-shifting strategies, which shift their operating period from peak to off-peak hours, or load-modulation strategies, which directly reduce or avoid energy use during peak hours. The Energy Smart Appliances (ESA) covered include cold and wet appliances; heating, ventilation, and air conditioning units (HVAC); battery storage; and smart electric vehicle charge points. But to unlock the full potential of ESA, designers and manufacturers must first address critical issues for effective DR as the following:

- **Cybersecurity**: The prevention of unauthorized access to ESA by third parties.
- **Data privacy**: The secure storage of personal data on the device or any controlling part.
- **Interoperability**: The ability of ESA to work seamlessly across any DR service operated by any system player.
- **Power quality**: The prevention of grid disturbances caused by the incorrect or simultaneous operation of ESA.

During the last few years, different IoT PQ sensors have been developed to be embedded in individual appliances. One of these versions is an advanced Grid-interactive Appliance Controller (Garrido-Zafra et al. 2022). It works as a Direct Load Controller as well as a PQ monitor with a wide range of alarms. Status information, configuration, PQ data, and even diagnostic data from the appliances can be recorded while in operation and transferred to the cloud for machine learning processing.

Symbols and Abbreviations

AC	alternating current
AMI	advance metering infrastructure
BRP	balancing responsible party
CAIDI	customer's average interruption duration index
CMP	capacity market programs
COP	conference of the parties
CPP	critical peak pricing
DBP	**Demand bidding/buyback programs**
DC	direct current
DER	distributed energy resources
DG	distributed generation
DLC	direct load control
DoE	Department of Energy
DR	demand response
DSM	demand-side management
DSO	distribution system operator
EE	energy efficiency
EMS	energy management system
ERDP	emergency demand response programs
ETP	European Technology Platform
EU	European Union
EV	electric vehicle
GEB	grid-interactive efficient buildings
GIAC	grid-interactive appliance controller
HC	hosting capacity
HVAC	heating, ventilation, and air conditioning
ICT	information and communication technologies
IEA	International Energy Agency
IEC	International Electrotechnical Commission
IEEE	Institute of Electric and Electronic Engineers
IoT	internet of things
PQ	power quality
PV	photovoltaic
RES	renewable energy sources
RTP	real-time pricing
SA	smart appliance
SAIDI	system average interruption duration index
SAIFI	system average interruption frequency index
TOE	tons of oil equivalent

TOU time of use
TSO transmission system operator
VoLL value of lost load
VPP virtual power plants
VRES variable renewable energy resources
VTP virtual transfer points

References

Akrami, A., Doostizadeh, M., and Aminifar, F. (2019). Power system flexibility: an overview of emergence to evolution. *Journal of Modern Power Systems and Clean Energy* 7 (5): 987–1007.

Babatunde, O.M., Munda, J.L., and Hamam, Y. (2020). Power system flexibility: a review. *Energy Reports* 6: 101–106.

Bhusal, N., Abdelmalak, M., Kamruzzaman, M., and Benidris, M. (2020). Power system resilience: current practices, challenges, and future directions. *IEEE Access* 8: 18064–18086.

Billinton, R. and Allan, R.N. (2003). Reliability of electric power systems: an overview. In: *Handbook of Reliability Engineering* (ed. H. Pham), 511–528. London, https://doi.org/10.1007/1-85233-841-5_28: Springer.

Bollen, M.H.J. (2003). What is power quality? *Electric Power Systems Research* 66 (1): 5–14.

Carr, J. (1996). North American and European distribution systems compared. *Power Technology International* 107–108, 111–112, https://www.osti.gov/etdeweb/biblio/272909.

Cecati, C., Mokryani, G., Piccolo, A., and Siano, P. (2010). An overview on the smart grid concept. In: *IECON 2010-36th Annual Conference on IEEE Industrial Electronics Society*, 3322–3327. IEEE.

Chan, J.Y. and Milanovic, J. (2010). Framework for assessment of economic feasibility of voltage sag mitigation solutions. University of Manchester, PhD thesis. https://pure.manchester.ac.uk/ws/portalfiles/portal/54595942/FULL_TEXT.PDF.

Chetan Sharma Consulting (n.d.). Correcting the IoT history. http://www.chetansharma.com/correcting-the-iot-history (accessed 26 August 2022).

Cochran, J., Miller, M., Zinaman, O. et al. (2014). Flexibility in 21st century power systems. National Renewable Energy Lab. (NREL), Golden, CO, USA.

Colak, I. (2016). Introduction to smart grid. In: *2016 International Smart Grid Workshop and Certificate Program (ISGWCP)*, 1–5. IEEE.

Colak, I., Bayindir, R., and Sagiroglu, S. (2020). The effects of the smart grid system on the national grids. In: *2020 8th International Conference on Smart Grid (icSmartGrid)*, 122–126. IEEE.

Dai, C., Cheng, K., Lei, Y., and Yang, Y. (2020, 2020). Research hotspots and evolution of energy prosumer: a literature review and bibliometric analysis. *Mathematical Problems in Engineering* 2020: 5703101.

Energy, Transport and Environment Statistics – Publications Office of the EU (2020). Energy, Transport and Environment Statistics. https://ec.europa.eu/eurostat/web/products-statistical-books/-/ks-dk-20-001 (accessed 7 June 2022).

ENTSO-E Transmission System Grid Map (n.d.). ENTSO-E Transmission System Map. https://www.entsoe.eu/data/map/ (accessed 7 June 2022).

Eto, J.H. (2017). The national cost of power interruptions to electricity customers – a revised update. *Lawrence Berkeley National Laboratory, Tech. Rep.* LBNL-58164.

European Technology Platform [ETP] SmartGrids (2010). Smart Grids: Strategic Deployment Document for Europe's electricity networks of the future. Strategic Deployment Document.

Flamm, A., P. Schell, U. Stougaard Kiil, *et al.* (2017). USEF: workstreams on aggregator implementation models – recommended practices and key considerations for a regulatory framework and market design on explicit demand response. *USEF, Tech. Rep.* https://www.usef.energy/app/uploads/2017/09/Recommended-practices-for-DR-market-design-2.pdf.

Garrido-Zafra, J., Gil-de-Castro, A.R., Savariego-Fernandez, R. et al. (2022). IoT cloud-based power quality extended functionality for grid-interactive appliance controllers. *IEEE Transactions on Industry Applications* 1. https://doi.org/10.1109/TIA.2022.3160410.

Gellings, C.W. (2017). Evolving practice of demand-side management. *Journal of Modern Power Systems and Clean Energy* 5 (1): 1–9.

Gholami, A., Shekari, T., Amirioun, M.H. et al. (2018). Toward a consensus on the definition and taxonomy of power system resilience. *IEEE Access* 6: 32035–32053.

Green Alliance (2020). A smarter way to save energy: using digital technology to increase business energy efficiency (2020). https://green-alliance.org.uk/wp-content/uploads/2021/11/A_smarter_way_to_save_energy.pdf (accessed 7 June 2022).

Guan, Y., Vasquez, J.C., Guerrero, J.M. et al. (2017). An open virtual neighbourhood network to connect IoT infrastructures and smart objects—vicinity: IoT enables interoperability as a service. In: *2017 Global Internet of Things Summit (GIoTS)*, 1–6. IEEE.

de Heer, H. (2015). USEF position paper. The independent aggregator. *Position Paper, Jun.*

Heffron, R.J., Körner, M.F., Schöpf, M. et al. (2021). The role of flexibility in the light of the COVID-19 pandemic and beyond: contributing to a sustainable and resilient energy future in Europe. *Renewable and Sustainable Energy Reviews* 140: 110743.

IEA (2021). Renewables 2021 – analysis and forecast to 2026. *International Energy Agency (IEA) Publications International.*

IEC (n.d.-a) IEC 61000-3-2 Ed. 5.1 en:2020|Electromagnetic Compatibility (EMC) – Part 3-2: Limits – limits for harmonic current emissions (equipment input current 16 A per phase). https://webstore.iec.ch/publication/28164 (accessed 7 June 2022).

IEC (n.d.-b) IEC 61000-3-3:2013+AMD1:2017+AMD2:2021 CSV|Electromagnetic Compatibility (EMC) – Part 3-3: Limits – limitation of voltage changes, voltage fluctuations and flicker in public low-voltage supply systems, for equipment with rated current <16 A per phase. https://webstore.iec.ch/publication/68776 (accessed 7 June 2022).

IEC (n.d.-c) IEC 61000-4-7:2002+AMD1:2008 Electromagnetic compatibility (EMC) Part 4-7: Testing and measurement techniques—general guide on harmonics and interharmonics measurements and instrumentation, for power supply systems and equipment connected thereto. https://webstore.iec.ch/publication/4228 (accessed 8 June 2022).

IEEE (2003). C62.41.1-2002 – IEEE guide on the surge environment in low-voltage (1000 V and less) AC power circuits.

IEEE (2012) '1366-2012 – IEEE guide for electric power distribution reliability indices – Redline.

IEEE (2015) '1453-2015 – IEEE recommended practice for the analysis of fluctuating installations on power systems.

IEEE 519-2014 – IEEE Recommended Practice and Requirements for Harmonic Control in Electric Power Systems (n.d.). IEEE Recommended Practice and Requirements for Harmonic Control in Electric Power Systems. https://ieeexplore.ieee.org/document/6826459 (accessed 16 November 2021).

IEEE Recommended Practice for Monitoring Electric Power Quality (2019). IEEE Std. 1159-2019 (Revision of IEEE Std. 1159-2009). pp. 1–98. https://doi.org/10.1109/IEEESTD.2019.8796486.

International Energy Agency (2019). *Status of Power System Transformation 2019: Power System Flexibility*. OECD Publishing.

Jeschke, S., Brecher, C., Meisen, T. et al. (2017). Industrial internet of things and cyber manufacturing systems. In: *Industrial Internet of Things* (ed. S. Jeschke, C. Brecher, H. Song and D.B. Rawat), 3–19. Springer, https://link.springer.com/chapter/10.1007/978-3-319-42559-7_1.

Jordehi, A.R. (2019). Optimisation of demand response in electric power systems, a review. *Renewable and Sustainable Energy Reviews* 103: 308–319.

Larsen, P., Kristina SH, Eto, J., & Sanstad, A. (2019). Frontiers in the economics of widespread, long-duration power interruptions: Proceedings from an Expert Workshop. https://eta-publications.lbl.gov/sites/default/files/long_duration_interruptions_workshop_proceedings.pdf.

Li, S., Da Xu, L., and Zhao, S. (2018). 5G Internet of things: a survey. *Journal of Industrial Information Integration* 10: 1–9.

Loughran, D.S. and Kulick, J. (2004). Demand-side management and energy efficiency in the United States. *The Energy Journal* 25 (1): 19–43.

Mapa del sistema eléctrico ibérico (n.d.). Mapa del Sistema eléctrico ibérico. https://www.ree.es/sites/default/files/downloadable/maptra2005.pdf (accessed 7 June 2022).

Mendes, G., Loew, A., and Honkapuro, S. (2019). Regional analysis of the economic impact of electricity outages in US healthcare facilities. In: *2019 IEEE Power & Energy Society General Meeting (PESGM)*, 1–5. IEEE.

MeriTalk (n.d.) Rx: ITaaS + Trust – MeriTalk. https://www.meritalk.com/study/rx-itaastrust (accessed 7 June 2022).

Mohandes, B., El Moursi, M.S., Hatziargyriou, N., and El Khatib, S. (2019). A review of power system flexibility with high penetration of renewables. *IEEE Transactions on Power Systems* 34 (4): 3140–3155.

Neukomm, M., Nubbe, V., and Fares, R. (2019). Grid-interactive efficient buildings. US Department of Energy (USDOE), Washington, DC, USA; Navigant Consulting, Inc.

Nikoobakht, A., Aghaei, J., Niknam, T. et al. (2019). Electric vehicle mobility and optimal grid reconfiguration as flexibility tools in wind integrated power systems. *International Journal of Electrical Power & Energy Systems* 110: 83–94.

Office of Electricity (2011). Title XII – Smart grid SEC. 1301–1308 statement of policy on modernization of electricity grid. US Department of Energy.

Palacios-Garcia, E.J., Moreno-Muñoz, A., Santiago, I. et al. (2017). PV hosting capacity analysis and enhancement using high resolution stochastic modeling. *Energies* 10 (10): https://doi.org/10.3390/en10101488.

Qdr, Q. (2006). Benefits of demand response in electricity markets and recommendations for achieving them. *US Department of Energy, Washington, DC, USA, Tech. Rep.* https://www.energy.gov/sites/default/files/oeprod/DocumentsandMedia/DOE_Benefits_of_Demand_Response_in_Electricity_Markets_and_Recommendations_for_Achieving_Them_Report_to_Congress.pdf.

Sharma, A., Rajpurohit, B.S., and Singh, S.N. (2018). A review on economics of power quality: impact, assessment and mitigation. *Renewable and Sustainable Energy Reviews* 88: 363–372.

Shrouf, F., Ordieres, J., and Miragliotta, G. (2014). Smart factories in Industry 4.0: a review of the concept and of energy management approached in production based on the Internet of Things paradigm. In: *2014 IEEE International Conference on Industrial Engineering and Engineering Management*, 697–701. IEEE.

Silverstein, A., Gramlich, R., and Goggin, M. (2018). A customer-focused framework for electric system resilience. Natural Resources Defense Council and the Environmental Defense Fund.

Statista (2022). Number of Internet of Things (IoT) connected devices worldwide from 2019 to 2021, with forecasts from 2022 to 2030. https://www.statista.com/statistics/1183457/iot-connected-devices-worldwide (accessed 25 August 2022).

Targosz, R. and Manson, J. (2007). Pan-European power quality survey. In: *2007 9th International Conference on Electrical Power Quality and Utilisation*, 1–6. IEEE.

UNE-EN 50160:2011/A2:2020 (n.d.). Voltage characteristics of electricity supplied by public electricity networks. https://www.une.org/encuentra-tu-norma/busca-tu-norma/norma/?c=N0064209 (accessed 15 November 2021).

UNFCCC (2015). ADOPTION OF THE PARIS AGREEMENT – Paris Agreement text English.

Vegunta, S.C., Watts, C.F.A., Djokic, S.Z. et al. (2019). Review of GB electricity distribution system's electricity security of supply, reliability and power quality in meeting U.K. industrial strategy requirements. *IET Generation Transmission and Distribution* 13 (16): 3513–3523. https://doi.org/10.1049/IET-GTD.2019.0052.

Zhai, Z.J. and Salazar, A. (2020). Assessing the implications of submetering with energy analytics to building energy savings. *Energy and Built Environment* 1 (1): 27–35.

2

A Deep Dive into the Smart Energy Home

Neomar Giacomini

Senior Engineering Manager for Electronics Hardware Development at Whirlpool Corporation, Benton Harbor, Michigan, USA

2.1 Smart Home Ecosystem

The time has come to think, and most importantly experience, our houses as a cohesive integrated ecosystem of devices that work together to make our lives easier. It is also time to see houses becoming part of a digital ecosystem that extends far beyond the traditional home, both from an energy perspective and other living needs such as faster communications enabling a multitude of services, data access, command, and control.

From a Smart Energy perspective, this integration has started with the publication of the Energy Independence and Security Act of 2007 (Congress.gov 2007), where the first official definition of Smart Grid was presented. Since then, a variety of studies and products became available in the areas of Smart Energy, Smart Cities, Smart Buildings and Electric Vehicles (EVs).

As for in-house, since Amazon Echo was launched on 6 November 2014 (Welch 2014) and subsequently Google Home was announced on 4 November 2016 (Chen 2016), the market has seen significant expansion in their use to enable consumers to build Smart Home ecosystems.

This book is timewise related to the COVID-19 pandemic, formally classified as such by the WHO (World Health Organization) on 11 March 2020 (UN 2020), which has quarantined millions of people in their homes and with the easiness of online shopping, the work-from-home period and other factors, the expansion in the Smart Homes adoption became inevitable.

Traditionally speaking for the pre-pandemic and pandemic timeframe, the definition of a Smart Home is somewhat simplistic. Figure 2.1 shows a diagram of a Smart Home comprised of a house with internet access, a smart thermostat, a connected camera, connected light bulbs, an Intelligent Virtual Assistant (IVA)

Energy Smart Appliances: Applications, Methodologies, and Challenges,
First Edition. Edited by Antonio Moreno-Munoz and Neomar Giacomini.

Figure 2.1 Traditional smart home in the year of 2020. Source: Neomar Giacomini (co-author).

to play music and enable basic commands such as alarms and timers. That is a reasonable visualization of what a Smart Home is considered in this timeframe.

This scenario, although simplistic, is not limited by current device's capabilities as of 2022, but instead by the consumers' limited knowledge and desire to live connected.

As briefly aforementioned, this scenario started changing drastically in recent years. With the pandemic lockdown, consumers started to get informed, to hear more from their inner circle on the benefits of certain connected devices, and to understand more about their capabilities and advanced functions.

Users of Smart Devices that on early states of adoption were just using their Virtual Assistants to play music and set timers, have taken the time to learn how to interconnect the Smart Home ecosystem to additional devices, learned how to create connected functional recipes through services such as IFTTT (If This Than That), and also became familiar with new gadgets and the benefits, whether minor or major, these devices can provide.

The growth in the adoption of Smart Homes worldwide will reach nearly 50% from mid-2022 to the end of 2025 according to the Statista Inc. study (Lasquety-Reyes 2021) shown in Figure 2.2. And it is worth noting this is related to the number of houses, not the number of devices. The forecasted amount of 478.22 million households represents a penetration rate forecast in the world

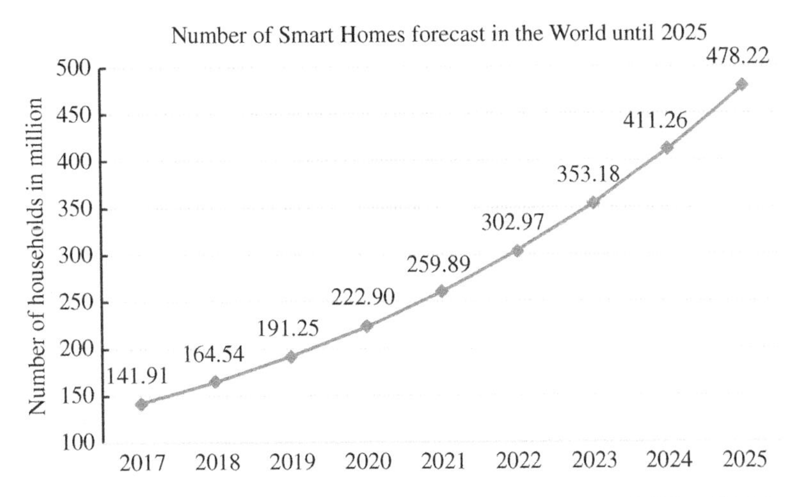

Figure 2.2 Number of Smart Homes forecast in the World from 2017 to 2025 (in millions). Source: Adapted from Lasquety-Reyes (2021)/https://www.statista.com/forecasts/887613/number-of-smart-homes-in-the-smart-home-market-in-the-world last accessed 18 November 2022.

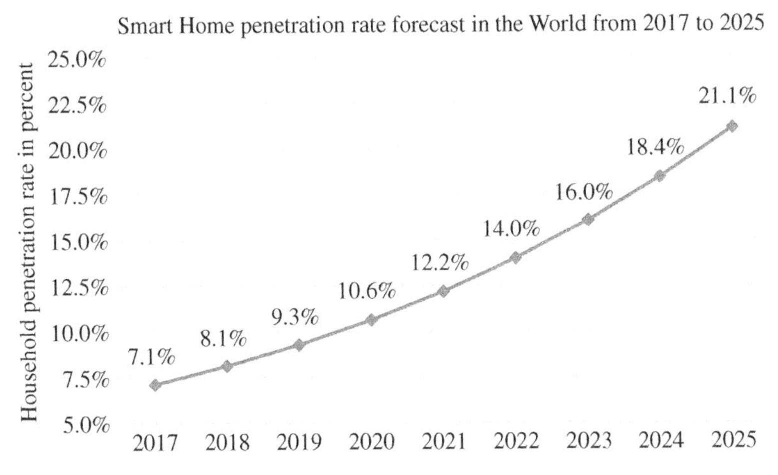

Figure 2.3 Smart Home penetration rate forecast in the World from 2017 to 2025. Source: Adapted from Statista (2021a)/https://www.statista.com/forecasts/887636/penetration-rate-of-smart-homes-in-the-world/ last accessed 18 November 2022.

reaching 21.09% in 2025 (Statista 2021a). Figure 2.3 shows the details on that forecast. These statistics do not consider devices whose primary function is not the automation or remote control of household equipment, e.g. smartphones and tablets. Similarly, devices that relate to household connection and remote control only to a limited extent, such as smart TVs, are not included either.

Table 2.1 Number of Internet of Things (IoT) connected devices worldwide from 2019 to 2030.

Number of Internet of Things connected devices worldwide from 2019 to 2030, by use case (in millions)					
	Connected vehicles	White goods	Environmental monitoring	HVAC[b]	Lighting
2019	355	6	36	48	109
2020[a]	408	8	40	56	131
2021[a]	499	12	49	72	181
2022[a]	604	18	60	91	237
2023[a]	724	29	75	112	311
2024[a]	853	48	92	136	407
2025[a]	991	76	115	163	530
2026[a]	1130	113	144	195	683
2027[a]	1270	164	180	232	871
2028[a]	1411	232	224	275	1096
2029[a]	1551	323	278	325	1358
2030[a]	1684	441	345	381	1651
Projected growth from 2022 to 2030	179%	2295%	473%	317%	597%

a) Forecast.
b) HVAC (Heating, Ventilation, and Air Conditioning).
Source: Adapted from Statista (2020).

Along with the growth in the number of Smart Homes, the number of Internet of Things (IoT) devices worldwide will also increase significantly. A few examples of the projected growth of IoT devices related to the Smart Energy Homes (SEH) topic are shown in Table 2.1. This table is a subset and adaptation of a study published by Statista Inc. (Vailshery 2020).

Worth stating that this table includes Connected Vehicles which are counted outside many of the Smart Home statistics since not all Connected Vehicles are EV. EVs are important in the scope of this book due to the energy collaboration aspects with the Smart House.

Besides the growing number of IoT devices, there are key enabling technologies around smart homes that are building the ecosystems to a point where one device can interact with many, and vice versa. These technologies and their penetration into the Smart Homes are enabling heavy adopters to reach over a hundred devices under a collaborative Smart Home ecosystem.

Technologies like Wi-Fi™, Bluetooth, Zigbee, and many networking protocols applied to these devices, are being implemented so that centralized control via the ecosystem management is possible, therefore also enabling cross-device collaboration. The collaboration is not entirely seamless as of 2022, but significant efforts by manufacturers and alliances to reach seamless integration are taking place and will be discussed later.

By no means does 100 devices in a single Smart Home, as aforementioned, are all important or even relevant to the Smart Energy Homes discussion, but it is worth depicting what a heavy connected house looks like in the year 2022. This is important to elaborate on the Smart Home scenario and expand the possibilities one can find while thinking on the different devices and their possible link to smart energy use cases. A new set of eyes with a different background is key for breakthroughs in any line of science.

From a simple Virtual Assistant, all to way to a robotic cat litter box, cameras, home appliances, smart beds, energy monitors, connected faucets, door locks, garage door openers, meat smokers, thermostats, and many other devices. It does not look like there is a limit to the creativity technology companies are showing to design connected devices around the traditional home use cases. Figure 2.4 depicts a scenario with nearly 120 devices, and although some of the icons may look not from this era in terms of connectivity, they are. All these possibilities are either already available in the market or announced during major technology shows.

If a Smart Home owner in the year 2022 takes the time to count device by device available in his or her house, this count will go near 20 connected devices very easily. Six connected light bulbs, two assistants, four cameras, two televisions, a thermostat, an alarm system, two computers, one tablet, and two smartphones linked to the ecosystem get to a total of 21 devices already. This is an exercise the author strongly recommends, take your time and start listing your connected devices, every other day you will find another device you forgot to list. Please note that the request is to list all the connected devices, not just IoT devices, more on that later.

On a personal note, as of May 2022, the author of this chapter currently lives in a house with 102 connected devices, with very good coverage of what is shown in Figure 2.4. No connected t-shirts, shoes, and other apparel around yet though.

Important to state that a personal computer is connected, but not by definition a traditional IoT device, same for smartphones and tablets. One can argue a Smart TV is connected, but it is not a sensor or actuator in order to fit in the IoT definition. The relevance of this short debate is to assure the discussion around Smart Homes does not get limited or misguided by assuming this is only about IoT devices.

It is of common agreement in the technical community around this topic that IoT relates to a network of devices that are embedded with sensors and other technologies to enable connection and exchange of data over the internet. Although this is an important aspect of Smart Energy systems, it is not a limiting boundary to what can be achieved on a Smart Energy Home with regular connected devices contributing to the Smart Home ecosystem as well.

Figure 2.4 Diagram of what a heavily connected Smart Home looks like in the year of 2022. Source: Neomar Giacomini (co-author).

Any connected resource in the Smart Home can provide data to enable Smart Energy use cases. For a short example, a computer is not an IoT device, but it is connected and is able to provide the Smart Home with information on the user's location, like the customer presence in the office for instance by detecting the user is active in the computer, which may be useful to trigger a given Smart Energy use case at that moment. Or smartphones that just by being present in the Wi-Fi network at a given moment can provide a reasonable count of number of people in the house.

To foster the imagination around potential use cases, and to provide a better understanding of connected/IoT devices currently available and/or applied to Smart Homes in the year 2022, a list is provided in Table 2.2.

Table 2.2 List of connected devices available for Smart Homes in the year of 2022.

Connected devices commonly found at, or available for Smart Homes in the year of 2022. Includes both connected and IoT devices		
Aquariums	Health monitors	Smoke and gas sensors
Air purifiers	Heaters	Sound sensors
Alarms	Humidifiers	Stoves, ovens, and cooktops
Assistant robots	HVAC	Streaming receivers (Cast feature)
Audio systems	Indoor exercise bike	Tablet computers
Bed frames	Kitchen small appliances	Televisions
Blinds and shades	Leakage detection sensors	Thermostats and thermometers
Cameras	Light bulbs	Toasters
Cars	Microwaves	Toys
Cat litter robots	Smartphones	Tracking devices
Coffee machines	Modems and routers	Treadmills
Desktop computers	Mop robots	Uninterrupted power supplies
Dishwashers	Pet feeders	Vacuum cleaners
Door locks	Pillows	Vacuum robots
Doorbell cameras	Power banks	Video games
Dryers	Power strips	Virtual assistants (multiple forms)
Energy monitors	Power tools	Virtual reality goggles
Faucets	Refrigerators	Washing machines
Garage door opener	Robotic lawn mowers	Water detection sensors
Gardening sensors	Sleep sensors	Water filtration systems
Grills and smokers	Smart power switches	Weather sensors
Haptics sensors	Smart watches	

Source: Neomar Giacomini (co-author).

The list provided is comprehensive but should not be considered absolute as this is an evolving market. Although at first look it may be perceived containing random connected devices irrelevant from the Smart Energy Home perspective, further reflection on each device can provide interesting insights.

The aquarium, for instance, is basically a large mass of water that has its temperature affected both by heaters and ambient temperature, and impacts the humidity of the house, therefore impacting air conditioning, ice formation in the air conditioner evaporator unit, and so on. An interesting fact, depending on the size of the aquarium, heating, air flow, and shape is that it will lose significant amount of water per day due to evaporation. No reliable scientific references were found to cite on this topic, but information collected from multiple discussion forums showed cases where water loss went from 1, all the way to 15 l/day depending on aquarium size. The author of this chapter experienced a water loss of 7 l/day on a 280 l aquarium, which is a significant amount of water that may end up in the air conditioner evaporator unit and impact performance.

Based on this example, it is strongly recommended to reflect on every device available and its impact on the house in terms of energy or any other data points they can provide. Also, the aquarium could be present in the house but not be connected to the internet, therefore unknown to a smart energy system. But since it is connected and that information is available, what can be done with it? That is an interesting point to debate and expand as a system is designed.

To provide a view of the current ownership of such devices in a US Smart Home, Figure 2.5 is provided. This chart shows the results of an online survey performed in early 2021 by Statista Inc. (2021d) with 3792 respondents between 18 and 64 years of age. This data does not provide direct basis for understanding the deployed volume, but instead the percentages and spread of Smart Home devices as an effect of consumer budget and preferences.

The device at the top of the list is the Smart TV, and in the context of Smart Energy Homes that is not a major enabler for use cases. The list follows with speakers, streaming devices and find the first potential candidate for the Smart Energy context at the fifth position, Smart Bulbs at 20%, then connected thermostats at 17%, smart plugs at 14%, and those plugs may be in use with a lamp, heater, Christmas tree lighting, who knows.

The importance of this data is in the fact that the devices considered as a foundation for Smart Energy use cases have considerably low adoption according to that 2021 research. Large appliances for instance are still below 10%, showing consumers are still at a slow adoption rate on critical items for Smart Energy Home use cases.

In relation specifically to Smart Appliances, the Smart Home Report 2021 publication from Statista (2021c) shows that the country with the highest Smart Appliance adoption rate is South Korea at 15.7%, Australia at 10%, and the United States at 7.4%. The data for the United States shows slight variation from Figure 2.5 due to data aggregation and sampling, but still with less than 2% deviation.

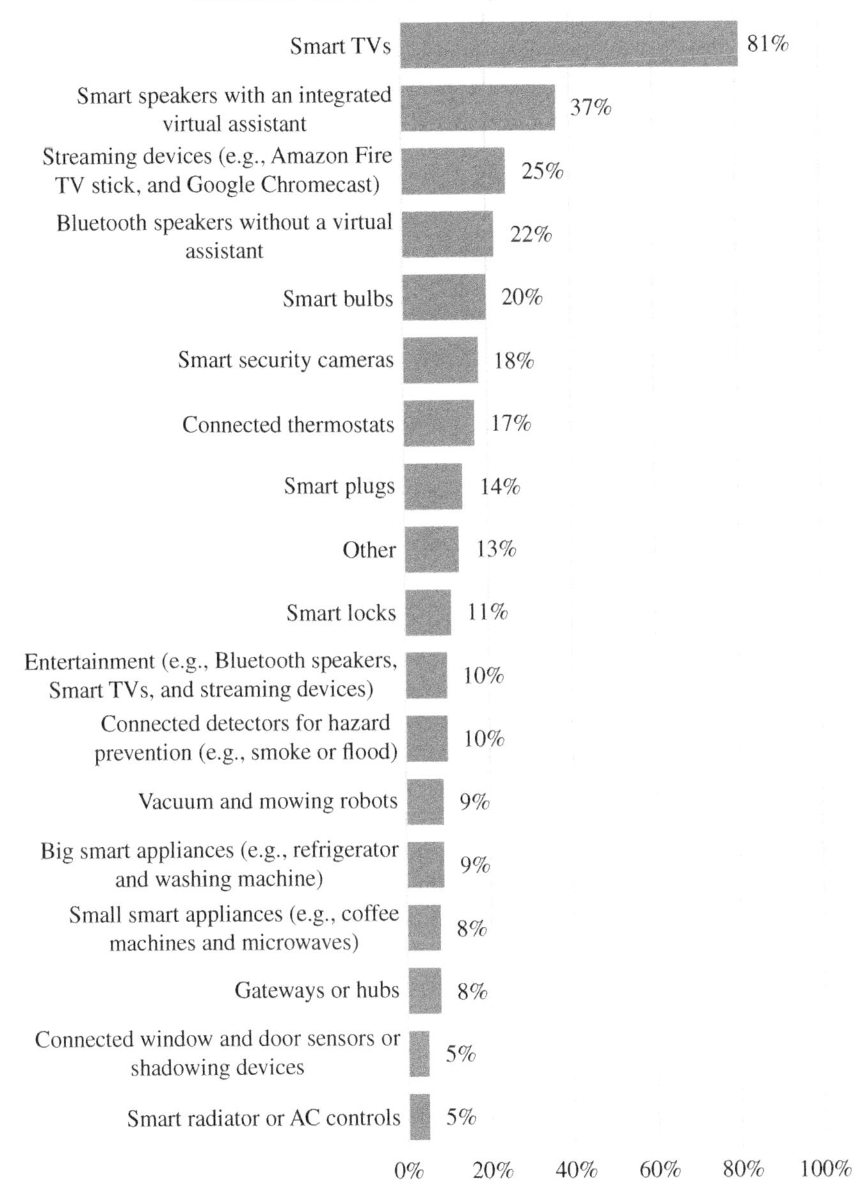

Figure 2.5 Ownership rate of Smart Home devices in the United States 2021. Which Smart Home devices does your household own? Source: Adapted from Statista (2021c)/https://www.statista.com/statistics/1253083/consumer-adoption-concerns-smart-home-technologies-in-the-united-states/ last accessed 18 November 2022.

With an understanding of connected devices vs. IoT devices, a broad grasp on the types of devices comprising and/or available for a Smart Home, and the insight on how important it is to deep dive on any type of connected device that can become a source of information either directly or indirectly, it is time to deep dive on the definitions, uses, benefits and restrictions of enabling technologies in this space.

2.2 Enabling Technologies

In this session, a short list of technologies currently available in the Smart Home scenario will be provided and briefly explained. The focus is here in on giving the macro information, benefits, and challenges associated with their use.

➢ **Wi-Fi**, without a doubt this is currently the most disseminated wireless communication technology for Smart Homes. It is based on the IEEE 802.11 family of standards and is available in all connected devices such as laptops, tablets, and smartphones, and in the Smart Home scenario is still the go to technology most companies are using to launch new IoT products. Although this statement is true, Wi-Fi is losing ground to other technologies due to power consumptions and other limitations.

Wi-Fi is currently broadly available in 2.4 and 5 GHz, with the newest generation Wi-Fi 6E also operating at 6 GHz. Due to the cost of the advanced versions, 2.4 GHz is still the preferred path for many IoT device manufacturers. This is an important aspect of this scenario as there is a significant larger number of devices in the 2.4 GHz band than the others.

Although taken as granted that most Wi-Fi routers would allow up to nearly 250+ devices to connect, an assumption from unaware association to Ethernet capabilities, that is not necessarily the case. Many Wi-Fi router vendors limit the maximum number of devices allowed to connect to 16, 32, or 64 devices. Newer models are in-fact advertising they enable up to 200 devices as an additional benefit of their product. That limitation was mostly due to memory costs and consumers historically not necessarily requiring large amounts of devices connected at the same time, and the newer devices able to connect more than 200 devices are currently costing well over a $1000.

This limitation is one of the main headaches most consumers going heavy connected encounter. Their current Wi-Fi routers are limited to a somewhat small number of concurrent devices, and unexpected behaviors will happen when reaching the limit until additional network capability is added. Unexpected behaviors such as random disconnection, lag in responsiveness are the most common effects.

Considering the year 2022 as a baseline, Wi-Fi in Smart Home devices is being broadly used on most types of devices. Applications with difficult access to a power outlet, however, such as door locks, window blinds, and some light bulb ecosystems are tending to use other technologies such as Zigbee and Z-Wave, but in number of devices, Wi-Fi is still at a much larger position. Additional information about Wi-Fi can be found in the Wi-Fi Alliance® website at https://www.wi-fi.org.

- **Bluetooth®** is a short-range wireless technology that in the Smart Home scenario is currently used mostly for Wi-Fi provisioning, computer peripherals, fitness devices, and audio systems. It operates in the 2.4 GHz band and its application is limited since it requires a gateway to reach the internet and cloud applications. Fitness scale for instance connects to a Smartphone and uses its internet connection to update a cloud database. Additional information about Bluetooth can be found in the Bluetooth Special Interest Group website at https://www.bluetooth.com.
- **Zigbee** is a low-power mesh network used for low data rate applications. Since Zigbee has a defined maximum data rate of 250 kbps, it is suitable for sensing, command, and control. Light switches, motion sensors, smart locks, certain lighting, and home sensing applications are common using this technology. Zigbee has applications in 2.4 GHz, but also 915 MHz in North America, and 868 MHz in Europe. A big advantage of this technology is its ability to support more than 65,000 devices on the same network with up to 30 hops, making it also well suitable for industrial applications. Additional information about Zigbee can be found in the Connectivity Standards Alliance (CSA) website at https://csa-iot.org/all-solutions/zigbee.
- **Z-Wave** is a sub 1 GHz mesh network that operates in similar fashion to Zigbee, resulting in one of the common question marks for consumers: Which one to get, Zigbee or Z-Wave? The main difference from Zigbee is the number of nodes and speed, this one is limited to 232 nodes, and a maximum data rate of 100 kbps. The number of hops is also much lower, four in this case. Z-Wave is commonly used for window blinds, smoke and gas detectors, smart locks, lighting controls, and so on. Z-Wave its well suited for home applications since the limit of 232 nodes would cover most command-and-control application needs. Additional information about Z-Wave can be found in the Z-Wave Alliance website at https://z-wavealliance.org.
- **Thread** is a low-power mesh networking technology focused on IoT, it uses 6LowPAN (IPv6 over Low-Power Wireless Personal Area Network) and competes with both Zigbee and Z-Wave. Thread is limited to 250 kbps in the 2.4 GHz band and is currently backed by over 60 sponsors and contributors, and due to the large effort to create a seamless connectivity standard that would work for all partners, in contrast to Zigbee and Z-Wave, Thread was developed to be independent of proprietary gateways like those that do exist for Zigbee and Z-Wave. One major benefit is that a Smart House already

having one Thread enabled Wi-Fi border router, allows the user to buy devices from whichever preferred provider without the need to add more proprietary hubs. You only add additional border routers if the network became physically too large and you need faster reach to the internet with less data hops, or for network reliability to avoid depending on a single link to the outside world. Thread networks are limited to 32 routers and 511 End Devices per router (OpenThread 2022). Additional information about Thread can be found on the Thread Group website at https://www.threadgroup.org.

- **Matter**, in contrast to the previous technologies, is not a physical wireless networking technology, but instead an application framework standard with Internet Protocol (IP) as its backbone. Its aim is to unify the best of the smart home technologies into a single application framework to make it easier for manufacturers to create products that will work with a variety of ecosystems and voice assistants (Kennis 2021). The cross-collaboration between different ecosystems such as Amazon Alexa, Apple Homekit, and Google Assistant to name a few, is not entirely seamless as of 2022, but significant effort on "matter" will enable future products to display the "matter" certification logo to reinforce its compatibility with multiple Smart Home ecosystems. Currently while purchasing devices customers need to be cautious to verify if the product is compatible with their needs and smart home ecosystem, which quite often may not be the case. Matter has launched over both Wi-Fi for high-bandwidth use cases like cameras and Thread for low-power, low-bandwidth devices.

2.3 Limitations

As of 2022, through the application of aforementioned technologies, Smart Homes have reached a sufficient level of maturity to start enabling Smart Home Energy systems, but still presenting limitations to any uninformed user that decides to implement the ecosystem by itself.

Unfortunately, the technologies have evolved significantly but presenting segmentation and limitations that make any larger smart home ecosystem contain a multitude of configurations and fine-tuning to be done. To depict this issue, a brief list of limitations faced during the implementation of Smart Homes is provided.

- **Smart Home Ecosystems** cross compatibility, as briefly mentioned while discussing "matter," is currently still an issue in the market. Consumers still need to be cautious during the purchase process to assure compatibility with pre-installed devices. Different vendors also approach the link to the ecosystem in different ways, some allowing direct integration with all functions available, others having the full set of features only available through a mobile application and basic ones through the integrated ecosystem, and others even offering solutions where the connected features are only available via smartphones.

- **Wi-Fi Routers** are another source of frustration for many users. The currently installed base of routers in consumers' homes have aged to some extent, with routers that are 5, even 10 years old in some cases. Such routers were designed in an era where the Wi-Fi network had to support a handful of connected devices, and for cost and complexity reasons many of these routers were designed with limitations on the maximum number of connections. Whether that maximum number is 16 or 32 devices, as soon as the user reaches those barriers the network will misbehave, and the latest added device will be the one to blame, while actually, the router is the cause of the issue.
- **Multiple networks** seem to be a solution for the router's maximum number of devices limitation. At first, the simple addition of another low-cost router and creation of a secondary Wi-Fi appears to solve the problem. However, many mobile applications and direct device connection solutions require device–device or device–mobile direct connection under the same Wi-Fi network. In other words, investing on expensive routers capable of hundreds of devices may be required.
- **Signal congestion**, although not a common issue, is definitely a concern, especially when a Smart Home is re-establishing itself after a power loss for instance. As the modem used to reach the internet provider takes significant more time than most smart devices in the house take to be ready for the service, all devices will start seeking connection through other networks, or even set themselves on pairing mode for reconfiguration. A consumer that is not well-versed in the wireless networking space will likely get confused through this process. Close proximity to high power repeaters or too many nearby networks is also an issue, quite common in neighborhoods where residences are close to each other, apartment buildings, and so on.
- **Construction** techniques have not changed over recent years, and the current approach is not a problem. But unfortunately, homeowners rarely know what is inside the walls. This is an important factor as installation points for routers, repeaters, and devices should avoid areas with materials that can cause signal degradation and other sources of noise. Some examples of issues are large televisions installed on an adjacent room right behind a wireless router, therefore blocking direct signal propagation to other devices; Wireless repeaters or devices installed by a wall with metal air ducts running inside of it, also degrading signal quality.
- **Credentials and provisioning** to get a Smart Home Ecosystem configured are a nuisance as of 2022. Setting up a Smart House running the ecosystem of one single vendor (i.e. Google Home, Amazon Alexa, etc.), and having over 100 devices from a couple of dozen different manufacturers over Wi-Fi is a challenge. This scenario is realistic as of 2022 and very similar to the personal

use case the author experienced in his personal Smart Home experiment. The consumer is required to download the ecosystem mobile application to start the process, download also a couple of dozen apps one for each manufacturer, and provide Wi-Fi network provisioning including password and other details to all 100 devices, in most cases individually. After that, a tedious process to link each device to the ecosystem central application and configure room placement, access control, and other details is also required. On top of this complexity, the configuration process for all these vendors varies significantly, but fortunately, the efforts around matter as a standard to solve these challenges are already becoming a reality confirmed by multiple vendors pledging adoption of the new standard during the 2022 Consumer Electronics Show (CES).

➢ **Privacy and security** are still concerns of many users. Whether to install a thermostat that knows if you are at home, or a camera that provides not just presence but audio and visuals of the house, consumers express all different levels of concern while joining the Smart Home trend. This is a limiting factor as it slows the adoption of IoT technologies that would help drive Smart Energy Homes use cases.

A study from the Continental Automated Buildings Association published by Statista Inc. (CABA 2021) and presented in Figure 2.6, shows statistics on the concerns that affect the United States consumer adoption of smart home technologies. Personal Information concerns are at 36%.

➢ **Debugging** issues in Smart Homes is a challenge as tools and applications are not yet centralized and easy at hand. Statistics on device activity, data usage, speed, network addresses, and other useful debugging information are scattered across a multitude of tools, and likely for safety reasons behind multiple different passwords. Even the lack of a log regarding when each device was added, using what credentials, to which network, what app it uses, and how it was linked and configured is an issue, as it is rarely available. Also, when physical issues arise such as poor signal propagation, excess noise, spectrum congestion due to apartment complexes and other factors, measurement equipment such as spectrum analyzers will be required and a regular consumer will not have access to one.

➢ **Power and internet dependency** is another concern, especially regarding security systems and access control. Certain commands like a request to close the window shades, even those being battery operated, may not get to the devices if there is a power outage and the internet modem and wireless routers are inoperative. That could be an issue for a user during vacation and using automatic control recipes to simulate presence in the house.

Very short-term power interruption off an entire house can also cause inconvenient situations. Many lighting systems as of 2022 still face issues when the power goes off and back on quickly, as that is the same behavior of a user flipping a light switch off and on, which will turn the lights on. This example is recently being solved by some vendors by coordinating data across all light

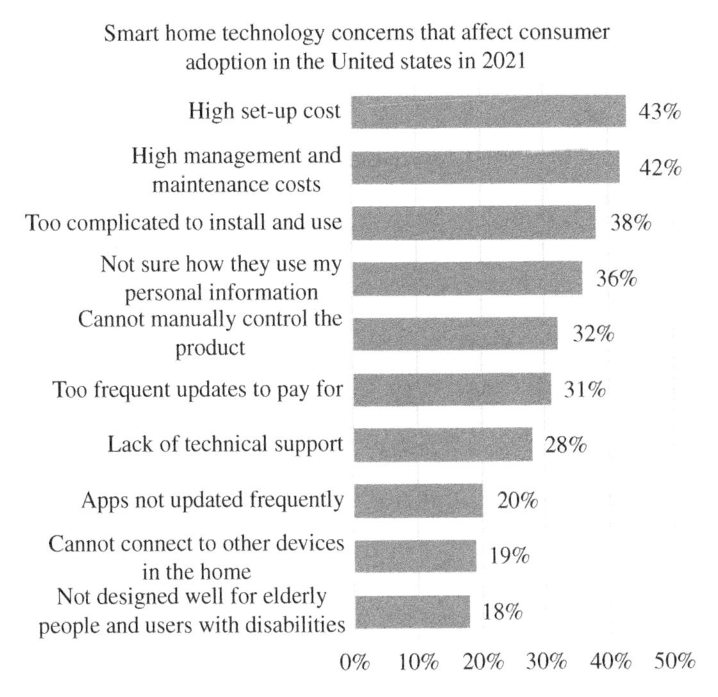

Figure 2.6 Smart home technology concerns that affect consumer adoption in the United States in 2021. Source: Adapted from CABA (2021)/https://www.statista.com/statistics/1253083/consumer-adoption-concerns-smart-home-technologies-in-the-united-states/ last accessed 18 November 2022.

bulbs in the house, but it shows how the consumer may confuse a quick power outrage with unauthorized access for instance. The author experienced such situation when a short power outage during the night turned on nearly 40 lights in the house. After a few seconds of confusion, a beep coming from a home appliance gave away that the reason was a power interruption, not some unauthorized access. Scary moment waking up to all of that trying to figure out who was invading the house, well, it was just technology.

One important point, however, is that most conventional devices that depend on power would not be available to the consumer anyway whether the device is connected or not. Garage door openers, camera systems, thermostats, lighting, and others would be inoperative on their non-connected versions the same.

Unfortunately, this book cannot accommodate providing statistics for all countries, but to provide a better idea on power disruption and how that may affect different consumers a brief example for the United States is provided.

The US Energy Information Administration (EIA) has collected and published interesting data on how many average hours of power interruption each consumer experiences in the United States per year (Hoff and Lindstrom 2021), Figure 2.7 shows this information from the year 2013 up to 2020.

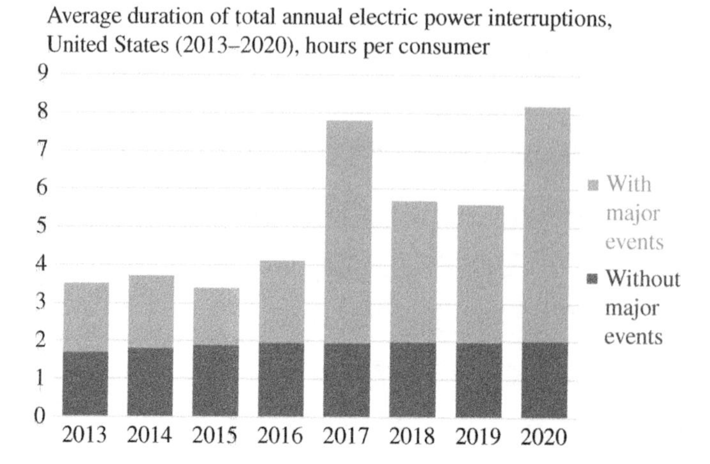

Figure 2.7 Average duration of total annual electric power interruptions, United States (2013–2020), hours per consumer. Source: Hoff and Lindstrom (2021)/Public Domain.

Another aspect that is very relevant to this scenario is the overall availability and reliability of your energy provider. The same study from EIA (Hoff and Lindstrom 2021) also shows the average number of interruptions per state vs. the total duration of the interruptions in 2020 per state in the United States. This data is plotted in Figure 2.8 and is worth noting that the United States has eight states that had more than 15 hours of accumulated interruptions.

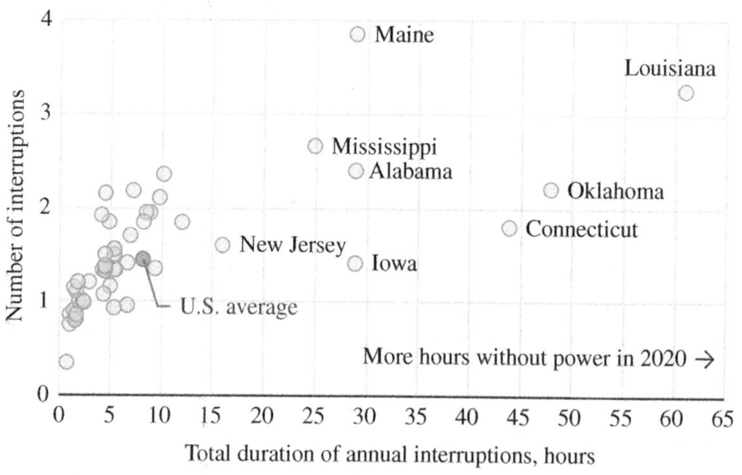

Figure 2.8 Average total annual electric power interruption duration and frequency per customer, by US state (2020). Source: Adapted from Hoff and Lindstrom (2021).

2.4 A Look into a Future Anchored in the Past

A lot has happened since the Energy Independence and Security Act of 2007 formally structured the Smart Grid's needs in a way that the overall industry could collaborate on a common ground, it triggered significant improvements in energy management and demand response, therefore enabling broader initiatives like Smart Cities to get implemented. With the release of Amazon Echo in 2014 and Google Home in 2016, Smart Homes were finally made broadly available for consumers to experiment with and the deployment of the additional infrastructure needed for Smart Energy Homes finally started.

Important to remember that both in-house and out-of-the-house infrastructure are required for advanced Smart Energy Home use cases. It is a must therefore for research to consider both inside and outside the house to understand where the future is taking us. This chapter has focused on Smart Homes and referred to outside-home factors such as the energy grid as a source of energy, but the Smart Energy Home scenario also considers local power generation through Solar Power and other means, and also how EVs will play an important role. EVs are discussed in more detail later, having Chapters 9 and 10 focused on this topic.

The statistics showed that even with the storyline extending for more than 15 years regarding Smart Grid and 8 years for Smart Home assistants, the acceptance and implementation are still quite low, and scattered in terms of types of devices and combinations thereof. Scattered in the sense that research cannot assume all Smart Homes are, or will be, enabling complex Smart Energy implementations anytime soon. A home with smart energy meter is not necessarily a Smart Energy Home; A home with a Smart Thermostat does not necessarily have Smart Appliances; A Smart Home that has just one smart light bulb cannot be compared to another with 40, and so on.

Smart thermostats are around a little longer than Smart Home Assistants. The Nest Learning Thermostat launched in October 2011 (Hu 2011) and is a prominent example. Even this technology is over 10 years old, according to the Smart Home Report 2021 – Energy Management from Statista (2021b), Smart Thermostats are present in less than 3% of global Smart Homes, not to confuse with the 17% specific to the United States installed base.

Ideally, to enable a comprehensive Smart Energy Home scenario where energy providers and consumers cooperatively flatten and reduce the demand in the most cost and energy-efficient way benefiting both sides of the equation, it would be required that manufacturers, governments, cities, and consumers build a somewhat utopic scenario given the current deployment.

Utopic by making an analogy to the automobile that was first invented and perfected in Germany and France in the late 1800s (Onion et al. 2018), but still these days has not reached, or become financially accessible, to all consumers worldwide. It is also an industry that had not agreed and converged to a single fuel source, to mention a few, gasoline, ethanol, electricity, diesel, propane, natural gas, biodiesel, and hydrogen now also showing up as a possible future. This reference for comparison is quite a stretch, but it shows how difficult it is to scale up initiatives that cross borders both commercially and geographically.

The Smart Home forecast presented in Figure 2.3 indicates a penetration rate forecast in the world reaching 21.09% in 2025, indicating that the years ahead will just scratch the surface in terms of installed base of smart thermostats, smart appliances, smart lightbulbs, and other devices needed to drive holistic Smart Energy Home, or even city-wide Smart Energy ecosystems. One should not expect complex city-wide Smart Energy uses cases including Smart Energy Homes to be broadly deployed anytime soon. Smart cities will indeed enable that in a select few cities at a time. But the coordination of Energy Availability, Demand Response, and inter-appliance collaboration extending beyond the house in a seamless way to optimize consumption and flatten the demand is a long way ahead in terms of high-volume deployment.

Before presenting more statistics and projections, let us depict an ideal Smart Energy Home scenario.

- ➢ A multitude of Smart Homes is connected to a smart grid and able to cooperate for energy cost and demand optimization.
- ➢ A smart energy grid provides energy and data related to cost and availability. Including historical data, current status, and forecasts. Also capable of receiving back into the grid energy produced by individual houses.
- ➢ The Smart Homes having smart energy meters capable of receiving and interpreting incoming energy data.
- ➢ Smart Homes having Energy Management System, acting as a central control to manage energy use cases within the house. (e.g., Demand Control, energy allocation, priority requests, etc.)
- ➢ Smart Homes have broadband internet access.
- ➢ Smart Homes have a Home Area Network (HAN) to enable connectivity to all devices.
- ➢ Smart Homes have a Smart Home Ecosystem capable of interoperability across all its devices.
- ➢ Said Ecosystem has self-healing capabilities to recover from power outages and other adversities.
- ➢ Smart Homes have means for energy storage (e.g. battery banks).
- ➢ Smart Homes potentially have local power generation (e.g. Solar panels).
- ➢ Smart Homes have bidirectional power capability (e.g. for regular consumption, to store and provide locally generated energy back to the grid).

- Consumers have an EV capable of bidirectional energy flow. Either charging at home or bringing energy back home as needed and acceptable based on customer's transportation and energy needs.
- Smart Homes' internal network (e.g. wired or wireless) and Energy Management System have means to provide device–device connectivity and coordination even in case the internet is not available. In other words, the system is robust against internet and cloud systems being unavailable.
- Smart Homes have HVAC systems controlled via smart thermostat and are capable of thermal energy recovery, storage, or repurposing (e.g. laundry dryer warm air repurposed for other applications).
- Smart Homes have home appliances with demand coordination capability, therefore enabling cycles delay, concurrent activation avoidance, etc. (e.g. avoid motors starting at the same time for washer and dryer, avoiding refrigerator compressor starting up at the same time the dishwasher heater is active). Said appliances also have a multitude of sensors to provide data to Energy Management System, such as user's presence, home temperature and other variables.
- Smart Homes have entertainment systems capable of cooperating with the Smart Energy House to aggressively optimize energy savings based on house and user behavior. (e.g., Smart Home detects user left the living room to go to the kitchen, therefore just listening to the TV, not watching it, and the TV display backlight brightness can be set to zero.).
- Smart Homes have devices to improve natural light and thermal usage, such as connected blinds, smart air vents, and so on.

A visual representation of this scenario is shown in Figure 2.9. As one can realize through this figure and description provided, there is a long way to go to develop Smart Cities with infrastructure and Smart houses as advanced as that.

Although this list may sound like news to professionals starting into the Smart Energy Home research and development, a good part of it is not. A portion of this list is already known since Title XIII – Smart Grid Sec. 1301–1308, Statement of Policy on Modernization of Electricity Grid was published by the Department of Energy (DoE) Office of Electricity in 2011 (DoE 2011).

The Smart Grid initiative in the US dates back more than 10 years and has documented most elements required for full Smart Energy Home operation and automation. However, since the Smart Grid initiative started driving the installation of Smart Energy Meters, or Advanced Metering Infrastructure (AMI) as referred by the DoE, it has succeeded, but partially.

The smart meters deployed in the field today in the United States use two-way communication, but data itself flows just one way, from the home to the energy provider. Data regarding demand, price, system capacity, etc., is not shared back

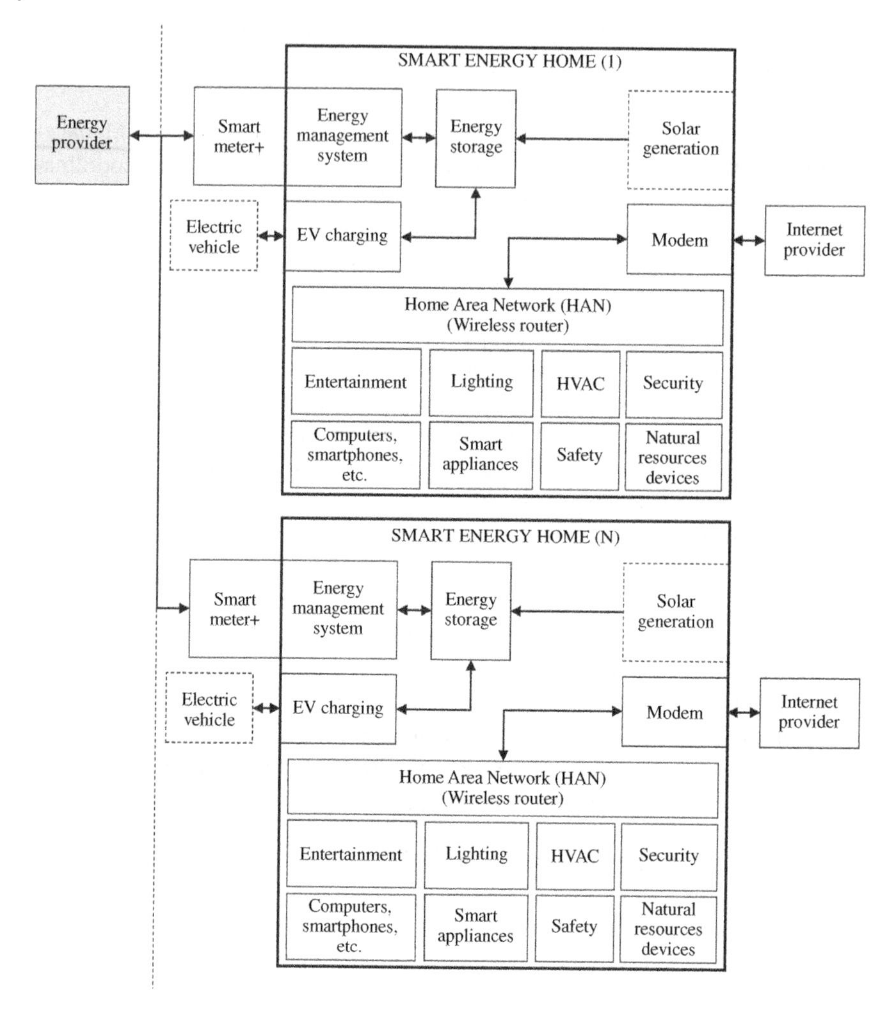

Figure 2.9 Smart Energy Home ecosystem showing extended scenario. Source: Neomar Giacomini (co-author) (2022).

to the house through the meter directly, therefore the house cannot rely on it as a source of data for energy use cases, but there are other solutions for that. These meters can also receive commands to disconnect or reconnect service of a unit in case of lack of payment for instance.

The solution for the lack of direct data sharing from the smart meter to the house is by closing the loop with the energy provider using the internet. The energy provider captures data from all relevant Smart Meters, compile such data to create the relevant indicators for each area, and makes this data available in the cloud, where through regular internet access and Application Programming Interfaces (API) the Smart Energy Homes can get the data and make operational decisions.

The US Energy Information Administration publishes every year the Electric Power Annual report (EIA 2021b) containing a great amount of information. Although being focused only on the US territory, this data is presented to exemplify that Smart Energy Homes initiatives are heavily dependent on the infrastructure available outside the house, which is not yet fully deployed yet. Figure 2.10 shows the data provided regarding metering penetration by technology type in the United States. As of 2020, 65% of the US households were equipped with smart meters. The trend in the past five years is an increase of 4–5% per year.

While looking at this data, it is also important to consider whether that 65% is evenly spread across residential neighborhoods, cities and states, or unevenly scattered providing no statistically significant coverage for certain areas to consider energy collaboration at the moment.

Figure 2.11 shows the percentage of Smart Meters in residential units per state in the United States, this data is from the 2020 Annual Electric Power Industry Report from the U.S.A. Energy Information Administration (2021a). Washington, DC is at 100% coverage through smart meters but relates to just about 300,000 units, while New York with 7,253,485 units is at just 13% coverage, but adds up to nearly a million units.

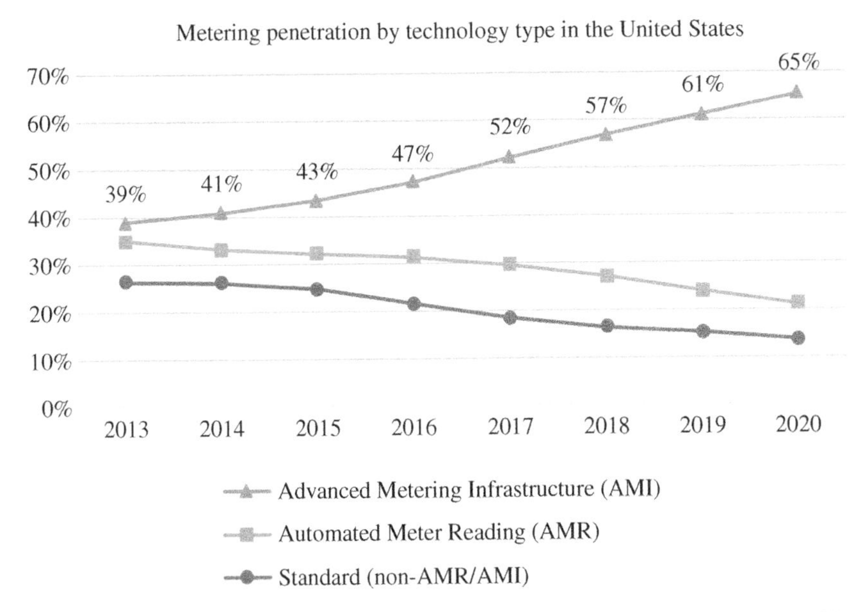

Figure 2.10 Metering penetration by technology type in the United States. Source: EIA (2021a).

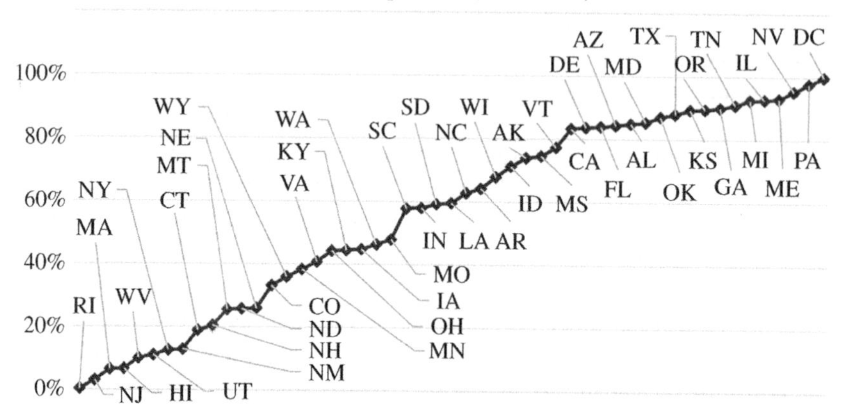

Figure 2.11 Percentage of Advanced Metering Infrastructure (AMI) per state as of December 2020. Source: EIA (2021a).

This breakdown intends to reinforce that the space of inference for new Smart Energy Home initiatives must be focused on where efforts are coming from all fronts including energy providers, industry, government, and consumers. Planned Smart Cities are a great example of that.

Moving inside the house, the first important pieces to analyze the maturity and penetration ratio are the Energy Management System and Energy Storage.

As of 2022 the Energy Management Systems are in majority available only in houses with local generation, solar being the main option for consumers opting for that feature. Such controllers are also in majority capable of bidirectional power exchange, meaning they have the capability to receive energy and/or provide it back to the grid.

Such bidirectional approach is a feature also published in Title XIII – Smart Grid Sec. 1301–1308, Statement of Policy on Modernization of Electricity Grid was published by the Department of Energy Office of Electricity in 2011 (DoE 2011), where it reads "The ability to develop, store, send and receive digital information concerning electricity use, costs, prices, time of use, nature of use, storage, or other information relevant to device, grid, or utility operations to or from a computer or other control device," but note it mentions bidirectional data as well which currently is quite limited on what the house receives from the grid. The same publication also mentions EVs in this context of storage as well, "Deployment and integration of advanced electricity storage and peak-shaving technologies, including plug-in electric and hybrid electric vehicles, and thermal storage air conditioning."

Unfortunately, there is still one issue impacting the deployment of technologies like solar panels, which if solved would drive broader availability of storage banks and controllers. Currently, the appeal for the consumer is still more on ecological than financial terms. Even in the United States after a decade of Smart Grid push and multiple financial incentives for Solar Energy implementation, only a few locations in the country result in a profitable investment for the consumers (Rogers et al. 2019).

On top of that, the residential energy storage attachment rate is very low (e.g. battery backup for the home), in the United States only 8.1% of the residential units with solar panels in 2020 had energy storage means (Barbose et al. 2021). The same study also shows that the average US residential solar installation is 6.5 kW, but it can vary widely from roughly 4–10 kW. The importance of this data, regarding any installation whether in the Unted States or any other country, is that incentives for the addition of energy storage should be increased, otherwise solar-powered homes will only be able to collaborate in terms of energy during daytime when generating in excess, and the installed capacity planned per unit will likely be specific to its power needs. The result at best would be the unit working on a direct revenue or credit basis (e.g. provide energy during daytime to recollect at nighttime, or get discount on a future date), but not storing energy to support the grid during high demand including nighttime, or emergencies.

From Energy Management System perspective, the current units are focused on energy management at house level only. However, data availability to enable Smart Energy use cases across all energy-related devices and appliances inside the house is still a major gap across the industry.

The development of devices capable of full interoperability depends on vendors collaborating to standardize data, open interfaces and use cases, both at Energy Management System level and appliances inside the house. This scenario will require all vendors to open information to a certain extent to assure that energy consumption, production and management in a structured manner can be done across the entire Smart Energy Home and beyond, such as in the case of a smart city.

To enable that level of interoperability, the European Telecommunications Standards Institute (ETSI) and European Standards Organization (ESO) that supports the timely development, ratification, and testing of globally applicable standards for Information and Communication Technologies (ICT), has developed the Smart Applications REFerence (SAREF).

The SAREF ontology is a shared model of consensus that facilitates the matching of existing assets in the smart applications domain and has its own dedicated chapter in this book. Chapter 7 refers to the SAREF ontology and its relationship with existing IoT ontologies focusing on the Smart Home and Smart Appliances domains.

Although initiatives such as SAREF provide a great foundation for vendors to build products enabling energy cooperation, without regulatory efforts requiring its implementation to be mandatory the deployment of a common language between all devices will continue at the slow pace we currently see in the market.

Not to speculate and diverge from data and facts too much, but from a rewards perspective the famous question "What is in it for me?" is key in this scenario. Will a comprehensive Smart Energy implementation pay back to consumers and vendors enough to justify the development funding, asset costs, risks, and other factors? The benefit is clear to the energy providers, energy grid, and overall future of reliable energy, but at the end of the day it will be either mandated by regulations, or the business case will have to be attractive to all parties, otherwise, this scenario will not become a reality. Just picture the scenario of a consumer having to invest thousands of dollars in the current date and having a 5–12 years payoff to equip a Smart Energy Home with solar panels (Warren 2022), but unable to even be sure if the family will still reside in that specific house after that period? The wide range for the payoff period is due to the differences in solar exposure, local energy cost, system size, type, having energy storage is available or not, and so on.

In January 2022 according to research from Pew Research Center with over 10,000 respondents (Tyson et al. 2022), the share of homeowners who have considered installing solar panels within the last 12 months in the United States was 39%, with only 8% already having it installed.

With all that in mind, one can wonder whether the technical solutions will converge to enable a full implementation and deployment in 5, 10, or even 15 years from now. Such prediction is difficult to make in face of constrains that show up along this process.

One example of a constrain that recently showed up in the United States is the newest version of the Net Energy Metering (NEM) program in California. The state has been a successful example in recent years with incentives to get consumers installing solar power, but the new NEM 3.0 is raising big concerns.

Net metering in its standard approach pays back one-for-one for energy delivered back to the grid from residential solar panels, but that changed in 2016 when NEM 2.0 was introduced. Homeowners started receiving two cents less per kWh delivered back to the grid. The newest version of the NEM program may cut the one-for-one to a half-to-one, or even more, making the investment in solar panels even less attractive. The California NEM 3.0 rulemaking process can be followed through the California Public Utilities Commission website (CPUC 2022).

Given this scenario, at the same time the technical community craves for regulatory efforts to enforce adoption of common protocols and practices to enable Smart Energy homes, the willingness from a consumer's perspective to adopt it, and invest considerable budget in it, is being challenged.

Knowing that other major initiatives such as Smart Grid are already taking more than a decade to become a reality with a large footprint, especially when considering global deployment, if one had to predict when fully equipped self-managed

Smart Energy Homes will be implemented at scale globally, it should not be a reality is less than 10 years. Hopefully, given the increasing global energy challenges and increase in EVs deployment, the push for fully integrated Smart Energy Homes will increase.

2.5 Conclusion

These are exciting times, the technologies needed both in terms of hardware and software are out there and just need to converge to speak a common language in order to enable a promising home energy management future.

Incentives that provide financial benefits and reliable energy to the consumer, and result in commercial advantage to vendors will be key to drive momentum on the Smart Energy Homes space. Whether this is driven through regulatory efforts, governmental funding programs or any other front, that is not really a concern as long as it happens.

The digitalization that is happening all around the globe, increase in the volume of EVs, and connected home appliances even still operating isolated from Energy Management System will continue building the foundation needed for the energy scenario depicted in this chapter. A great advantage of many of these appliances, even still operating isolated from Energy Management System is that many of them have over-the-air (OTA) software update capability, which means these devices may not be compliant with a common language to enable interoperability today, such as the SAREF ontology described on Chapter 7, but they can still receive OTA updates pushed through the internet by the vendor and enable such functionalities.

Aside from the regular appliances included in Table 2.2, the biggest challenge seems to be in the Energy Management System. This device will require self-adapting capabilities to understand each different household, build a reasonable model and database of their capabilities and be able to interact with the energy grid controllers even with the houses not containing the same device types, quantities, features, and energy capabilities. The need to be capable to analyze the house digitally enabled devices and build its digital twin will be a challenging task, but necessary. As IBM defines, "A digital twin is a virtual representation of an object or system that spans its lifecycle, is updated from real-time data, and uses simulation, machine learning and reasoning to help decision-making" (IBM 2022).

In summary, the next decade will be a waiting game for the consumers, and a push–pull in the technology community to find the right middle ground where everyone operates with competitive advantage in this scenario, but sooner or later it will come to reality, and with the right push a 10-year timeframe seems reasonable.

Symbols and Abbreviations

6LowPAN	IPv6 over Low-Power Wireless Personal Area Network
AMI	Advanced Metering Infrastructure
API	Application Programming Interfaces
CABA	Continental Automated Buildings Association
CSA	Connectivity Standards Alliance
DoE	Department of Energy
EIA	Energy Information Administration
ESO	European Standards Organization
ETSI	European Telecommunications Standards Institute
EV	Electrical Vehicles
HAN	Home Area Network
HVAC	Heating, Ventilation, and Air Conditioning
ICT	Information and Communication Technologies
IFTTT	If This Than That
IoT	Internet of Things
IVA	Intelligent Virtual Assistant
OTA	Over-the-air
SAREF	Smart Applications REFerence
WHO	World Health Organization

Glossary

Communication technologies Electronics apparatus that enable different devices to exchange information.

Connected devices Products (end-user devices) that use communication technologies to maintain a link to an online service or another product.

Credentials Username and password to enable access to a service.

Congestion In communication technologies it relates to a moment in time when more data is flowing through a communication channel than it is capable of handling, therefore causing delays.

Debugging Action of monitoring data and actions to identify possible issues.

Ecosystem The macro scenario is created while all parts of a system and multiple systems are in place operating and cooperating.

Home Area Network The communications infrastructure of a house, popularly usually expressed as "the house WiFi™."

Internet of Things The interconnection via the internet of devices enabling them to send and receive data.

Mesh A mesh network is a network in which devices (nodes) are linked together and branch further using each other as communication bridges.

Metering The act of measuring energy used by a house.

Ontology A set of concepts and categories that shows their properties and relations.

Penetration rate A percentage that indicates the deployment of technology in a given region.

Provisioning The process through which a communications technology accepts a new device in the network and assigns an address, access type, and other information.

Smart Home A home which contains an IoT ecosystem.

Smart Energy Grid A Energy grid with enhanced capabilities in terms of energy measurement, distribution, and communication with the end users.

Standard A formal agreement in the industry to enable multiple companies to design towards the same set of requirements.

Virtual assistant A device/service that enables users to request/assign tasks from devices in the Smart Home Ecosystem.

Wireless Related to communication technologies that use over-the-air communication means.

References

Barbose, G., Darghouth, N., O'Shaughnessy, E., and Forrester, S. (2021). Tracking the Sun. https://emp.lbl.gov/sites/default/files/2_tracking_the_sun_2021_report.pdf (accessed 3 June 2022).

CABA (2021). Smart home technology concerns that affect consumer adoption in the United States in 2021. https://www.statista.com/statistics/1253083/consumer-adoption-concerns-smart-home-technologies-in-the-united-states (accessed 25 May 2022).

Chen, B.X. (2016). Google Home vs. Amazon Echo: a face-off of smart speakers. https://www.nytimes.com/2016/11/04/technology/personaltech/google-home-vs-amazon-echo-a-face-off-of-smart-speakers.html (accessed 4 April 2022).

Congress.gov (2007). H.R.6 – Energy Independence and Security Act of 2007. https://www.congress.gov/bill/110th-congress/house-bill/6 (accessed 15 May 2022).

CPUC (2022). R2008020 – Proceeding. https://apps.cpuc.ca.gov/apex/f?p=401:59:0::::: (accessed 9 June 2022).

DoE (2011). Title XIII – Smart Grid Sec. 1301–1308 Statement of Policy on Modernization of Electricity Grid. https://www.energy.gov/oe/downloads/title-xiii-smart-grid-sec-1301-1308-statement-policy-modernization-electricity-grid (accessed 1 June 2022).

EIA (2021a). Annual Electric Power Industry Report, Form EIA-861 detailed data files. https://www.eia.gov/electricity/data/eia861 (accessed 2 June 2022).

EIA (2021b). Electric Power Annual. https://www.eia.gov/electricity/annual (accessed 1 June 2022).

Hoff, S. and Lindstrom, A. (2021). EIA – Energy Information Administration. https://www.eia.gov/todayinenergy/detail.php?id=50316 (accessed 26 May 2022).

Hu, R. (2011). The Nest learning thermostat. https://www.core77.com/posts/20860/the-nest-learning-thermostat-20860 (accessed 30 May 2022).

IBM (2022). What is a digital twin? https://www.ibm.com/topics/what-is-a-digital-twin#:~:text=A%20digital%20twin%20is%20a,reasoning%20to%20help%20decision%2Dmaking (accessed 9 June 2022).

Kennis, J. (2021). Thread, matter, and CHIP – this glossary will quickly bring you up to speed. https://www.threadgroup.org/news-events/blog/ID/287/Thread-Matter-And-CHIP--This-Glossary-Will-Quickly-Bring-You-Up-To-Speed#.YowdPqjMIuU (accessed 23 May 2022).

Lasquety-Reyes, D.J. (2021). *Number of Smart Homes Forecast in the World from 2017 to 2025*. s.l.: Statista.

Onion, A., Sullivan, M., and Mullen, M., 2018. Automobile history. https://www.history.com/topics/inventions/automobiles (accessed 26 May 2022).

OpenThread (2022). Node roles and types|OpenThread. https://openthread.io/guides/thread-primer/node-roles-and-types (accessed 23 May 2022).

Rogers, M., Kane, S., and Finkelstein, J. (2019). How residential energy storage could help support the power grid. https://www.mckinsey.com/industries/electric-power-and-natural-gas/our-insights/how-residential-energy-storage-could-help-support-the-power-grid (accessed 3 June 2022).

Statista (2021a). Smart Home penetration rate forecast in the world 2017–2025. https://www.statista.com/forecasts/887636/penetration-rate-of-smart-homes-in-the-world (accessed 25 May 2022).

Statista (2021b). Smart Home report 2021 – energy management. https://www.statista.com/study/36297/smart-home-report-energy-management (accessed 26 May 2022).

Statista (2021c). Smart Home report 2021 – smart appliances. https://www.statista.com/study/50587/smart-home-report-smart-appliances (accessed 26 May 2022).

Statista (2021d). Which smart home devices does your household own? https://www.statista.com/statistics/1124290/smart-home-device-ownership-us (accessed 25 May 2022).

Tyson, A., Funk, C., and Kennedy, B. (2022). Americans largely favor U.S. taking steps to become carbon neutral by 2050. https://www.pewresearch.org/science/

wp-content/uploads/sites/16/2022/02/PS_2022.03.01_carbon-neutral-2050_REPORT.pdf (accessed 8 June 2022).

U.N. (2020). Coronavirus Outbreak (COVID-19): WHO update (11 March 2020). https://www.youtube.com/watch?v=Btlzrwl9Lcw (accessed 4 April 2022).

Vailshery, L.S. (2020). *Number of Internet of Things (IoT) Connected Devices Worldwide from 2019 to 2030, ByUuse Case*. s.l.: Statista.

Warren, T. (2022). The average payback period for home solar panels in the US. https://www.energybot.com/blog/payback-period-for-solar-panels.html (accessed 9 June 2022).

Welch, C. (2014). Amazon just surprised everyone with a crazy speaker that talks to you. https://www.theverge.com/2014/11/6/7167793/amazon-echo-speaker-announced (accessed 4 April 2022).

3

Household Energy Demand Management

*Esther Palomar, Ignacio Bravo, and Carlos Cruz**

Department of Electronics, University of Alcalá, 28871, Madrid, Spain

After reading this chapter you should be able to:

- Understand the demand-side management (DSM) umbrella term for the current energy efficiency of the demand side.
- Identify the technical components and applications related to the organization and management of energy demand at home.
- Understand the key elements for the development of demand response (DR) programs to address energy efficiency challenges.

3.1 Introduction

Energy efficiency is emerging as a key element in tackling climate change and reducing our carbon footprint (Iralde et al., 2021). Towards the transition to a more sustainable energy-efficient reality, government, research institutions, and industry are orchestrating a number of measures that could inspire consumers' lifestyle choices such as, for instance, the use of hybrid/electric vehicles, food from sustainable agriculture and/or energy-efficient household's tools and appliances. Public financial and funding instruments are also providing programmed support, e.g. equipment retrofitting, building insulation, or renewable generation for self-consumption that can contribute to significantly mitigate greenhouse emissions and foster energy communities (Caramizaru and Uihlein, 2020; Gruber et al., 2021). Moreover, smart appliances and other connected devices are raising

*Corresponding author email: carlos.crut@uah.es

Energy Smart Appliances: Applications, Methodologies, and Challenges,
First Edition. Edited by Antonio Moreno-Munoz and Neomar Giacomini.

expectations for the potential of Information and Communication Technologies (ICTs) in deploying Demand-Side Management (DSM) and Demand Response (DR) programs to reduce demand peaks and induce a behavioral change in consumers (Wilhite et al., 2000).

DSM and DR are considered as reference tools for encouraging reductions in energy use and resource optimization (Hu et al., 2013; Rajasekhar et al., 2019; Chen et al., 2021). Due to technology advancement of smart grids such as intelligent energy management systems (EMSs), Advanced metering infrastructure (AMI), and information technologies compatible with smart grids, power system operators in several developed countries have adopted DR strategies and developed a variety of technical instruments (Waseem et al., 2020; Sezgen et al., 2007; Siano, 2014; Pfeifer et al., 2018). For instance, Waseem et al. (2020) present an exhaustive analysis of the methods and strategies for air conditioners' load control and the implications of implementing DR programs over their total energy utilization in residential buildings. Simulated results revealed obvious effects in load control during peak demand hours at several locations, e.g. reducing about 9% of the total peak demand under certain situations in Australia and United States where a price reward strategy is deployed and customers are provided with smart devices to control ACs events and usage hours.

Technology has a significant role to play in providing an effective market framework for consumer engagement in DR programs. When deploying price-based strategies (also known as indirect load control), consumers have to shift their 'controllable' loads (swimming pool pump, electric water heater, dishwasher, clothes washer, dryer, AC, and electric vehicle) to off-peak periods for the reduction of electricity bill, while it also provides a reduction in peak demand of the total system (Vivekananthan et al., 2014). Therefore, customers must have access to simple methods of forecasting prices. To encounter this challenge, technology can have a role in facilitating an efficient response to DR signals. From the perspective of consumers, this can be achieved by i) providing information about the use of electricity at a particular point in time, ii) providing information about the cost/price of electricity at a particular point in time, and iii) assisting consumers to respond to price signals (or any other type of event). For the latter, home automation technologies together with Home Energy Management Systems (HEMS) help consumers to assess the information and make an informed decision, i.e. load curtailment, load shifting, and fuel substitution. Such automation can be undertaken based on the preferences of consumers, while it avoids the need for constant decision-making by consumers (Sezgen et al., 2007).

This chapter overviews the current landscape of the technical characteristics of the home automation environment that allow consumers to engage in DSM and DR programs and systems. The focus is on describing the main communication infrastructures and protocols as well as the software and hardware platforms used for different smart grid and Internet-of-Things (IoT) applications at household level. Cost-effectiveness, compatibility, scalability, security, configuration, and lightweight design properties are reviewed within the most relevant demonstrable pilots and marketable solutions deploying DSM and DR programs. Energy preservation and its benefits from both individual and aggregated approaches to DSM are also analyzed among the clean energy initiatives (Abi Ghanem and Crosbie, 2021).

3.2 Technical Opportunities and Challenges for DSM

Under the IoT umbrella concept, research and development conducted on smart homes have raised a variety of opportunities to increase energy efficiency at home by extending the network assets as a cyber-physical smart system and IoT-enabled sensor data (Li et al., 2011). Appliances and user devices can be accessed, managed, and controlled by autonomous hardware platforms according to a day-ahead schedule or by receiving commands from anywhere in the world. Moreover, web-based and mobile-based technologies have gained their importance in this cutting-edge technology as enablers of user-friendly applications (Lobaccaro et al., 2016).

Indeed, smart appliances and connected devices are increasingly offering benefits in terms of energy and bill savings. However, to find a single application that helps consumers to configure, manage, and monitor all these devices and so their information is not an easy task (Dusparic et al., 2017). One of the main challenges relates to the number of different standards and communication protocols that are being used for device and sensor communication. Scalability is also at stake (Plantevin et al., 2017). This section outlines the advances on software, hardware, and communication networks to counter these challenges.

3.2.1 Software Solutions

Consumer electronic device control, energy management, and efficiency home automation have become a reality through a number of industry and research projects, solutions and services proposed (Han and Lim, 2010). Consumers nowadays are capable of obtaining real-time appliance-specific breakdown of energy consumption by deploying sensors (e.g. smart plugs) that monitor the consumption of each appliance. Though deploying such a sensing infrastructure could be costly and not that easy to configure in some cases (Li et al., 2018), the home energy management system market size is experiencing healthy growth and a broad market of technologies and services that consumers use to better manage and control their home energy consumption and production is emerging as suitable for any budget (Mahmoudi et al., 2017).

Appliance load monitoring (ALM) and remote control are popular services within HEM that provide adaptive information and applications to consumers anywhere anytime using different devices. For instance, consumers can obtain and visualize appliance-specific energy consumption statistics that can further be used to devise load scheduling strategies for optimal energy utilization (Zoha et al., 2012). There are two major approaches to ALM, namely intrusive load monitoring (ILM) and non-intrusive load monitoring (NILM). ILM method is more accurate in measuring appliance-specific energy consumption compared with NILM since it requires one or more sensors per appliance, whereas NILM requires only a single meter per house or a building that is to be monitored.

High costs and multiple sensor configurations, installation complexity as well as the large-scale smart meter deployments worldwide favored the use of NILM, especially for the case of large-scale deployments. However, the complex task of energy disaggregation is then needed to estimate the energy consumed by every individual appliance in a house from a single energy measurement device like a smart meter (Tabatabaei et al., 2016).

The emergence of low-cost platforms has fostered the proliferation of smart home energy controllers and application programming interface (API) (Mhanna et al., 2016). Smart energy controllers or HEMS are the technologies that can respond to altered conditions independently with minimum human intervention. HEMS can shift or curtail the demand in response to the electricity price and/or according to the resident comfort to optimize the electricity consumption at peak hours.

Home energy controllers can be developed through open source and locally hosted platforms such as *Mozilla WebThings* (Stark et al., 2020) or *OpenHAB* (Parocha and Macabebe, 2019) amongst others. For instance, *Home Assistant* (Ali et al., 2020) offers reliable, secure, and flexible HEMS on a dedicated board to control and monitor home devices through an app. The system shows at a glance how much it is consuming, producing and where this energy is going. Automation rules can be triggered by a variety of events such as when a specific message is received on given Message Queuing Telemetry Transport (MQTT[1]) topic. This feature is particularly interesting when participating in automated DR (Jia et al., 2018; Jahić et al., 2021) where an automation DR server facilitates the consumer response to a certain strategy like the real-time price, the proximity to periods of high demand, or the availability of supply from local renewables. In fact, scheduling optimization in HEMS is an NP-hard problem involving a large number of variables and constraints, i.e. the arbitrary dynamics of renewable energy, consumer demand, consumer behavior, and electricity price.

---

1 MQTT is a popular IoT messaging protocol, offering publish-subscribe, one-to-many messaging mechanisms across standard Internet connections (see Section 3.2.3).

Automated DR strategies and algorithms have been mostly addressed by machine learning and optimization scheduling techniques including artificial neural network (ANN) control, fuzzy logic (FL) control, control by genetic algorithms (GAs), and other evolutionary methods discussed in Logenthiran et al. (2012) amongst other meta-heuristics such as particle swarm optimization (PSO), linear optimization and/or Markov decision process (MDP) (Tan et al., 2020). The optimized schedule is devoted to set an optimal policy of the set of appliances to be controlled according to a given objective, e.g. bill reduction so minimizing the total cost of energy consumption at the consumer's house. Laboratory testbeds and living-lab pilots have demonstrated that usage awareness and scheduling optimization alone have the potential to reduce consumption by 15% in private households (Cabras et al., 2015).

Aggregation algorithms emerged aiming at ensuring the optimal utilization of energy resources and the stability of the system (Cruz et al., 2019). Andruszkiewicz et al. (2021) proving that large residential consumers aggregated by tariff incentives may have a significant impact on the power system's load. The roles and business models that are taken on by aggregators are quite varied, e.g. the energy provider, retailer, or the balance responsible party. Generally, the aggregator represents the energy-sharing coordinator of various technological components in the energy community aiming at balancing generation and load as well as at optimizing some goal like minimal price or maximal self-consumption. This is often done with the help of demand-side management, day-ahead operating scheduling, and/or flexibility transaction (Good and Mancarella, 2017).

Scalability and interoperability issues are trying to be solved through technical specifications in consensus and, for instance, the Smart Appliance REFerence ontology (SAREF) (Daniele et al., 2015) that seeks semantic interoperability of smart devices. Most of the aforementioned application-layer APIs allow IoT service developments and assist in embedding the system nodes with minimal effort as well as hiding network-layer functions (Koshizuka and Sakamura, 2010). Dealing with scalability, they have adopted the ubiquitous ID (uID) architecture that defines a semantic knowledge back-end database providing a unique identifier (called ucode) that is separate from network addresses. Unlike network addresses, ucodes are assured to be always unique, static, and never reused, all of which are necessary for the knowledge data using them to remain useful even after the network topology and connected object change.

3.2.2 Hardware Platforms

The aforementioned strategies or algorithms may lose efficacy in real projects due to the interdependence and interplay between the cyber system and physical system (Jia et al., 2018). Many smart devices such as smart meters, smart appliances, and smart plugs have been used to support smart buildings so as to assist homeowners to control the electrical appliances remotely from their smart phone and make better decisions about energy consumption (Ahmed et al., 2015). In addition to controllable appliances, plug-in hybrid electric vehicle (PHEV) technologies will expectedly penetrate into smart homes because of their environmental advantages. Apart from real-time monitoring of controllable loads, these devices can offer sufficiently advanced electronics on-board to support automated DR systems such as load scheduling and instantaneous load interruption (Rastegar et al., 2016). For example, washing machines, dishwashers, electric vehicle (EV) charging, electric storage heating, and AC systems among others, are loads that could be intelligently scheduled to stagger instantaneous load through the day Jahić et al. (2021).

Proposed HEMS architectures locate a central unit as a primary piece of hardware for the HEMS acting as the central point for all the communications and data transmission between energy management devices, the consumer, and household appliances (Solomita et al., 2004). In 2006, Whirlpool corporation's HEM controller (Ghent, 2006) could manage the energy usage in home-based devices which received information regarding the off-peak time slots for the day. Current research and developments are applying cost-efficient platforms for HEMS implementations paying special attention to cost-effectiveness, compatibility, and versatility (Saleem et al., 2019). Table 3.1 compiles a state-of-the-art on these developments. For instance, the emergence of inexpensive microcontrollers such as Arduino (2021) has also enabled the implementation of low-cost energy services so as to generate energy demand profiles and predictive pattern (Amer et al., 2015). Similarly, current models of the Raspberry Pi (Qureshi et al., 2017) are integrating communication protocols of interest in the smart home such as Bluetooth, WiFi™ and Zigbee (Baraka et al., 2013). The BeagleBone Black (Skeledzija et al., 2014) is another open-source hardware platform dedicated for devices' control and the Libelium Waspmote (Quintana-Suárez et al., 2017) that has built-in sensor monitors to develop remote monitoring system.

3.2.3 Communication Infrastructures

A number of architectures for HEMS have been conceptualized, designed, and built upon various networking and communication protocols along with specific hardware implementation (Mahapatra and Nayyar, 2019). Proposed architectures are generally divided into two environments, i.e. home and the outside world, which communicate through an access point, gateway, or home energy controller. This gateway, which is one of the critical components of the HEMS, is regarded as a middleware between the cloud layer and the sensor layer. It reports the data acquired from sensors to the cloud while delivering commands generated by the

Table 3.1 Hardware platforms (a benchmark study – May 2022).

Hardware	Features	Communication transceivers	Operating system	Power consumption	Strengths/weakness
Raspberry Pi 3	1.2 GHz Quad Core BCM2837 64bit CPU 1GB	4 USB, Wi-Fi, Bluetooth, optional ZigBee and Z-Wave	Raspbian Ubuntu Windows 10	1.8 W	Open-source platform; Use Python or C++;
Arduino	32 MHz Micro controller based on ATmega2560 32 kB	WiFi, Bluetooth, ZigBee, GSM	Processing-based	0.2W	Open-source platform hardware/software; High flexibility. Appliances compatibility
BeagleBone	720 MHz MR Cortex-A8 processor 512 MB	1 USB port, PLC, Bluetooth, Ethernet	Angstrom Linux	1 W	Open-source platform similar to Raspberry; Easy setting up.
RADXA	ROCK Pi 4 is a Rockchip RK3399 based SBC	WiFi, Bluetooth 5.0, USB Port, GbE LAN	Linux	2.3 W	Open-source platform; High flexibility.
Libelium Waspmote	14.7 MHz ATmega1281 28 kB	1USB, 802.15.4/ZigBee LoRaWAN,WiFi PRO GSM/GPRS,4G modules	Linux	2 W	High flexibility; Starter kit:200; ZigBee,WiFi and LoRaWAN support

(Continued)

Table 3.1 (Continued)

Hardware	Features	Communication transceivers	Operating system	Power consumption	Strengths/weakness
Xilinx Spartan	16 Mb SPI flash memory, 100 MHz	Ethernet, USB port	Linux	2 W	SH, Deep Learning, Autonomous System
PYNQ	Embedded systems Xilinx Zynq Systems on Chips (SoCs)	Bluetooth, Ethernet, USB port	Linux	2.3 W	IoT hardware development in Python
Control4Home Automation	Control4Home owners enjoy personalized smart living experiences	Bluetooth, WiFi Z-Wave and ZigBee	Licensed	–	Operation with Internet connection; Not user installation
Nexia	Smart home automation system	Z-Wave	Licensed	–	No knowledge of installation required/ Only Z-Wave support; Low compatibility
LG smart appliance	Control key features on LG smart appliances from your smartphone	WiFi	Licensed	–	No knowledge of installation required/ Only for LG appliances

cloud layer to actuators (Jia et al., 2018). Gateway, smart sockets, and sensors, which are physically connected to the smart appliances, support appliance-based metering, and direct load control. Besides, the consumer's household can be controlled from anywhere and at any time by a smartphone.

Within the Home Area Network (HAN), ZigBee and IEEE 802.15 Wireless Personal Area Network (WPAN) are generally adopted as communication methods for the smart appliances integration and data transmission (Andreadou et al., 2016). For instance, ZigBee (Bilgin and Gungor, 2012) offers an adequate communication range up to 100 m with a low data rate (up to 250 kbps) and a low power consumption. Z-Wave (Mahmood et al., 2015) is used for short-range communication due to the low communication latency in small data packets. Others like Bluetooth, WiFi, and 6LoWPAN (Leithon et al., 2020) are also widely used in HEMS over short distances (Collotta and Pau, 2015). By contrast, Power Line Communications (PLC) are commonly applied to wired smart meters for remote monitoring and load disaggregation.

For prosumers, the energy generation of solar and/or wind power system can be monitored through a renewable energy gateway (Zafar et al., 2018). The solar power system generally comprises solar panels, PLC modems, a solar inverter, and a Renewable Energy Gateway (REG). A PLC modem is deployed on the back side of each photovoltaic (PV) module. PLC modems have both sensing and communication capability. They measure the voltage, current, and temperature of the attached PV module and communicate with the REG through the DC power line. The solar inverter converts DC power to AC power; it also monitors accumulated energy and transient power and reports it to the gateway. The wind power system comprises a similar architecture. This renewable energy gateway transfers the gathered data to the home energy controller through Ethernet in most cases. Furthermore, the REG has both wired and wireless communication capabilities. The home energy controller analyses the data and produces the energy and power generation profile (Han et al., 2014).

Some architectures include aggregation logic into the neighbourhood area network (NAN), which collects data from multiple local HANs' gateways and sends it to a data concentrator. In some cases, Thread, a low-power IPv6-based mesh networking technology for IoT, provides security and reduces complexity within the NAN environment. This technology uses 6LoWPAN under the IEEE 802.15.4 wireless protocol (Aradindh et al., 2017). The use of fiber optics is also justified when a high data transmission rate is required (Shakerighadi et al., 2018). Cellular networks such as 5G/4G LTE wireless service can also be used for higher performance and speed, as well as a lower latency (Huang et al., 2012).

For instance, cellular network standards based on WiMax™ and GSM are suitable for the Wide Area Network (WAN), which connects to the service or utility provider (Saleem et al., 2019). GSM is a low-cost communication system with an excellent signal quality that is also implemented between several home energy controllers. Low Power Wide Area Network (LPWAN) and 5G protocols demonstrate high speed and responsiveness, and operate in various licensed and unlicensed frequency bands, though their application to IoT communications remain slow and other technologies seem more promising at present when several devices are involved. For example, long range (LoRa) LPWAN meets most of the IoT challenges and applications (Han et al., 2015). Table 3.2 highlights the main characteristics of the examined technologies and includes recommendations on the most appropriate areas of application.

3.2.4 Communication Protocols

Current communication architectures within the smart home could design the device interaction through (i) user datagram protocol (UDP) sockets, where there is no connection between client and server; (ii) MQTT (Light, 2017), which is created to connect devices that allow data encryption; and/or (iii) Constrained application protocol (CoAP) (CoAP, 2020) over Datagram Transport Layer Security (DTLS), which guarantees the confidentiality and integrity of the content for the data transmission. Simulated experiments such as in Venckauskas et al. (2019) and Chen and Kunz (2016) have studied and compared these protocols and others according to the restrictions given by the application and transport layer requirements (see Figure 3.1). Relevant conclusions are as follows:

- UDP Sockets: Each clientserver application operates on a channel communication based on UDP over the IEEE 802.11 standard. The channel descriptor, i.e. *socket*, indicates the communication protocol deployed, the *socket* network address, the local and remote IP network address, and the port number. Each client-server application is therefore uniquely identified. Figure 3.2 illustrates the interaction of clientserver communication. It is not necessary to establish or release the connection under UDP, since the data is directly sent indicating (within the data structure) the destination address.
- MQTT: Widely used among IoT devices due to its high portability and reduced consumption in terms of memory and power, MQTT protocol also assures secure communication with Transport Layer Security (TLS) on the port 8883. It implements a publish/subscribe communication mechanism for one-to-many data transmission as shown in Figure 3.3 and can establish three different levels of quality of service (QoS): Level 0 sends the message at most once following the message distribution flow and does not check if the message reaches its destination; level 1 sends the message at least once and guarantees that the message is received, but causes inconsistencies to the server if the message is delivered more than once; and level 2 sends the message exactly once.

Table 3.2 Communication infrastructures (benchmark study – May 2022).

Technology	Standard	Max. data rate	Frequency band	Power consumption	Transmission range	Strengths	Application areas	Encryption/ authentication
Bluetooth	IEEE802.15.1	24 Mbps (v3.0)	2.4 GHz	Low	10 m typical	Small networks Security, speed Easy access Flexibility	HAN	Challenge response scheme/ CRC32
WiFi	EEE802.11x	11,54 to 300 Mbps outdoor	2.4 GHz 5 GHz	Very high	Up to 100 m	Popular in HAN Speed, flexibility	HAN	4-Way handshake/ CRC32
Z-Wave	802.11	100Kbps	2.4GHz 868.42 MHz (EU)	Low	30 m indoor; 100 m outdoor	No interferences	HAN, NAN	AES128/32bit home I.D
ZigBee	IEEEE802.15.4	256 Kbps	2.4 GHz	Very low	10–100 m	Low cost Low consume Flexible topology	HAN,NAN	ENC-MIC-128 Encrypted key/CRC16
LPWAN	SigFox LoRaWAN NB-IoT	0.3 to 50 kbit/s per channel	915 MHz	Low	10 km in open space	Low power Low cost	NAN,WAN	Symmetric key cryptography/ AES 128b

(Continued)

Table 3.2 (Continued)

Technology	Standard	Max. data rate	Frequency band	Power consumption	Transmission range	Strengths	Application areas	Encryption/ authentication
6LoWPAN	IEEEE802.15.4	250 Kbps	2.4 GHz	Low	Up to 200 m	Low energy use	HAN, NAN	Symmetric key cryptography/ AES 128b
GSM/GPRS	ETSI GSM EN 301349 EN 301347	14.4 Kbps (GSM) 114 Kbps (GPRS)	935 MHz Europe 1800 MHz	Low	Several Km	Low cost Signal quality	HAN, NAN WAN	64 bit A5/1 encryption/ Session key generation
WLAN	IEEE 802.11	150 Mbps	2.4 GHz Europe	Low	250m	Robustness	HAN, WAN	WEP, WPA, WPA2/Open, Shared EAP
5G	5G Tech Tracker	20 Gbps	3400-3800 MHz awarding trial licenses (EU)	Very Low	46 m indoor; 92m outdoor	High speed Low latency	HAN, WAN	Symmetric key encryp- tion/Mobility management entity
3G/4G	UMTS	14.4 Mbps	450,800 MHz 1.9 GHz	Low	Up to 100 m	Fast Data Transfer	HAN,WAN	CDMA2000 /Authentica- tion and Key Agreement

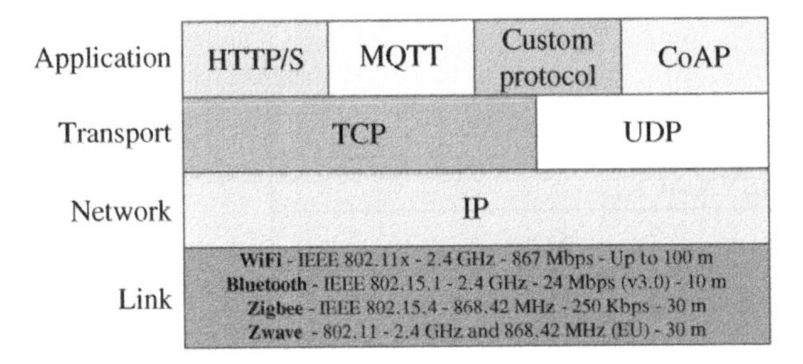

Figure 3.1 Protocols for IoT systems, adapted from Cruz et al. (2020).

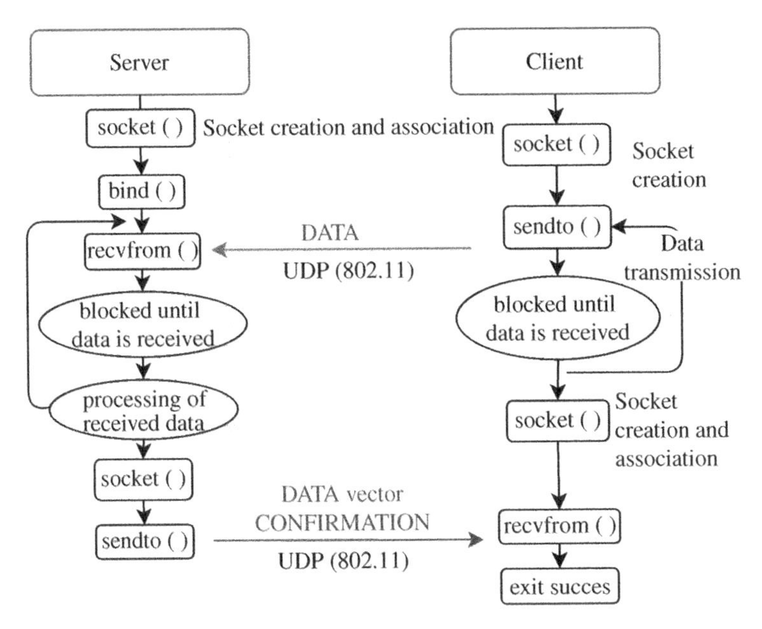

Figure 3.2 UDP communication clientserver, adapted from Cruz et al. (2020).

- CoAP: Specially designed for Machine-to-Machine (M2M) applications such as smart energy and building automation, and IoT environments (devices and networks with limitations, e.g. low power, and/or low data rate), the communication is based on UDP for reducing the communication cost, and its request/reply packet structure is made very compact to be used on low-power lossy networks like 6LoWPAN over IEEE 802.15.4 (Yashiro et al., 2013). It is based on the Representational state transfer (REST) model: Servers make resources available under a Uniform Resource Locator (URL), and clients access these resources using web transfer methods. It also provides strong

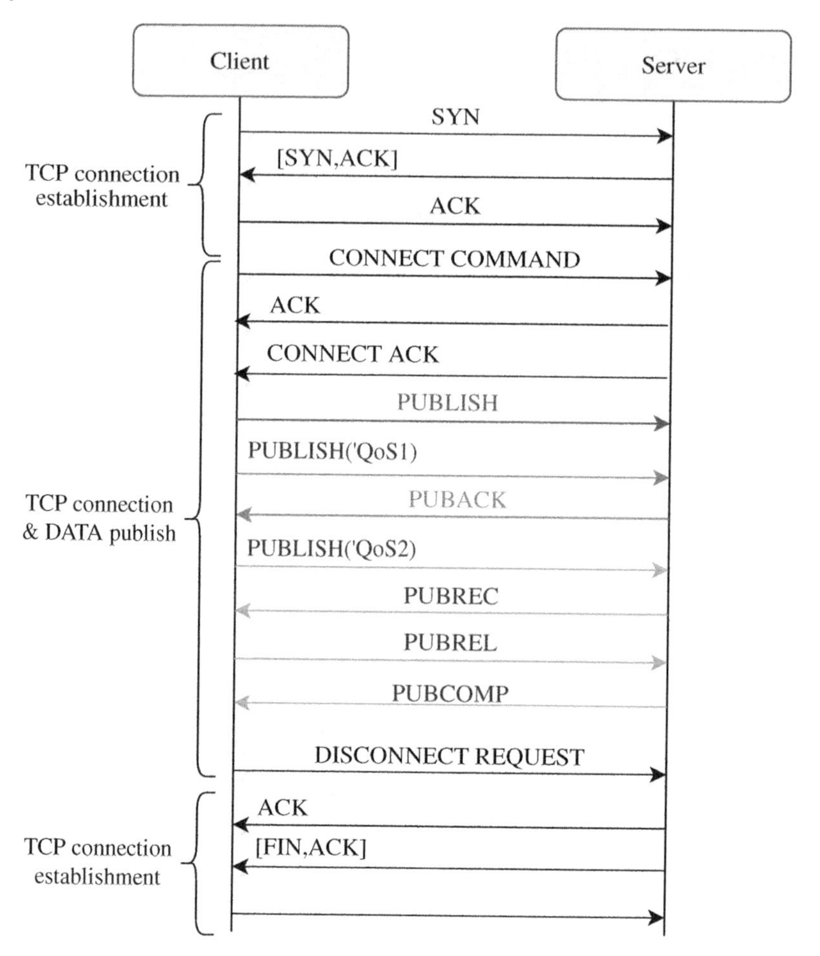

Figure 3.3 MQTT *Handshake* communication clientserver, adapted from Cruz et al. (2020).

security over DTLS: CoAP's default choice of DTLS parameters is equivalent to 3072-bit RSA keys. In particular, the implementation can be performed by DTLS v1.2 under FreeCoAP, a C-library developed for GNU/Linux devices using GnuTLS. Figure 3.4 illustrates a DTLS session procedure where both, client and server, are equipped with certificates and private keys. Experiments (Martí et al., 2019; Iglesias-Urkia et al., 2017; Kothmayr et al., 2013) show performance efficiency and interoperability as key advantages of CoAP in comparison with other protocols such as MQTT.

Further empirical analysis of these communication protocols on a DR pilot is described in Section 3.3.

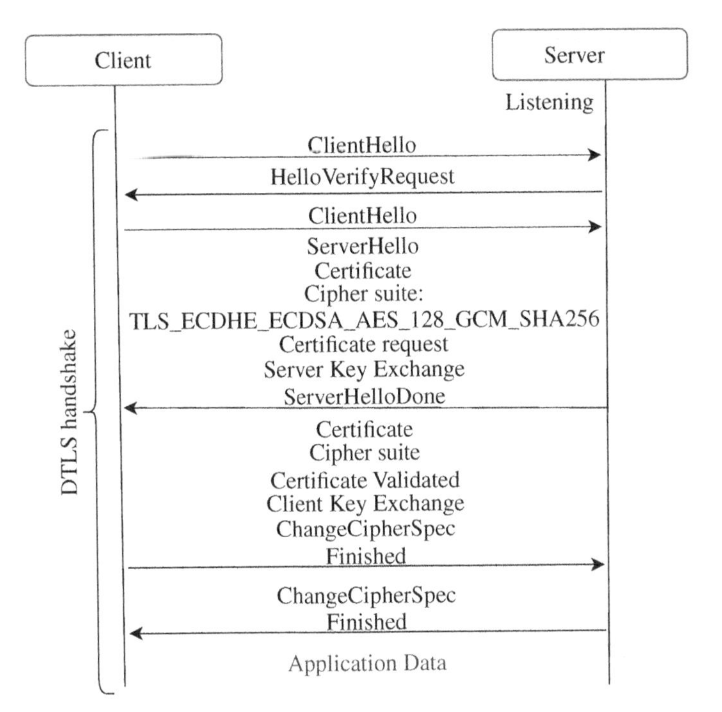

Figure 3.4 CoAP DTLS *Handshake* communication clientserver, adapted from Cruz et al. (2020).

3.2.5 Security Concerns

HEMS and DR systems are becoming part of the critical infrastructure for electric power since they can be often seen as virtual generators (Mohan and Mashima, 2014). The two-way communication capability of the smart grid and the recent cloud-based energy management services (Chen and Chang, 2015, 2016) are enabling command-response exchange between DR servers, aggregators and home energy controllers through the Internet. For instance, some cloud-based deployments take advantages of big data analytics for energy that can be leveraged for improving the performance of DR programs.

However, the 3-domain architectures for energy management that count on aggregators include additional security concerns in terms of data and communication protection as well as critical assets protection. On one hand, DR clients or home energy controllers interact with the physical electric equipment to shed load or to programmatically alter their loads. Legitimate[2] HAN devices are

2 A HAN device is considered as legitimate if the digital certificate provided by the HAN device is signed by a trusted certification authority. The digital certificate of the HAN device needs to be stored in a tamper-resistant hardware to ensure its integrity.

non-malicious devices equipped with sufficient functions for energy services and well-tested for interoperability with energy controllers, and service providers (Tanaka et al., 2012). On the other hand, electric utilities provide their customers with metering data, pricing information, and demand response signals as commercial energy services through their smart meters or dedicated APIs. Communication channels between these assets should therefore comply with cybersecurity frameworks for critical infrastructures such as NIST (Tibbals and Dolezilek, 2006) and domain-specific standards like OpenADR 2.0 (OpenADR Alliance, 2013).

Most of these security specifications and standards for DR communications require TLS with client authentication. This mechanism is an effective solution for communications security even when the deployments migrate to cloud. The use of Extensible Markup Language (XML) Signatures are also proposed for non-repudiation as an optional measure, which is an effective solution. Moreover, in non-cloud deployments, cyber vulnerability assessment would require evaluation of the deployment infrastructure of the DR provider, but migration into cloud changes the attack surface of the DR system, which requires the reconsideration of requirements for deployment of intrusion detection systems, network monitors, etc. The protection of electronic access points and security perimeters becomes the responsibility of the cloud provider.

Different types of attacks and/or risks (Park et al., 2019; Hennebert and Santos, 2014) could be mounted to assess the degree of the data, system, and communication security and privacy within the HEMS under the relevant security properties, i.e. availability, confidentiality, and integrity, as follows:

1. **Eavesdropping attacks:** The adversary tries to intercept packets, to find credentials (client IDs, usernames, or passwords), and to connect to the server in case these credentials are found. The server manages to block all anonymous connections using authentication mechanisms. When DTLS/TLS is implemented, an attacker does not listen to the transmitted messages. Message reply attacks are also prevented as the attacker is not able to replay the session with the previously exchanged information.
2. **Message modification attacks:** The adversary tries to generate packets as to raise errors on the server that affect the hardware platform performance. A secure channel prevents the modification of messages. The attacker exploits the attack, e.g., by sending manipulated preferences to the server or can also delete some of the exchanged messages if sufficient control over the network infrastructure is achieved.
3. **Spoofing attacks:** Messages exchanged between trusted peers are safe from spoofing if they are encrypted by authenticated public keys. Therefore, the attacker cannot complete a correct authentication.

4. **Attacks against the public key authentication process:** DTLS/TLS suites are negotiated between server and client and both exchange information that allows them to agree on the same secret key. For example, with RSA, the client uses the server's public key, obtained from the public key certificate, to encrypt the secret key information. The client sends the encrypted secret key information to the server. Only the server can decrypt this message because the server's private key is required for this decryption. Data transmission is then encrypted in Galois/Counter Mode (GCM) mode. The design of *Mac – then – Encrypt* is applied to authenticate the Elliptic Curve Digital Signature Algorithm (ECDSA) key exchange to prevent attacks.
5. **Ensuring data integrity:** Any modification to DTLS/TLS messages can be detected.
6. **Spoofing of IDs:** The so-called IP spoofing is a technique in which TCP/IP or UDP/IP data packets are sent with a spoofed sender address. TLS/DTLS add a cookie sent by the server as protection against spoofing. This mechanism forces the attacker/client to be able to receive the cookie.
7. **Brute-force attacks:** The adversary tries to break the password-based authentication by creating a list of words and attempts to guess valid credentials by brute force. In order to prevent password cracking, passwords and default user accounts should be disabled. Additionally, the secure communications protocol SSH must be secured by allowing the access to machines with authorized secure shell (SSH) keys only.
8. **Denial of service attacks:** The TLS stateless cookie is a protection against possible denial of service (DoS) attacks. This mechanism does not provide any defence against DoS attacks from valid IP addresses.

In case of a private server hosted in-house, the entry point for the adversary is limited to the open network ports. However, on the cloud, physical isolation of the hardware is not always guaranteed, whereas some functionality or resource could be shared with other tenants. Data provided by HEMS offers more rich information making it possible to identify consumers' usage habits and socio-economic status, appliances/devices, energy sources, and automation routines, which may not be available with other electricity meters. As a result, it provides a larger surface area for privacy invasions than data from other smart home systems and devices. Data has to be protected against other tenants, external entities, and a cloud service provider. While implementing data encryption and access control may address the first two, privacy protection against cloud service providers would require one step further. Ideally, data has to be encrypted even when it is processed on memory (Ramokapane et al., 2022). It is important therefore to help and provide consumers with an understanding of what data is collected by HEMS and how they can share data without the risk of unwanted disclosures and/or processing.

3.3 Pilots and Experimental Settings

DR technology and demand flexibility is a market still under development. Some key elements for the development of DR programs are: (i) definition of independent aggregators, (ii) participation in DR and (iii) implementation of DR. Portugal, Spain, Italy or Croatia have neither yet actively setup a DR policy nor yet adjusted their structures to allow demand-side resources to participate in the markets. The process of defining the role of an independent aggregator have not been started. On the contrary, Denmark, Germany, Austria, Finland, or Sweden have enabled DR through the energy retailer. They offer their demand-side solutions as a package with their electricity bill. Pilot implementations such as in Belgium (Dahulst et al., 2015), Shanghai (Zhang et al., 2016), or the Netherlands (Gercek et al., 2019) have measured the smart meters flexibility, the feedback of household electricity consumption, pattern recognition, and the effectiveness of price incentives. The study in Belgium comprised a total of 418 programmable appliances across 186 households. Their results showed the positive effects of incentive payments in participants offering flexibility every 40 hours. Studies (Stamminger and Anstett, 2013) in Germany focus their results on different households to monitor consumption where customers are exposed to price modification over a period of two years. Moreover, Kobus et al. (2015) found a consumption reduction in 77 households by controlling appliances operation (e.g. washing machines) that could reduce peak-time consumption (an average of 48% reduction if compared to other control groups with non-smart appliances).

All these pilots and testbeds are of great importance to validate and encourage the final introduction and acceptance of HEMS technology at our homes. Reliable internal communication and proper interoperability of smart appliances are very important issues to be addressed for future large-scale deployment.

3.4 Conclusions

This chapter has outlined the technologies that can be feasibly utilized to support the adoption of demand-side management tools and demand response programs at the domestic level reviewing the technical components (hardware platforms, software tools, and communication networks) related to the organization and management of energy demand at home. The analysis of the relevant literature on Internet-of-things and cyber-physical systems application in the smart home has determined promising system and network architectures for an efficient, cost-effective, and secure deployment of automated demand response systems, which in general comprised three domains and roles: the home area network and energy controller, the service provider or aggregation server at the neighborhood/district, and the utility. For instance, data transmission between appliances has been reviewed and tested over the current marketable solutions, i.e. ZigBee (features low data rate and power consumption), Z-Wave (used for short-range communication due to its low latency of small data packet communication in scalable environments), and WiFi (which has a high data rate and long range). Protocols such as 3G/5G or LPWAN represent the best strategy to configure the connection between the most promising DR service provider business models and the local energy service substations. Pilots also evidence the suitability of cost-effective platforms such as Raspberry Pi boards for demand and consumption control and monitoring at home. Security and privacy are also validated on a series of network configuration scenarios as to envision the security challenges to consider when migrating a DR system model into the cloud.

Symbols and Abbreviations

AC	air conditioning
AC	power alternating current power
AMI	advanced metering infrastructure
ALM	appliance load monitoring
ANN	artificial neural network
CoAP	constrained application protocol
DC	power direct current power
DoS	denial of service
DTLS	Datagram Transport Layer Security
DR	demand response
DSM	demand-side management
EMS	energy management system
EV	electric vehicle
FL	fuzzy logic
GA	genetic algorithm
HAN	home area network
HEMS	home energy management system
HTTPS	hypertext transfer protocol secure
ICT	information and communication technology
ILM	intrusive load monitoring
IoT	Internet of Things
IP	internet protocol
M2M	machine-to-machine
MQTT	Message Queuing Telemetry Transport
NAN	neighborhood area network

NILM	Non-intrusive load monitoring
PLC	Power line communications
PHEV	Plug-in hybrid electric vehicle
PSO	Particle swarm optimization
PUBACK	MQTT Publish acknowledgement
PUBREC	MQTT Publish Received
PUBREL	MQTT Publish Release
PUBCOMP	MQTT Publish Complete
QoS	quality of service
SAREF	Smart Appliance Reference Ontology
SSH	Secure SHell
SYN/ACK	synchronization and confirmation TCP messages
SoC	systems on chips
TLS	transport layer security
TCP	transmission control protocol
uID	ubiquitous ID
UDP	User Datagram Protocol
WAM	wide area network
WPAN	wireless personal area network
XML	extensible markup language
6LoWPAN	IPv6 Low Power Wide Area Network

Glossary

Keep track and create a list with all important term definitions that must be added to the Glossary. This will be included in the final version of the manuscript (end of the book) and each author is responsible for your own list.

Advanced metering infrastructure Integrated system of smart meters, data management systems, and communication networks that enable two-way communication between the utilities and the customers.

Appliance load monitoring Process for analyzing changes in the voltage and current going into a house and deducing what appliances are used in the house as well as their individual energy consumption.

Automated demand response Strategy that expands electricity consumers' energy management capabilities by participating in Demand Response programs using automated electric controls and management strategies.

Constrained application protocol Specialized Internet application protocol for constrained devices, as defined in RFC 7252.

Consumer flexibility Capacity of consuming electricity without having a loss of comfort or utility.

Datagram transport layer security Communications protocol providing security to datagram-based applications by allowing them to communicate in a way designed RFC 4347 to prevent eavesdropping, tampering, or message forgery.

Day-ahead schedule Daily programming schedule corresponding to the day after the deadline date for the reception of bids for the session.

Demand aggregation Key efficiency lever leading to higher buying power for the participants resulting in higher discounts and cost savings.

Demand response Change in the power consumption of an electric utility customer to better match the demand for power with the supply.

Demand response signal A signal sent by the local utility, Independent System Operator (ISO), or designated curtailment service provider or aggregator, to a customer, indicating a price or a request to modify electricity consumption, for a limited time period.

Demand side management Strategy used by electricity utilities to control demand by encouraging consumers to modify their level and pattern of electricity usage.

Energy community A wide range of collective energy actions that involve citizens' participation in the energy system.

Energy disaggregation The problem of separating an aggregate energy signal into the consumption of individual appliances in a household.

Energy efficiency The use of less energy to perform the same task or produce the same result.

Home energy management system A combination of its hard- and software components that work together to efficiently manage the energy usage of a home.

Internet-of-things The network of physical objects – "things" – that are embedded with sensors, software, and other technologies for the purpose of connecting and exchanging data with other devices and systems over the internet.

Load curtailment The deliberate reduction in output below what could have been produced in order to balance energy supply and demand or due to transmission constraints.

Load shifting A load management technique that aims to move demand from peak hours to off-peak hours of the day.

Machine-to-machine Direct communication between devices using any communications channel, including wired and wireless.

Message Queuing Telemetry Transport A lightweight, publish-subscribe, machine-to-machine network protocol.

Neighborhood area network An offshoot of Wi-Fi hotspots and wireless local area networks (WLAN), which enable users to connect to the Internet quickly and at very little expense.

Optimization algorithm A procedure, which is executed iteratively by comparing various solutions until an optimum or a satisfactory solution is found.

Prosumer Someone that both produces and consumes energy.

Smart appliance An appliance becomes a smart appliance when it is connected to a central system and able to be programmed or controlled remotely or operate autonomously, based on input from sensors detecting things like temperature, light levels, or activity.

Smart grid An electricity supply network that uses digital communications technology to detect and react to local changes in usage.

Smart meter An electronic device that records information such as consumption of electric energy, voltage levels, current, and power factor.

Wide area network A collection of local area networks (LANs) or other networks that communicate with one another.

Wireless local area networks A wireless computer network that links two or more devices using wireless communication to form a local area network (LAN) within a limited area such as a home, school, computer laboratory, campus, or office building.

References

Abi Ghanem, D. and Crosbie, T. (2021). The transition to clean energy: are people living in island communities ready for smart grids and demand response? *Energies* 14 (19): 6218.

Ahmed, M.S., Mohamed, A., Homod, R.Z. et al. (2015). Smart plug prototype for monitoring electrical appliances in home energy management system. *2015 IEEE Student Conferences on Research and Development (SCOReD)*, pp. 32–36. IEEE.

Ali, N., Radzi, F., Jaaafar, A. et al. (2020). Home automation monitoring system based on Internet-of-Things application. *Journal of Physics: Conference Series* 1502: 012041.

Amer, M., El-Zonkoly, A.M., Aziz, N., and M'Sirdi, N. (2015). *Smart Home Energy Management System for Peak Average Ratio Reduction*. Annals of the University of Craiova.

Andreadou, N., Guardiola, M.O., and Fulli, G. (2016). Telecommunication technologies for smart grid projects with focus on smart metering applications. *Energies* 9 (5): 375.

Andruszkiewicz, J., Lorenc, J., and Weychan, A. (2021). Price-based demand side response programs and their effectiveness on the example of tou electricity tariff for residential consumers. *Energies* 14 (2): 287.

Aradindh, J., Srevarshan, V.B., Kishore, R., and Amirthavalli, R. (2017). Home automation in IoT using 6LoWPAN. *International Journal of Advanced Computational Engineering and Networking*, p. 3. https://doi.org/2320-2106, Volume-5.

Arduino (2021). Arduino. https://www.arduino.cc (accessed 28 March 2021).

Baraka, K., Ghobril, M., Malek, S. et al. (2013). Low-cost Arduino/Android-based energy-efficient home automation system with smart task scheduling. *Proceedings of the 2013 5th International Conference on Computational Intelligence, Communication Systems and Networks*, CICSYN '13, pp. 296–301. Washington, DC, USA: IEEE Computer Society.

Bilgin, B.E. and Gungor, V.C. (2012). Performance evaluations of ZigBee in different smart grid environments. *Computer Networks* 56 (8): 2196–2205.

Cabras, M., Pilloni, V., and Atzori, L. (2015). A novel smart home energy management system: cooperative neighbourhood and adaptive renewable energy usage. *2015 IEEE International Conference on Communications (ICC)*, pp. 716–721. IEEE.

Caramizaru, A. and Uihlein, A. (2020). *Energy Communities: An Overview of Energy and Social Innovation*, vol. 30083. Publications Office of the European Union Luxembourg.

Chen, Y.-W. and Chang, J.M. (2015). EMaaS: Cloud-based energy management service for distributed renewable energy integration. *IEEE Transactions on Smart Grid* 6 (6): 2816–2824.

Chen, Y.-W. and Chang, J.M. (2016). Fair demand response with electric vehicles for the cloud based energy management service. *IEEE Transactions on Smart Grid* 9 (1): 458–468.

Chen, Y. and Kunz, T. (2016). Performance evaluation of IoT protocols under a constrained wireless access network. *2016 International Conference on Selected Topics in Mobile Wireless Networking (MoWNeT)*, pp. 1–7.

Chen, J.J., Qi, B.X., Rong, Z.K. et al. (2021). Multi-energy coordinated microgrid scheduling with integrated demand response for flexibility improvement. *Energy* 217: 119–387.

CoAP (2020). CoAP. http://coap.technology/ (last accessed 29 June 2021).

Collotta, M. and Pau, G. (2015). A solution based on bluetooth low energy for smart home energy management. *Energies* 8 (10): 11916–11938.

Cruz, C., Palomar, E., Bravo, I., and Gardel, A. (2019). Towards sustainable energy-efficient communities based on a scheduling algorithm. *Sensors* 19 (18): 3973.

Cruz, C., Palomar, E., Bravo, I., and Gardel, A. (2020). Cooperative demand response framework for a smart community targeting renewables: testbed implementation and performance evaluation. *Energies* 13 (11): 2910.

Dahulst, R., Labeeuw, W., Beusen, B. et al. (2015). Demand response flexibility and flexibility potential of residential smart appliances: experiences from large pilot test in Belgium. *Applied Energy* 155: 79–90.

Daniele, L., Hartog, F., and Roes, J. (2015). Created in close interaction with the industry: the smart appliances reference (SAREF) ontology. In: *Formal Ontologies Meet Industry. FOMI 2015 Internatinal Workshop Formal Ontologies Meet Industries, Lecture Notes in Business Information Processing*, vol. 225 (ed. R. Cuel and R. Young), 100–112. Cham: Springer.

Dusparic, I., Taylor, A., Marinescu, A. et al. (2017). Residential demand response: Experimental evaluation and comparison of self-organizing techniques. *Renewable and Sustainable Energy Reviews* 80: 1528–1536.

Gercek, C., Schram, W., Lampropoulos, I. et al. (2019). A comparison of households energy balance in residential smart grid pilots in the netherlands. *Applied Sciences* 9: 2993.

Ghent, B.A. (2006). Energy management system for an appliance. US Patent 7,110,832, 19 September 2006.

Good, N. and Mancarella, P. (2017). Flexibility in multi-energy communities with electrical and thermal storage: a stochastic, robust approach for multi-service demand response. *IEEE Transactions on Smart Grid* 10 (1): 503–513.

Gruber, L., Bachhiesl, U., and Wogrin, S. (2021). The current state of research on energy communities. *e & i Elektrotechnik und Informationstechnik* 138 (8): 515–524.

Han, D.-M. and Lim, J.-H. (2010). Design and implementation of smart home energy management systems based on ZigBee. *IEEE Transactions on Consumer Electronics* 56 (3): 1417–1425.

Han, J., Choi, C.-S., Park, W.-K. et al. (2014). Smart home energy management system including renewable energy based on ZigBee and PLC. *IEEE Transactions on Consumer Electronics* 60 (2): 198–202.

Han, S.N., Cao, Q.H., Alinia, B., and Crespi, N. (2015). Design, implementation, and evaluation of 6LoWPAN for home and building automation in the Internet of Things. *2015 IEEE/ACS 12th International Conference of Computer Systems and Applications (AICCSA)*, pp. 1–8.

Hennebert, C. and Santos, J.D. (2014). Security protocols and privacy issues into 6LoWPAN stack: a synthesis. *IEEE Internet of Things Journal* 1 (5): 384–398.

Hu, Z., Han, X., and Wen, Q. (2013). *Integrated Resource Strategic Planning and Power Demand-Side Management*, vol. 80. Springer.

Huang, J., Qian, F., Gerber, A. et al. (2012). A close examination of performance and power characteristics of 4G LTE networks. *Proceedings of the 10th International*

Conference on Mobile Systems, Applications, and Services, MobiSys'12, pp. 225–238. Low Wood Bay, Lake District, UK: Association for Computing Machinery.

Iglesias-Urkia, M., Orive, A.., and Urbieta, A. (2017). Analysis of CoAP implementations for industrial Internet of Things: a survey. *Procedia Computer Science* 109: 188–195. From Proceedings of the 8th International Conference on Ambient Systems, Networks and Technologies, ANT-2017 and the 7th International Conference on Sustainable Energy Information Technology, SEIT 2017, 16–19 May 2017, Madeira, Portugal.

Iralde, N.S.I., Pascual, J., and Salom, J. (2021). Energy retrofit of residential building clusters. A literature review of crossover recommended measures, policies instruments and allocated funds in Spain. *Energy and Buildings* 252: 111409.

Jahić, A., Hastings, J., Morrow, D.J., and Laverty, D. (2021). Hardware-in-the-loop demonstration of automated demand response for distribution networks using PMU and MQTT. *IET Smart Grid* 4 (1): 107–120.

Jia, K., Xiao, J., Fan, S., and He, G. (2018). A MQTT/MQTT-SN-based user energy management system for automated residential demand response: formal verification and cyber-physical performance evaluation. *Applied Sciences* 8 (7): 1035.

Kobus, C.B.A., Klaassen, E.A.M., Mugge, R., and Schoormans, J.P.L. (2015). A real-life assessment on the effect of smart appliances for shifting households' electricity demand. *Applied Energy* 147: 335–343.

Koshizuka, N. and Sakamura, K. (2010). Ubiquitous ID: standards for ubiquitous computing and the Internet of Things. *IEEE Pervasive Computing* 9 (4): 98–101.

Kothmayr, T., Schmitt, C., Hu, W. et al. (2013). DTLS based security and two-way authentication for the Internet of Things. *Ad Hoc Networks* 11 (8): 2710–2723.

Leithon, J., Werner, S., and Koivunen, V. (2020). Cost-aware renewable energy management: centralized vs. distributed generation. *Renewable Energy* 147: 1164–1179.

Li, X., Lu, R., Liang, X. et al. (2011). Smart community: an Internet of Things application. *IEEE Communications Magazine* 49 (11): 68–75.

Li, Y., Sheng, W., Yang, G. et al. (2018). Home assistant-based collaborative framework of multi-sensor fusion for social robot. *2018 13th World Congress on Intelligent Control and Automation (WCICA)*, pp. 401–406. IEEE.

Light, R.A. (2017). Mosquitto: server and client implementation of the MQTT protocol. *The Journal of Open Source Software* 2 (13): 265.

Lobaccaro, G., Carlucci, S., and Löfström, E. (2016). A review of systems and technologies for smart homes and smart grids. *Energies* 9 (5): 348.

Logenthiran, T., Srinivasan, D., and Shun, T.Z. (2012). Demand side management in smart grid using heuristic optimization. *IEEE Transactions on Smart Grid* 3 (3): 1244–1252. https://doi.org/10.1109/TSG.2012.2195686.

Mahapatra, B. and Nayyar, A. (2019). Home energy management system (Hems): concept, architecture, infrastructure, challenges and energy management schemes. *Energy Systems* 13: 643–669.

Mahmood, A., Javaid, N., and Razzaq, S. (2015). A review of wireless communications for smart grid. *Renewable and Sustainable Energy Reviews* 41: 248–260.

Mahmoudi, N., Heydarian-Forushani, E., Shafie-khah, M. et al. (2017). A bottom-up approach for demand response aggregators' participation in electricity markets. *Electric Power Systems Research* 143: 121–129.

Martí, M., Garcia-Rubio, C., and Campo, C. (2019). Performance evaluation of CoAP and MQTT_SN in an IoT environment. *Multidisciplinary Digital Publishing Institute Proceedings* 31: 49.

Mhanna, S., Chapman, A.C., and Verbic, G. (2016). A distributed algorithm for demand response with mixed-integer variables. *IEEE Transactions on Smart Grid* 7 (3): 1754–1755.

Mohan, A. and Mashima, D. (2014). Towards secure demand-response systems on the cloud. *2014 IEEE International Conference on Distributed Computing in Sensor Systems*, pp. 361–366. IEEE.

OpenADR Alliance (2013). OpenADR 2.0 Profile Specification, A Profile. *v. 0.9, Final Draft*, pp. 3–18.

Park, M., Oh, H., and Lee, K. (2019). Security risk measurement for information leakage in IoT-based smart homes from a situational awareness perspective. *Sensors* 19 (9): 2148.

Parocha, R.C. and Macabebe, E.Q.B. (2019). Implementation of home automation system using OpenHAB framework for heterogeneous IoT devices. *2019 IEEE International Conference on Internet of Things and Intelligence System (IoTaIS)*, pp. 67–73. IEEE.

Pfeifer, A., Dobravec, V., Pavlinek, L. et al. (2018). Integration of renewable energy and demand response technologies in interconnected energy systems. *Energy* 161: 447–455.

Plantevin, V., Bouzouane, A., and Gaboury, S. (2017). The light node communication framework: a new way to communicate inside smart homes. *Sensors* 17 (10): 2397.

Quintana-Suárez, M.A., Sánchez-Rodríguez, D., Alonso-González, I., and Alonso-Hernández, J.B. (2017). A low cost wireless acoustic sensor for ambient assisted living systems. *Applied Sciences* 7 (9): 877.

Qureshi, M.U., Girault, A., Mauger, M., and Grijalva, S. (2017). Implementation of home energy management system with optimal load scheduling based on real-time electricity pricing models. *2017 IEEE 7th International Conference on Consumer Electronics - Berlin (ICCE-Berlin)*, pp. 134–139.

Rajasekhar, B., Pindoriya, N., Tushar, W., and Yuen, C. (2019). Collaborative energy management for a residential community: a non-cooperative and evolutionary approach. *IEEE Transactions on Emerging Topics in Computational Intelligence* 3 (3): 177–192.

Ramokapane, K.M., Bird, C., Rashid, A., and Chitchyan, R. (2022). Privacy design strategies for home energy management systems (HEMS). *CHI Conference on Human Factors in Computing Systems*, pp. 1–15.

Rastegar, M., Fotuhi-Firuzabad, M., and Zareipour, H. (2016). Home energy management incorporating operational priority of appliances. *International Journal of Electrical Power & Energy Systems* 74: 286–292.

Saleem, Y., Crespi, N., Rehmani, M.H., and Copeland, R. (2019). Internet of Things-aided smart grid: technologies, architectures, applications, prototypes, and future research directions. *IEEE Access* 7: 62962–63003.

Sezgen, O., Goldman, C.A., and Krishnarao, P. (2007). Option value of electricity demand response. *Energy* 32 (2): 108–119.

Shakerighadi, B., Anvari-Moghaddam, A., Vasquez, J.C., and Guerrero, J.M. (2018). Internet of Things for modern energy systems: state-of-the-art, challenges, and open issues. *Energies* 11 (5): 1252.

Siano, P. (2014). Demand response and smart grids: a survey. *Renewable and sustainable energy reviews* 30: 461–478.

Skeledzija, N., Cesic, J., Koco, E. et al. (2014). Smart home automation system for energy-efficient housing. *2014 37th International Convention on Information and Communication Technology, Electronics and Microelectronics (MIPRO)*, pp. 166–171. IEEE.

Solomita, M., Ewald, A., Hebert, J. et al. (2004). Demand-response energy management system. US Patent Appl. 10/601,399, 19 February 2004.

Stamminger, R. and Anstett, V. (2013). The effect of variable electricity tariffs in the household on usage of household appliances. *Smart Grid and Renewable Energy* 4 (4): Article ID: 34030, 13 pages.

Stark, E., Schindler, F., Kučera, E. et al. (2020). Adapter implementation into Mozilla webthings IoT platform using Javascript. *2020 Cybernetics & Informatics (K&I)*, pp. 1–7. IEEE.

Tabatabaei, S.M., Dick, S., and Xu, W. (2016). Toward non-intrusive load monitoring via multi-label classification. *IEEE Transactions on Smart Grid* 8 (1): 26–40.

Tan, Z., Yang, S., Lin, H. et al. (2020). Multi-scenario operation optimization model for park integrated energy system based on multi-energy demand response. *Sustainable Cities and Society* 53: 101973.

Tanaka, Y., Terashima, Y., Kanda, M., and Ohba, Y. (2012). A security architecture for communication between smart meters and HAN devices. *2012 IEEE 3rd International Conference on Smart Grid Communications (SmartGridComm)*, pp. 460–464. IEEE.

Tibbals, T. and Dolezilek, D. (2006). Communications technologies and practices to satisfy NERC critical infrastructure protection (CIP). *2006 Power Systems*

Conference: Advanced Metering, Protection, Control, Communication, and Distributed Resources, pp. 454–479. IEEE.

Venckauskas, A., Morkevicius, N., Jukavicius, V. et al. (2019). An edge-fog secure self-authenticable data transfer protocol. *Sensors* 19 (16): 3612.

Vivekananthan, C., Mishra, Y., Ledwich, G., and Li, F. (2014). Demand response for residential appliances via customer reward scheme. *IEEE Transactions on Smart Grid* 5 (2): 809–820.

Waseem, M., Lin, Z., Ding, Y. et al. (2020). Technologies and practical implementations of air-conditioner based demand response. *Journal of Modern Power Systems and Clean Energy* 9 (6): 1395–1413.

Wilhite, H., Shove, E., Lutzenhiser, L., and Kempton, W. (2000). The legacy of twenty years of energy demand management: we know more about individual behaviour but next to nothing about demand. In: *Society, Behaviour, and Climate Change Mitigation, Advances in Global Change Research*, vol. 8 (ed. E. Jochem, J. Sathaye, and D. Bouille), 109–126. Dordrecht: Springer.

Yashiro, T., Kobayashi, S., Koshizuka, N., and Sakamura, K. (2013). An Internet of Things (IoT) architecture for embedded appliances. *2013 IEEE Region 10 Humanitarian Technology Conference*, pp. 314–319. IEEE.

Zafar, R., Mahmood, A., Razzaq, S. et al. (2018). Prosumer based energy management and sharing in smart grid. *Renewable and Sustainable Energy Reviews* 82: 1675–1684.

Zhang, X., Shen, J., Yang, T. et al. (2016). Smart meter and in-home display for energy savings in residential buildings: a pilot investigation in Shanghai, China. *Intelligent Buildings International* 11 (1): 4–26.

Zoha, A., Gluhak, A., Imran, M.A., and Rajasegarar, S. (2012). Non-intrusive load monitoring approaches for disaggregated energy sensing: a survey. *Sensors* 12 (12): 16838–16866.

4

Demand-Side Management and Demand Response

Neyre Tekbıyık-Ersoy

Energy Systems Engineering, Faculty of Engineering, Cyprus International University, Nicosia, North Cyprus via Mersin 10, Turkey

4.1 Introduction

Due to the continuous increase in world's population, and the adoption of the energy-intensive lifestyle, global electricity consumption has increased dramatically in the last few decades. As seen in Figure 4.1, this increase is a cumulative effect of the rise in electricity demand in almost all sectors. Figure 4.1 has been constructed by using the data provided by the International Energy Agency, IEA (2021); based on IEA data from: https://www.iea.org/data-and-statistics/charts/world-electricity-final-consumption-by-sector-1974-2019. All rights reserved; as modified by Neyre Tekbıyık-Ersoy.

Usually, supply is designed to meet the general demand requirements of an area by using historical data. However, adding too much capacity in order to guarantee a quality of service increases expenses for utility companies. On the other hand, insufficient capacity may cause supply failures which may result in a significant amount of reduction in consumer satisfaction due to blackouts. As there should always be a balance between demand and supply, the aim should be obtaining a properly sized capacity that will supply reliable power at low rates (Butler 2019). Any extra investment that will be done due to increased demand, will need to be reflected to the customers' energy bills. In order to provide a balance, it is important to understand what is contributing to the energy demand, and if possible to vary the demand according to the available resources. This can be done via Demand Response (DR) and DSM (Demand Side Management). As DSM/DR strategies are very important both for the electrical grids and for the consumers willing to achieve savings, this book chapter will be devoted to these two concepts. The book chapter will also help the readers to have an insight into the in-home systems and requirements for DSM.

Energy Smart Appliances: Applications, Methodologies, and Challenges,
First Edition. Edited by Antonio Moreno-Munoz and Neomar Giacomini.

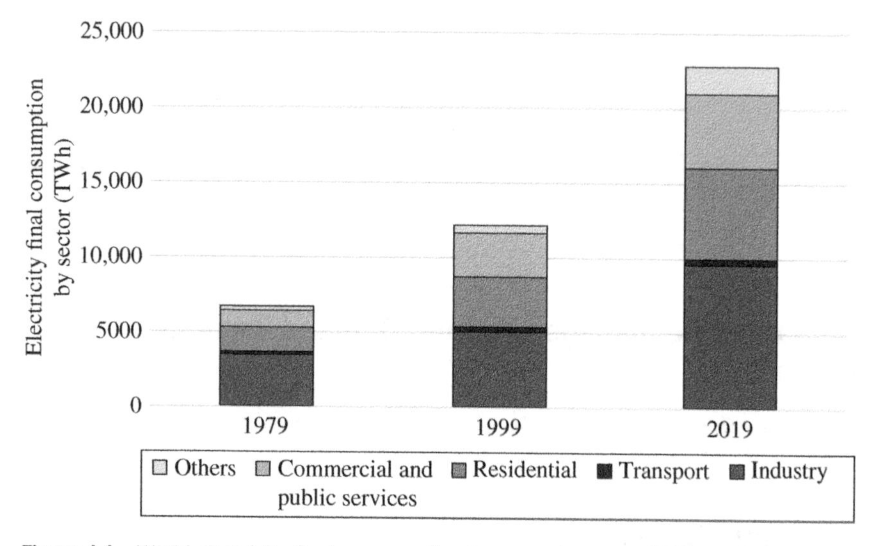

Figure 4.1 World electricity final consumption by sector between 1979 and 2019. Source: IEA (2021). Based on IEA data from: https://www.iea.org/data-and-statistics/charts/world-electricity-final-consumption-by-sector-1974-2019. All rights reserved; as modified by Neyre Tekbıyık-Ersoy.

4.2 Demand Response vs. Demand-Side Management

Although the concepts of DR and DSM are often used interchangeably, they are not the same. DR is about altering the energy demand in response to the available energy supply by financially incentivizing users to make short-term reductions in energy demand. On the other hand, DSM covers any program that encourages the end user to be more energy efficient (Fadlullah and Kato 2015). It considers the planning, implementation, and monitoring of utility activities that are designed to affect the use of electricity. DSM includes both DR and longer-term or permanent energy efficiency measures such as building automation upgrades, lighting retrofits, and Heating, Ventilation and Air Conditioning (HVAC) related improvements. Therefore, DR is just a tool for DSM, and it offers direct economic benefits. Indeed, (Milligan 2015) supports this claim by stating that it is cheaper to pay load to turn off (DR) for the 89 problem hours (1%) than to increase spinning reserve for 8760 hours/year.

4.3 The Need for Demand Response/Demand-Side Management

DSM, and more specifically DR programs are needed both for the utility companies and for the consumers. The utility companies need DSM strategies to optimize the management of the power grid. This way they can make sure that the energy consumption takes place at those times that are better for the energy producers and for the power grid. In traditional power grids, the balance is obtained by adjusting the supply and demand, by using forecasting, scheduling, and large-scale interconnections (Boudec 2019). However, especially in the last decades, the concerns about greenhouse gas emissions and depletion of fossil fuels have increased. This resulted in a shift toward renewables. But, renewables are stochastic, which means that the variability of the resource (such as wind speed, solar irradiation, etc.) has to be considered. This variability requires the grid to be more flexible and to have increased amount of regulation. DR can be used in providing that flexibility.

For example, in summer, there is more AC usage, and therefore energy consumption increases. When there is high amount of solar energy production, this may not be a problem. However, in the afternoon, solar energy generation decreases, but the peak consumption may still be there. Therefore, in order to prevent the blackouts, either expensive peak plants should be used or the peak has to be moved with the utilization of pricing signals. If the electricity price is arranged in a way that it is more expensive at peak times, then, some of the demand will be automatically shifted to another time on the same day (consumers will choose to consume energy at cheaper times), lowering the energy consumption at peak time and avoiding blackouts and brownouts. This is also beneficial for the consumers, as they will not be experiencing problems due to lack of electricity. However, this is not the only reason that the DR is needed for the consumers. In DR, the consumers are actively involved in grid operations, because they can adjust their electricity consumption during peak hours and in turn, they may benefit through financial incentives.

DR is also a critical component of the smart grid (a modernized electricity transmission and distribution network that includes two-way communication systems that deliver real-time information and enable the near-instantaneous balance of supply and demand). In conventional power grids, consumers are passive, meaning that they cannot monitor or participate in controlling their adjustable devices intelligently. However, in smart grids, consumers are more involved in many aspects of the grid. Alotaibi et al. (2020) claim that DR will have a vital role in shaping the future power grid, communication infrastructure, storage technologies, and distributed generation.

4.4 DSM Strategies

4.4.1 Energy Efficiency/Energy Conservation

Utilities aim to maximize the resources they have, and usually, energy efficiency is the first choice as it is the cheapest option. At this stage, it should be noted that energy efficiency and energy conservation are different concepts. In energy efficiency, less energy is used to provide the same service. For example, a 100 W incandescent light bulb can be replaced with an 18 W Light Emitting Diode (LED) light bulb, and the same amount of light can be obtained. This would decrease energy consumption by more than 80%. More examples could be listed such as choosing refrigerators, washing machines, etc., based on their energy efficiency classes in order to reach further energy savings. The good side of energy efficiency is that the consumers do not need to change their behavior. In energy conservation, however, the consumers should change their behavior to save energy. Examples would be changing the thermostat settings to need less AC and turning off the lights when leaving the room, etc. Energy efficiency aims to decrease the load at all times (not only during the peak). This way, more cumulative savings are expected to be achieved.

4.4.2 Peak Demand Clipping

The power grid must maintain power balance at all times. However, during some hours of the day, the load is much larger than the average. This period is called the peak period, or peak hours. The load at those hours (the peak load) should also be matched by the utility, in order to prevent a blackout. However, at those times, the transmission is highly congested and basically, they need peakers (expensive power plants as they are operated only during over-consumption hours) to generate power.

Peak clipping aims at reducing the peak loads. It is usually used in case of insufficient power capacity on the generation side. Sometimes, high incremental costs in electricity production can also make peak clipping preferable. Peak clipping can be achieved by direct or indirect reduction of the power consumption of customers during peak hours. This can be done by means of price signals or control signals. Through price signals, higher energy price, or rate, is applied during the peak hours so that users will prefer not to use electricity during that period. Another way is to automatically control the devices, by sending control signals to reduce the peak. In such a case, the load is directly controlled to reduce the stress of demand during the peak period, but this affects the satisfaction and/or comfort level of the consumers.

4.4.3 Demand Valley Filling

In valley filling, the aim is to increase the power demand of consumers in time periods where long-run incremental costs are less than the average electricity price. Increasing the demand in those times helps the related authority to avoid startup costs and ramping costs (Logan 2020). Examples of valley filling include; designing buildings with thermal storage facilities (space heating, water heating, etc.) which can generate demand during off-peak period, and benefit from special electricity rates favoring load increase during off-peak hours. This way the consumers are encouraged to switch on their water heating appliances during off-peak hours.

4.4.4 Load Shifting

Load shifting is a part of DR, and it can be referred as the combination of peak clipping and valley filling. The main difference between load shifting and peak clipping is that during load shifting, overall average energy consumption stays the same. In load shifting, the main goal is to shift the power demand of the consumers from the peak period to the off-peak period. Postponing an industrial process to another time would serve as a nice example.

According to GridBeyond (2019), shifting the load to another time causes greater returns due to DR participation. It also provides more energy cost savings when compared to the loss of production. The main question here is "when," rather than "how much." Load shifting can be obtained by using energy storage systems or by means of pricing policies. Mostly, the latter one is preferred. Such a pricing policy encourages customers to defer non-critical loads such as operating a dishwasher, washing machine, or clothes drier to off-peak hours.

4.4.5 Flexible Load Shaping

Flexible load shaping refers to more flexible and more complex modifications of the power demand patterns of the consumers. It is actually the combination of peak clipping, valley filling, and load shifting. Although flexible load shaping is more complicated for the system operators in terms of planning and operation, it allows more elasticity depending on the consumers' needs.

In this scheme, the related electricity authority decides which load pattern has to be applied. The required load pattern may be shaped based on specific targets or limits during different times of the day. Reducing demand by using this scheme allows grid operators to interrupt loads, with consequent changes in the quality of the service and reliability (Domínguez-Garabitos et al. 2022).

4.4.6 Strategic Load Growth

Strategic load growth aims at increasing the power consumption of consumers following economic development policies. It is not a part of DR, but a part of DSM. Strategic load growth is actually the opposite of energy efficiency. It may be hard to understand why the government is supporting to increase the load. However, electric cars pose a direct motivation for such a support. Some countries may want to increase the use of electric cars and the time spent on charging the cars. This way, the consumers would avoid using petroleum-derived products, such as gasoline. This will be beneficial for the environment and it will help those countries to reach their emission reduction targets.

4.5 Demand Response Programs

4.5.1 Types of Loads: Elastic vs. Non-elastic

The main idea of DR is to control the consumer side of the problem, which is mainly about the load. There are two types of loads that are important for DR; elastic loads and non-elastic loads (Boudec 2019). Non-elastic loads cannot be controlled (by utilities) without causing service disruption for the consumers. Examples of such loads are; lighting, TV, hair dryer, etc. When the consumer wants to use these loads, they should be able to use them as they are needed at specific times. For example, lights are used mostly in the evening. TV and hair dryer are usually used based on request, and therefore delaying them would not be preferable. Elastic loads, on the other hand, are the loads that can be controlled or delayed. Elastic loads can react to the signals coming from a smart grid. They can be controlled by the utility/smart grid within a predetermined range (within some customer-defined interval and by considering minimum comfort level) to achieve overall load balancing. The general idea is to shift elastic loads from high-demand (peak) period to low-demand (off-peak) period. Examples can be listed as; washing machines, dishwashers, air conditioners (AC), freezers, etc. For example, if a consumer switches off his/her AC for 15–20 minutes that would most probably do not disturb the consumer. However, it can help to relieve the stress on the grid.

4.5.2 General Approaches to Demand Response

There are two general approaches to DR; direct load control (DLC) and indirect load control (ILC). In DLC, the utility has remote access to some loads of the consumers, such as; AC, water heater, etc. The utility can remotely turn on or off the load whenever needed. This is done by using relays on the consumer side, or a digital command in case of connected smart equipment. For example, the utility may interrupt the consumer's heating systems or other systems by using those approaches. In such a case, the service may be guaranteed by the utility for a certain amount of time (e.g. 10 hours/day). However, DLC programs require special equipment and maintenance. Moreover, DLC programs are mainly preferred in the residential sector as the industrial sector needs more predictability.

Indirect load control is an alternative to DLC. This concept is based on informing the consumers about the price changes so that they would naturally try to avoid the higher price hours, rather than controlling their load directly (Mohsenian-Rad 2012). This is a successful mechanism in terms of reducing the peak load. Usually, there are two peaks throughout the day; morning peak and evening peak. According to Boudec (2019), variable prices can be offered on those periods. For example; if the average price of electricity is 8 cents/kWh, it can be set to a factor higher than this price. This does not prevent the customers from using electricity during those hours, but it will encourage them. More detailed information about different types of pricing can be found in Section 4.5.3.

4.5.3 Smart Pricing Models for DR

DR requires price-based programs that offer different electricity prices at different times. Based on pricing information, the consumers are expected to use less electricity during the times that electricity prices are high. This would automatically reduce the demand at peak hours. However, the main focus may not always be on the peak hours. In some situations, the total energy consumption during some time can also be an issue of interest. In such cases, different pricing mechanisms or models are needed. Deng et al. (2015) categorized smart pricing models into four categories; Time of Use (TOU) pricing, Critical Peak Pricing (CPP), Real-Time Pricing (RTP), and Inclining Block Rate (IBR).

In TOU, when utility customers consume energy at different time intervals of the day, they pay different electricity prices. TOU may also change from one season to another, depending on the application preferred by the electricity authority. Typically, a day is divided into three or more intervals (on-peak, mid-peak, off-peak, etc.). The on-peak period is usually the most expensive one, to serve the purpose of shifting the demand to other hours. TOU pricing usually keeps unchanged for a long period of time and it is announced far in advance. Critical peak pricing is the same as TOU, except for those times when the grid reliability is jeopardized. In such times, the normal peak price is replaced by a pre-specified higher price to reduce the energy demand faster. The main goal of CPP is to provide a balance between demand and supply. Therefore, it is employed only for a limited number of hours per year.

In real-time pricing, which is also known as dynamic pricing, the electricity price changes every hour (or more often). RTP is usually announced either one hour or one day before it is applied. According to Deng et al. (2015), RTP is considered to be one of the most economic and efficient price-based programs. Finally, inclining block rate mechanism constitutes a two-level electricity rate structure (lower and higher blocks), such that the more electricity is used by a consumer, the higher the electricity price becomes per kWh. Electricity price changes are triggered by exceeding the pre-defined thresholds (set for the user's electricity consumption over a time period). IBR helps to reduce the peak-to-average ratio (PAR) of the power grid. It is desirable to have PAR close to 1.

DR can also be applied for reactive power by utilizing customized pricing/ incentive mechanisms. It has been reported by Shigenobu et al. (2017) that providing incentive to consumers based on the reactive power consumption can improve the voltage profile of the power system. That study investigates the technical impacts of customer-side real and reactive power flow management using DR incentive strategy in the smart grid system. Similarly, Valinejad et al. (2020) consider both the active and the reactive DR and provide a new DR based on the power factor. Their study proposes a two-stage model for DSM considering polynomial and induction motor loads. According to the results of Valinejad et al. (2020), the proposed method can successfully reach an optimal trade-off between four objectives; peak demand, customer costs, voltage security, and power losses.

4.6 Smallest Communication Subsystem Enabling DSM: HAN

Home Area Network (HAN) is a very important part of the DSM, as it makes in-home communication possible. It is also known as Building Area Network (BAN). It is the smallest sub-system (the consumer-related network) in the hierarchical chain of smart grid, and it allows proactive involvement of the consumers and power users. HAN facilitates the connection of smart devices to the smart meters or controllers to implement energy management techniques, by allowing communication with other load-control devices. It also offers the consumer an interface to interact with the market.

4.6.1 General Structure

HAN consists of smart devices with sensors and actuators, a smart meter, a network router, and in-home display for energy management system (EMS) (Niyato et al. 2011). EMS is used for monitoring and controlling various electrical appliances. HAN has capability to get connected to outer world by Neighborhood Area Network (NAN) interactions. The general components of a HAN can be seen in Figure 4.2.

HAN's may use different portals/gateways, such as smart meters, HAN devices, or Utility Smart Network Access Port (U-SNAP). According to Rajawat (2019), in case of using smart meter as a gateway, NAN connects to the smart meter, and the smart meter connects to HAN. However, once installed the smart meter is not replaced for a long time, such option is not future-proof. When a HAN device is

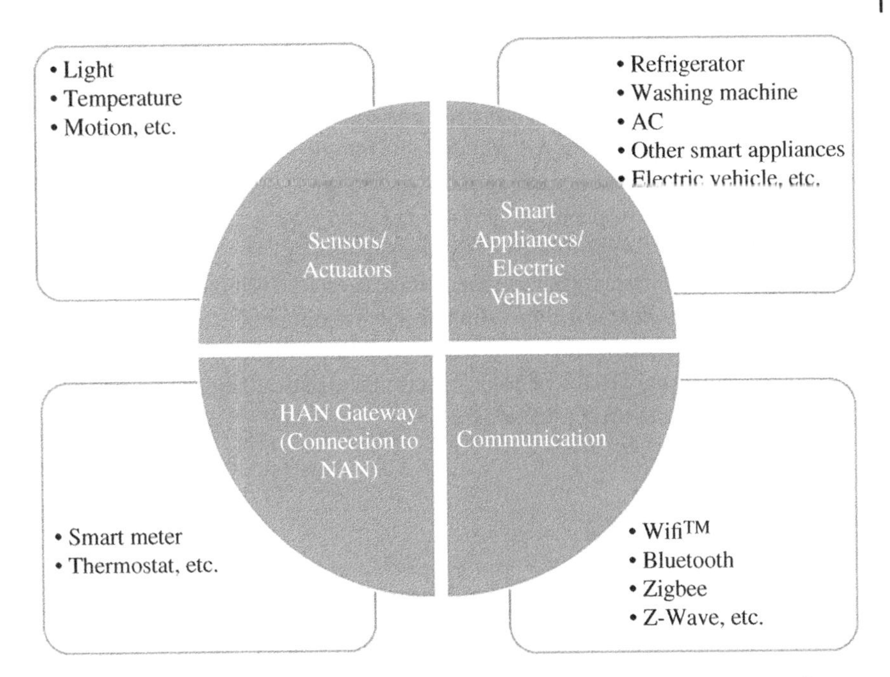

Figure 4.2 General components of a home area network. Source: Neyre Tekbiyik-Ersoy.

to be used as a gateway, a thermostat can be a good choice. In such a case, the HAN devices and the smart meter do not have to be located close to each other (they can be considerably far apart). However, when the meter serves as a gateway, in-home HAN devices should be located close to the meter due to connectivity issues. Another option is using U-SNAP. The U-SNAP standard provides interoperability between devices regardless of the communications technology and the appliance type. Thus, it creates both the ability to support any device on any network and to easily upgrade or change communications in the field by replacing the module.

4.6.2 Enabling Communication Technologies

HANs usually utilize short-range wireless technologies such as; Wi-Fi™, Bluetooth, ZigBee, and Z-Wave. But it may also encompass wired technologies such as; PLC (Power Line Communication) and Ethernet. The data rate is one of the essential criteria to be taken into account when selecting the appropriate communication technology. The maximum theoretical data rates of the above-mentioned wireless technologies are 2–600 Mbps for Wi-Fi, 721 kbps for Bluetooth, 250 kbps for Zigbee, and 40 kbps for Z-Wave (Rajawat 2019).

While the data rate differs from one technology to another, the coverage range is another essential factor to be considered. According to Niyato et al. (2011), Bluetooth has the lowest coverage range at 10 m, whereas the indoor range of Wi-Fi and Zigbee is approximately 70 m. Among these technologies, the ones allowing the maximum and the minimum number of nodes are Zigbee (with more than 64,000) and Bluetooth (with 8) respectively (Niyato et al. 2011). More detailed comparison of these technologies can be found in Niyato et al. (2011).

Wired technologies usually have higher data ranges. PLC (HomePlug) can reach up to 14–200 Mbps, whereas the maximum theoretical range of Ethernet is 10 Gbps (Rajawat 2019). Although Ethernet is faster, PLC has a higher coverage range of up to 200 m. However, wired technologies have the disadvantage of needing wire installations for setting up connections.

4.7 Smart Metering

4.7.1 Smart Meters vs. Conventional Meters

The most common types of meters are accumulation meter, interval meter, and smart meter. Accumulation meter is the oldest and most common type of meter. These meters record the total amount of energy used over a period of time, usually about a month. Due to lack of knowledge about the consumer's daily or hourly energy consumption patterns, accumulation meters enforce flat electricity rates. Also, for such meters, a meter reader must visit the property and physically read the meter, in order to determine the energy consumption and the bill amount. The physical meter reading is needed also for interval meters. However, in contrast to accumulation meters, interval meters can record energy use over short intervals, usually every 30 minutes. This makes interval meters advantageous over accumulation meters, as they can measure when and how much energy is used. This way, the utility company can offer prices and deals based on the time the consumer uses power.

Smart meters are the most advanced meters, as they can record when the consumer uses electricity as well as how much is used. Smart meters allow wireless communication between the energy provider and the meter. With this communication, connecting or disconnecting the electricity, reducing controllable loads or feeding electricity back into the grid from a renewable energy system, becomes much cheaper and easier. Smart meter also increases the amount of information that the consumer can access about his/her energy consumption. There are several differences between conventional meters and the smart meters. Some of these differences are listed in Table 4.1. In order to construct Table 4.1, the measure, recording, and communication-related information have been retrieved

Table 4.1 Comparison of smart meter to conventional meter.

Criteria	Conventional meter	Smart meter
Measure	Electricity consumption can be measured over a billing period (typically one or more months)	Electricity consumption can be measured at short intervals (typically hourly or more frequently)
Recording	Meter reading is manual. An employee of the distribution company should visit the premises to record the data	Meter reading is automatic. Meters send the data to the distribution companies through a wireless network
Billing accuracy	When the meter is behind the locked doors or inaccessible by the employee responsible for reading the meter, estimations may need to be made causing inaccurate bills	Smart meters record and send real-time data. Hence, the accuracy of the bill is higher
Time	Sending the employee to the consumer's house is time-consuming	Sending data automatically via network is very time-efficient
Communication	No communication capability	Two-way communication capability between distribution companies and the meters
Privacy	No privacy issues	Privacy concerns arise as the data readings are transmitted over wireless channels, and the energy provider has too much information about the energy consumption patterns of the consumer

Source: Adapted from Office of the Auditor General of Ontario (2014).

from Office of the Auditor General of Ontario (2014), while billing accuracy, time, and privacy details have been added for this book chapter.

As seen in Table 4.1, the smart meter is able to record more detailed information and send that information to the utility company more frequently. However, data recorded by smart meters must be detailed. Therefore, it may indirectly indicate what individual appliances a consumer is using at what time. This becomes possible via power usage to personal activity mapping. However, there are solutions to such privacy problems, such as; having local renewable energy (RE) generation, having a battery, etc. This way, the meter reading would be different. For example, as renewable power generation, such as solar power and wind power, has a strong dependence on weather conditions, the changes reflected in the meter readings could simply be due to reduced sunshine because of a cloud or shadow, or low/high wind speed observed for some time. This way, the consumers' personal activities cannot be linked to power usage (Bunder 2020).

4.7.2 What Should Consumers Know About the Advanced Metering Infrastructure

Advanced Metering Infrastructure (AMI) system supports two-way and real-time communication between utility companies and consumers. AMI includes a system that collects time-differentiated energy usage from meters. Hence, it is capable of providing information that enables utilities to provide DR products. AMI allows pricing based on total consumption, TOU pricing, critical peak pricing, and real-time pricing (King 2004). With AMI, the customers can obtain feedback via monthly bill, and can obtain detailed reports via Web or in-home displays. By using these data, consumers can choose to turn off energy-intensive appliances or they may shift the usage of some appliances to the off-peak period to provide energy and bill savings. Detailed data can also be used for personal electricity demand and bill prediction. Hence, AMI allows more consumer interaction.

In conventional power grid operating without AMI, power outages should be reported by customers, most of the time via phone calls. Otherwise, the utility company may not be aware of the problem. In these systems, the related electricity authority cannot respond to the outage until a customer calls them and warns them.

However, AMI infrastructure allows sending automatic notifications. There is an online notification system and the smart meter (the key component of AMI structure) notifies the related electricity authority automatically. Without AMI, most of the time, they do not know exactly where the outage is. So, a team needs to be sent to the area to check different lines and detect the exact location/source of the outage. But, this requires a lot of time and resources. Because of that, the energy expenses of the related electricity authority increase and this results in increased electricity prices, and hence consumer bills. AMI infrastructure with smart meters can detect where the outage is. Whenever it happens, the utility company is directly informed, and they can send the service trucks immediately to the area. This reduces the overall outage time, minimizes the interruptions to the daily routines of the consumers, and increases their comfort level.

4.8 Energy Usage Patterns of Households

DSM is about the management of consumption of loads, and in DR the users are financially incentivized to modify their consumption. Hence, the key to both DSM and DR is the electricity consumption profile of the consumers. With the emergence of smart meters and various technological devices, it became possible to obtain the hourly electricity consumption patterns of the consumers. These patterns can be used in designing better DSM and DR strategies. According to Logan (2020), the first two steps in DSM implementation are to conduct an end-user demand forecast, and to perform research on demand market (determining end-use patterns, and potential barriers), respectively. According to the data provided by the International Energy Agency (IEA 2021) (Based on IEA data from: https://www.iea.org/data-and-statistics/charts/world-electricity-final-consumption-by-sector-1974-2019. All rights reserved; as modified by Neyre Tekbıyık-Ersoy.) in 2019, the residential sector constituted approximately 26.6% of the world electricity final consumption. That indicates households constitute a considerable part of the end-use demand. Therefore, it is important to understand the energy usage patterns of households.

Among appliances available in households, the water heaters usually have high energy consumption, due to their high power rating. Usually, during the weekdays, water heating demand increases in the evening. However, this pattern may change during the weekends, as the house occupancy and the schedule changes significantly. Another device consuming a significant amount of energy is AC. Unlike water heaters, the ACs are usually operational in the afternoon in summer (in Northern Hemisphere) when the outside temperature is high and the building needs to be cooled. In winter (in Northern Hemisphere) however, AC usage can vary dramatically depending on user preferences and weather conditions. The other three obvious characteristics belong to the oven/stove, the tumble dryer, and the dishwasher. Oven usage can contribute to the peak in the morning (or before noon time) and in the afternoon around dinner time. This is due to the meal requirements of the people living in the house. Similar to the oven, tumble dryer causes routine peaks. This is because of the consumers' lifestyles. According to Timpe (2009), tumble dryer can be used before the noon time by those people who use the washing machine early in the morning (before going to work or school). However, most of the time, tumble dryer is used in the evening hours. When people come back from work, they wash their clothes and then, they want to dry them. Around the same time, the dishwasher also becomes active, usually after dinner. When people finish their meals, they load the dishwasher.

However, it should be noted that these patterns may change throughout the seasons. For example, in hot seasons, people may prefer to dry their clothes outside (or simply in a place inside their house) rather than using the tumble dryer, to reduce their energy bills. Moreover, in some countries with mild climates (where the temperatures are not so high or low throughout the year), people may have much less AC usage, as there would be less need to cool or heat their houses. Similarly, in some developed countries where many insulated houses exist,

AC usage would decrease due to reduced heat losses. When different climates are considered, it is also important to note those regions that temperatures change significantly during the day. In those regions, people may enjoy comfortable temperatures during the day, while in the evening when the temperatures drop significantly, the heating demand increases. Or contrary to this scenario, warm temperatures during the night can be followed by very hot temperatures during the day, increasing the cooling demand within the day.

Such patterns like the ones explained above should be investigated in those countries planning to apply DSM and DR strategies. Because they actually show what time of the day is the most appropriate to use load management options with smart appliances. However, in performing this load management, the shiftable and non-shiftable appliances should also be considered. There are some appliances that cannot be typically shifted. For example, when the consumer wants a hot meal around the noon time, DSM will not be able to act on that appliance. But there are some appliances that can easily be shifted, like washing machines.

4.9 Energy Consumption Scheduling

DR is heavily based on electricity prices and the consumers' reaction to these prices. The success of DR depends on many factors. One of the most important is informing consumers about the price changes, especially in indirect load control. The users can be informed via utility website, email, text message, automated voice call, a mobile app, an in-home display, indicators in the user interface and the smart meter. However, sometimes the consumers cannot properly react to smart pricing mechanisms. This can be due to the fact that there is too much information to follow, especially when more than one smart pricing mechanism is adopted at the same time. The solution can be the use of automated energy consumption scheduling (ECS), which can be a part of a smart meter or a separate device.

Usually, the consumer wants to minimize the energy cost, and the utility aims to manage the available energy with maximum profit. Therefore, the primary aim of ECS is to achieve these objectives and provide a balance between users' needs and the utility's requirements. ECS is based on optimization. The idea is to find the best operation schedule for the appliances in such a way that specific objectives will be achieved. In doing so, the constraints, which may change from one scenario, region, or user to another, are also considered. The main question is how the ECS should schedule the load based on the given price values. Mohsenian-Rad (2012) provides the general structure of an ECS problem as follows:

- Objective function (minimizing electricity cost)
  - The electricity cost (for a scheduling period) is calculated by considering the price of electricity and the energy consumption of each appliance (washing machine, dishwasher, etc.) in each time slot.

- The scheduling period considered by Mohsenian-Rad is 24 hours, and each time slot corresponds to one hour.
- The price for each time slot may change based on the smart pricing mechanism (TOU, RTP, etc.) adopted in the country of application.

- Constraints
  - The total energy consumption of an appliance between the defined beginning and ending times should be equal to the total energy needed for the operation of that appliance.
    - o The acceptable earliest beginning time (α_x) and the latest ending time (β_x) for appliance x are stated so that the scheduling of that appliance is done between α_x and β_x.
  - The energy consumption of appliance x in a time slot which is not between α_x and β_x, should be zero.
    - o Appliance should not be operational during the times not preferred by the consumer.
  - The energy consumption of an appliance in any time slot cannot be less than the minimum power level stated for that appliance and cannot be greater than the maximum power level stated for that appliance
  - The total energy consumption, due to various appliances operating at the same time, cannot exceed the predefined limit.
    - o In some cases, the total hourly load may need to be limited (mostly due to grid constraints, or the rules of the utility company). This constraint is used to model that scenario.

The stated problem formulation is very beneficial in terms of understanding the requirements of ECS. However, there are many studies considering various types of objectives, such as minimizing different types of costs, minimizing PAR, minimizing the peak load, maximizing user comfort, etc. Some studies also consider multiple objective functions and perform multi-objective optimization. The constraints also vary widely depending on the scenario considered for the study. Some studies in literature incorporate renewable energy or electric vehicles into ECS, as some households have solar or wind energy system installations and some utilize electric vehicles. However, the constraints are not limited to these. In some studies, the authors consider different scenarios in which some appliances need to operate after some other appliances (such as dryer being operated after the washing machine). This can also be arranged differently as some appliances may need to be operated at the same time (for example, TV and decoder).

4.10 Demand Response Options for Appliances

Section 4.9 has revealed the key characteristics of determining optimal appliance scheduling by shifting some appliances and changing energy consumption patterns. However, what kind of options do we have for shifting the appliances?

For how long an appliance can be shifted or interrupted? Timpe (2009) provided the results of a study giving very valuable information about smart timing of appliances and the interruption of the appliance schedule.

Let's consider shifting the time of operation of a washing machine or a dryer. People do not want their clothes to wait in the washing machine and stay wet for a long time. According to Timpe, people typically prefer a delay of less than three hours and maximum of nine hours for the washing machine and the dryer. The dishwashers, on the other hand, can typically be shifted for 6 hours, but this can be elongated until 12 hours. They may start their dishwasher in the morning before they leave the house, and then, when they come back from work in the evening, they can take the dishes out of the dishwasher.

An option to shifting the appliances can be interrupting the appliance cycles. In doing that, the technical constraints and the constraints from consumers should also be considered. For example, washing machines typically have hot cycles. In some cases, it may not be beneficial to allow the water to cool down and then try to heat it again. According to Timpe, the typical accepted interruption for the washing machine is about 10 minutes. Exceeding this duration may cause the water to cool down to such a point that re-heating would need more energy than it would require if the appliance was not interrupted. But, the dryer, on the other hand, can keep the heat for a longer time and can be interrupted for up to 30 minutes. The suggested interruption cycles for the dishwasher and other appliances are about 10 and 15 minutes respectively.

4.11 Bidirectional Effects of Demand Response

Most of the DR applications were based on load shifting, mainly moving the load from the peak times to the off-peak times. However, recently it has become more about flexible load shaping. This is basically arranging the demand in a way that the renewable energy generation (such as wind and solar) would be balanced. Hence, DR is important in balancing renewable energy generation.

4.11.1 Value of Demand Response for Balancing Renewable Energy Generation

The most important problems about renewables are their intermittency and uncertainty. This may cause the energy generation to be lower than expected sometimes, while at other times it may be more than the predicted amount. This uncertainty is reflected in the grid as an imbalance between demand and supply. For example, solar energy production increases around the noon time, however, most of the people are either at work or at school, etc., around that time.

This causes a reduction in residential energy demand. Similarly, when the wind speed considerably increases, causing wind turbines to produce more power, the generation increases. But this increase may not be in parallel with the demand at that time. Similarly, during the nighttime, solar panels cannot produce electricity, but the demand still exists. In all of these cases, as the demand cannot keep up with the pace of the increase/decrease in renewable energy generation, this causes an imbalance in the system.

Flexible load shaping can be used to maintain the balance between demand and supply, by simply increasing the demand when there is more generation and decreasing the demand when the production is insufficient. However, that is not the only way that DR helps the renewables and serves the purposes of using renewable energy. Without or with the low utilization level of DR; when the wind speed increases, some wind turbines may need to be shut down in order to preserve the demand–supply balance. However, this causes underutilization of those wind turbines, both technically and financially. In order to prevent such cases, DR can be used.

As DR balances the short-term variability caused by renewable energy, it also allows integrating more installed renewable energy capacity and diversifying the types of utilized renewable energy resources. When more renewable energy can be integrated into the system, more renewable energy support policies can be enacted (as one of the most important issues regarding high levels of renewable energy integration [mainly demand–supply imbalance] will be considerably resolved). Increase in the number and quality of renewable energy support policies are also expected to increase the interest and therefore the investment in renewable energy. This way, the costs related to renewable energy technologies are expected to reduce, which will also cause an increasing attention to renewable energy. This way, the countries with renewable energy targets can reach their targets comparably easier.

However, in order for DR to efficiently perform the balancing task, the consumers should effectively participate in the adopted DR options. If the announced pricing policies are not clear (or hard to follow), or the consumers are not willing to participate, the imbalances may continue. Hence, the consumers have a very important role in DR.

4.11.2 Value of Demand Response for Reducing Household Energy Expenses

DR enables the customers to track and adjust their energy usage in order to avoid costly peak demand charges and reduce their overall energy expenses. As seen from the ECS-related section provided earlier, the main goal of most of the optimization problems, designed for ECS, is to minimize the cost, or the electricity bill. This optimization helps the consumers in reducing their energy expenses by simply shifting the time period that they use some appliances. Most of the time, this scheduling is done by taking into account the comfort requirements of the customer, and the technical constraints about the grid and energy generation.

Especially TOU tariffs are an important enabler of DR as they incentivize the customers to shift their electricity use from high-demand periods to low-demand periods. Hence, the customer saves from energy expenses, and the power system benefits from this action.

4.12 Consumer Objections and Wishes Related to Smart Appliances and Demand Response

Consumers have a very important role in DR. The success of a DR program depends on a sufficient level of consumer participation. Hence, it is important to understand the factors that motivate or demotivate participation. Sloot et al. (2022) investigated potential motivations that underlie consumers' participation in DR programs. The authors report that they have conducted a nationally representative online questionnaire study among a commercial panel of German consumers. It was found in the study that both environmental and financial beliefs underlie participation. An interesting result of the study is that the participation was found to be generally higher when participation was described as voluntary compared to mandatory. According to the study, consumers tend to accept and participate in quota schemes when they perceive both environmental and personal benefits and low costs and risks of participating. They are also likely to participate when they feel that they are able to be flexible in their energy use.

Another study investigating the acceptance of DR noted (Tantau et al. 2021) that there is a need to improve communication between consumers and energy providers. It aims to determine the acceptance level of DR and its implementation due to the collaboration with an aggregator on the consumer side. The results of the study indicate that there is a willingness potential to implement DR programs with aggregators as intermediaries between energy providers and the consumers of electricity. The same study also reports that 81% of subjects said that they would like to participate in DR programs to contribute to the reduction of CO_2 emissions and of global warming. However, it should be noted that the results of the study are limited to three countries; Serbia, Hungary, and Romania.

Timpe (2009) provided the results of an intensive consumer research utilizing questionnaires, focus groups, and phone interviews, in terms of consumer objections/concerns and wishes. The consumers were concerned about the additional cost of smart appliances. They also did not like the idea of utility company to control if their appliances (such as washing machines or dishwashers) are running or not. They wanted to be able to override any automatic function or external control. The results also showed that there were doubts about the maturity level of the technology and skepticism about the ecological benefits that it would offer. In addition to these, concerns about health and safety issues were also reported. An appliance that is controlled automatically should deliver the service safely. Similarly, the operation of refrigerators or freezers should be kept within normal operating conditions for the intended functions.

Consumers also had some wishes related to DR. According to Timpe, these wishes are as follows: The consumers would like to have economic incentives for buying smart appliances. Enhanced safety functions are also required, such as being informed by the system in case if there is overloading on any appliance, the water is stopped in case if there is any water leakage, automatic detection of any technical faults, etc. In addition to these, any extra comfort and support given to the consumers will make DR more attractive, and therefore, it will increase the success of the DR programs.

At this stage, it should be noted that awareness is another important issue. There may be some people who are not aware of DR-related applications or even what DR is. Consumers will not participate in any program that they are not aware of, or do not have enough knowledge about. Hence, education and outreach are needed. The consumers should be informed about DSM, DR, smart appliances, and their contributions. This can be done via organizing informative seminars about smart appliances, DSM and DR, or adopting green labels so that the consumers can learn the life-cycle cost of the appliances they use. Doing so would focus on how much the appliance costs during its lifetime rather than focusing on how much they spend for buying the appliance. In order to plan these seminars or trainings, first, the awareness level of the country should be determined. This can be done by using consumer surveys or questionnaires. Children should also be educated about the need for energy savings and benefits both for themselves and for the country.

4.13 Costs and Benefits of Demand-Side Management

Logan (2020) claims that as more variable renewable energy generation (such as wind, solar, etc.) will become available, DSM will also need to support the management of distributed energy resources and the related flexibility needs. For instance, when the energy production from renewables is lower, there will be more need for thermal, nuclear, or other sources-based power generation in order to ensure the stability of the power grid. With DSM, high levels of flexibility can be achieved which in turn can prevent using additional energy sources for flexibility.

Logan states that energy efficiency programs, which constitute a part of DSM, provide more energy savings when compared to DR, but it is also noted that energy efficiency does not help the peak demand as much as DR. Logan also shows that, between 2014 and 2017, the highest energy savings in the United States have been achieved in 2017, as almost 30 TWh. When this amount is compared to the total demand of 4000 TWh, it does not seem to be high. However, it should be noted that energy efficiency offers cumulative savings. Hence, a high amount of energy savings can be achieved in the long term as energy efficiency may help avoid transmission and generation-related investments. However, it does not help the peak load considerably. DR, on the other hand, is very effective at saving peak demand but has very modest impacts on energy saving overall. Logan also states that in 2017, approximately 12 GW peak demand was saved through DR. Although when compared to 700 GW (the national peak load) it may not seem to be a considerable amount, it may still reduce the need for peaker power plants. Actually, DSM can be considered as the cheapest option to achieve a better demand profile.

Sajip (2018) reports that grid operators from around the world are creating DSM incentive programs as an alternative to expensive grid upgrades. For example, instead of adding a distribution capacity of 10 MW in an urban region, they prefer to lower down the peak demand by an equal amount. By using DSM, the costs of energy generation can be reduced, which may prevent an increase in electricity prices that should be paid by the consumers. The overall benefits of DSM can be summarized as follows:

It reduces:

- peak demand (due to DR, which is a part of DSM)
- investments related to new infrastructure
- fossil fuel and water usage (as the need for thermal and nuclear power generation decreases)
- emissions, pollution, and health-related effects (as the usage of fossil fuels decreases).

It improves:

- grid flexibility and reliability
- overall operation of electricity markets
- the competitiveness of providers due to lower energy costs
- distribution network's efficiency.

As it can be inferred from the provided information, demand-side management offers serious benefits. However, similar to every technological advancement, DSM comes with costs. Ladwig (2021) divided the costs of DSM into two groups; initialization costs and activation costs. These costs depend on several factors, such as economic situation, utilization of production, and consumer behavior.

Initialization costs consist of investment-related costs (investments related to the infrastructure of measurement, control, and communication technologies) and yearly fixed costs (related to the operation of these technologies). Activation costs, on the other hand, occur immediately when the electricity demand increases or decreases, as they are related to the activation of power plants. Activation costs mainly differ from one DSM category to another one (such as load shedding, load shifting, etc.). In difference to that, investment costs strongly vary among different sectors. The installation and the activation cost comparisons (with competing flexibility options) of DSM are presented in Ladwig (2021). According to the results of the study, load-shedding-related applications have the highest activation costs. The study also reveals that the activation costs of conventional power plants are considerably lower than the ones of load shedding. With regards to initialization costs, however, the cost of conventional power plants is considerably higher when compared to load shedding. Therefore, according to Ladwig, committing to load-shedding applications is only cost-efficient from a system perspective, when the generation capacity from power plants is not enough to fulfill the energy demand. For the detailed cost analysis of DSM, the reader is referred to Ladwig (2021).

Symbols and Abbreviations

AC	air conditioner
AMI	advanced metering infrastructure
BAN	building area network
CPP	critical peak pricing
DLC	direct load control
DR	demand response
DSM	demand-side management
ECS	energy consumption scheduling
EMS	energy management system
HAN	home area network
HVAC	heating, ventilation, and air conditioning
IBR	inclining block rate
IEA	International Energy Agency
ILC	indirect load control
LED	light emitting diode
NAN	neighborhood area network
PAR	peak-to-average ratio
PLC	power line communication
RE	renewable energy

RTP real-time pricing
TOU time of use
U-SNAP Utility Smart Network Access Port
α_x earliest beginning time for appliance x
β_x latest ending time for appliance x

Glossary

Bluetooth A short-range wireless technology used for exchanging data between mobile and fixed devices.
Fossil fuel A generic term for hydrocarbon-containing non-renewable energy sources such as coal, natural gas, and oil.
Grid flexibility The capability of a power system to preserve balance between power supply and demand during uncertainty.
Objective function The equation that needs to be minimized/maximized in an optimization problem.
Off-peak period The period of the day with the lowest total power demand.
Peak to average ratio The ratio between peak demand and average demand of the power system.
Peak period The period of the day with the highest total power demand.
PLC A technology that utilizes the existing electric power lines as a communication medium.
Wi-Fi A widely used wireless technology operating in the 2.4 GHz/5 GHz bands.
Zigbee A short-range, low-data rate, energy-efficient wireless protocol.
Z-Wave A short-range, low-data-rate wireless standard designed for home control automation.

References

Alotaibi, I., Abido, M.A., Khalid, M., and Savkin, A.V. (2020). Comprehensive review of recent advances in smart grids: a sustainable future with renewable energy resources. *Energies* 13 (23): 6269.
Boudec J-Y.L. (2019). Demand response [Presentation] [Online]. EPFL.
Bunder, R. (2020). We know when you are sleeping: The rise of energy smart meters [Presentation]. linux.conf.au.
Butler, J. (2019). Fundamentals of demand side energy management [Webinar]. Green Builder Media.
Deng, R., Yang, Z., Chow, M.-Y., and Chen, J. (2015). A survey on demand response in smart grids: mathematical models and approaches. *IEEE Transactions on Industrial Informatics* 11 (3): 570–582.

Domínguez-Garabitos, M.A., Ocaña-Guevara, V.S., Santos-García, F. et al. (2022). A methodological proposal for implementing demand-shifting strategies in the wholesale electricity market. *Energies* 15 (4): 1307.

Fadlullah, Z.M. and Kato, N. (2015). *Evolution of Smart Grids*. Springer Cham.

GridBeyond. (2019). The load shifting low-down: your quick guide. https://gridbeyond.com/load-shifting-low-down-guide (accessed June 2022).

IEA. (2021). World electricity final consumption by sector, 1974–2019. https://www.iea.org/data-and-statistics/charts/world-electricity-final-consumption-by-sector-1974-2019 (accessed December 2021).

King, C. (2004). Advanced metering infrastructure (AMI) overview of system features and capabilities [Presentation]. eMeter Corporation.

Ladwig, T. (2021). A techno-economic comparison of demand side management with other flexibility options. In: *The Future European Energy System* (ed. D. Möst, S. Schreiber, A. Herbst, et al.), 155–173. Springer.

Logan, J. (2020). Demand side management: From baseload reductions to flexibility provider. Enhanced Load Modeling and Forecasting Online Course, Module 2, Unit 3, National Renewable Energy Laboratory.

Milligan, M. (2015). Integrating variable renewable energy into the grid: Key issues and emerging solutions [Webinar]. Clean Energy Solutions Center.

Mohsenian-Rad, H. (2012). Topic 4: Demand response [Presentation]. ECE5333: Communications & Control in Smart Grid, Texas Tech University.

Niyato, D., Hu, R.Q., Hossain, E. and Qian, Y. (2011). Communications and networking for smart grid systems [Presentation]. IEEE GLOBECOM.

Office of the Auditor General of Ontario. (2014). Annual Report 2014. Queen's Printer for Ontario.

Rajawat, K. (2019). Smart grid communications [Presentation]. IIT Kanpur, QIP Short Term Course on Smart Grid Technology.

Sajip, J. (2018). Improving building energy performance with demand side management. https://www.ny-engineers.com/blog/demand-side-management (accessed June 2022).

Shigenobu, R., Adewuyi, O.B., Yona, A., and Senjyu, T. (2017). Demand response strategy management with active and reactive power incentive in the smart grid: a two-level optimization approach. *AIMS Energy* 5 (3): 482–505.

Sloot, D., Lehmann, N., and Ardone, A. (2022). Explaining and promoting participation in demand response programs: the role of rational and moral motivations among German energy consumers. *Energy Research & Social Science* 84: 102431.

Tantau, A., Puskás-Tompos, A., Fratila, L., and Stanciu, C. (2021). Acceptance of demand response and aggregators as a solution to optimize the relation between energy producers and consumers in order to increase the amount of renewable energy in the grid. *Energies* 14 (12): 3441.

Timpe, C. (2009). 'Smart domestic appliances provide flexibility for sustainable energy systems, SMART-A' [Webinar]. Leonardo Energy.

Valinejad, J., Marzband, M., Ansari, M., and Labonne, A. (2020). Demand response based on the power factor considering polynomial and induction motor loads. In: *2020 IEEE Texas Power and Energy Conference (TPEC)*, 1–6. https://doi.org/10.1109/TPEC48276.2020.9042519.

5

Standardizing Demand-Side Management: The OpenADR Standard and Complementary Protocols

Rolf Bienert

OpenADR Alliance, 111 Deerwoord Road, Suite 200, San Ramon, CA 94583, USA

5.1 History and Creation of OpenADR

The beginning of Demand Response (DR) and Demand-Side Management (DSM) goes back several decades to the first interruptible tariffs for large commercial and industrial customers. Utility operators would contact a customer to change power consumption on-site. At that time, demand response was primarily used to provide Energy (MWh) and/or Capacity (MW) when wholesale prices were unusually high, or when there were capacity shortages due to atypical consumer demand.

The way these early programs were implemented was rather simple in principle. Utilities would seek out customers with large energy consumption to onboard in order to minimize the number of participants while maximizing the possible curtailments. Most of the resources had no or little automation and any reduction in consumption had to be started and managed manually by a facility manager. During the onboarding, experts would evaluate the facilities together with the resident staff to determine the best possible response when a DR event was called out. Often, manual checklists were created for the operator on-site to create a more predictable response to a request. Note that these manual mechanisms at customer facilities are still around to this day. Factories for instance cannot be expected to automatically change their electricity load without any human intervention and approval.

Once the onboarding was completed, the customer was added to the DR Program of the respective area and provider. Utility staff would then call or later page a primary customer contact to manually change power consumption on-site with no immediate feedback in the utility control room. As the number of participants increased, the program operators had to get creative with the

Energy Smart Appliances: Applications, Methodologies, and Challenges,
First Edition. Edited by Antonio Moreno-Munoz and Neomar Giacomini.

notification mechanisms. A pager was already an easier way to contact more participants at once but not all customers would be interested in and attentive to a pager at that time. Telephone and fax were still the preferred methods early on, so the initial automation was focused on speeding up these methods. Automated dialing systems with recorded messages were installed to simplify the process and to keep the personnel need in check. Fun fact: Some of these automated dialing systems were still in place in recent years.

While the implementation of these communication mechanisms was rather simple from a technology standpoint, it was certainly not future-proof. The main issues can be summarized in three categories.

No feedback: While personal phone calls would at least provide a "message received" feedback, there was no way to tell how much energy consumption was really curtailed after an event was called. The use of one-way pagers and dialing machines even eliminated that initial feedback. It was also almost impossible to review the change in power consumption after the fact without Advanced Metering Infrastructure (AMI) and Smart Meters. Basically, the utilities had to rely on readings in their substations, overall power consumption on the grid, and data based on previous experiences. This might have been enough in these early days, but it would not be sustainable over time.

Scalability: There is a limited number of larger consumers that can be controlled using these manual processes. Even with dialing machines, fax machines, and pagers, the number of people you can reach during a limited time period is finite and not very large. Remember, at this time residential and small-medium businesses had very little automation and connectivity beyond the phone line. Onboarding was different for every customer and required significant effort.

No harmonization: Every utility essentially built its own program with different parameters and objectives. While this is in principle not a bad thing, it also meant that suppliers and integrators would customize solutions as needed which created higher costs and long implementation timelines despite the relatively low amount of technology involved.

Then the 2000/2001 energy crisis hit California and other Western States, and things changed. Due to market manipulations, mainly by service companies that provided capacity resources, California ran into supply shortages and increased energy prices – up to 800% in the year 2000. Rolling blackouts circulated around the state to make up for the shortages which affected businesses and private entities alike. Due to capped retail electricity prices, Pacific Gas & Electric Company had to declare bankruptcy and Southern California Edison only just escaped the same fate. The crisis is estimated to have caused $40–45 billion in losses.

Preliminary research from a California Public Utility Commission (CPUC) analysis of the blackouts determined that, as stated by the CPUC Deputy

Executive Director for Energy and Climate Policy Edward Randolph, even the still small amount of available DR "contributed quite a bit to grid support," and, "This is the first event in many years that required sustained demand response. The analysis of its performance will inform future decision-making." More emphatically, Gridworks Executive Director and former CPUC Energy Advisor Matthew Tisdale said that "If California had already seriously embraced flexible demand response, it would not have even come close to blackouts."[1]

Therefore, after the crisis was contained, the California Energy Commission (CEC) identified the need for a better, faster, and more scalable communication mechanism for demand response. In particular, to eliminate the need for rolling blackouts and instead replace them with targeted and automated DR events. The CEC's Public Interest Energy Research (PIER) program created a General Funding Opportunity (GFO) to entice companies to become creative and offer solutions to the state of California. The Demand Response Research Center (DRRC – https://buildings.lbl.gov/demand-response) which is managed by Lawrence Berkeley National Laboratory (LBNL) and the company Akuacom (later Honeywell) took on the project and worked with the California Investor Owned Utilities (IOUs – Southern California Edison – SCE, San Diego Gas & Electric – SDG&E and Pacific Gas and Electric – PG&E) to create the first standard for demand response communications – OpenADR 1.0. LBNL describes the standard as "a communications data model designed to facilitate sending and receiving DR signals from a utility or independent system operator to electric customers. The intention of the data model is to interact with the building and industrial control systems that are pre-programmed to take action based on a DR signal, enabling a demand response event to be fully automated, with no manual intervention. The OpenADR specification is a highly flexible infrastructure design to facilitate common information exchange between a utility or Independent System Operator (ISO) and their end-use participants. The concept of an open specification is intended to allow anyone to implement the signaling systems, providing the automation server or the automation clients."

OpenADR 1.0 offered an implementable way to quickly and effectively communicate between a management server at the utility (often referred to as a DRAS – Demand Response Access Server – a term later trademarked by Honeywell) and a client device at the energy customer facility end, often also referred to as the resource side. Implementation started between 2005 and 2006. The standard was however focused on existing or newly implemented DR programs in California and was also limited to Akuacom/Honeywell and their partner companies.

1 https://www.utilitydive.com/news/demand-response-failed-california-20-years-ago-the-states-recent-outages/584878

5.2 Re-development of OpenADR 2.0

As the Smart Grid efforts increased after 2005, more emphasis was put on interoperability and the creation of common standards. The Smart Grid Interoperability Panel (SGIP) was tasked by the US National Institute of Standards and Technology (NIST) to oversee standardization of the Smart Grid. SGIP collected use cases and market requirements from different utilities, areas, and interest groups to add to a framework for a demand response standard. With OpenADR 1.0 functionality as its basis, the task of creating this new framework was given to the OASIS Energy Interoperation Technical Committee (EITC). The EITC used the work of both the OASIS Energy Market Information Exchange (EMIX) and the OASIS Web Services Calendar (WS-Calendar) to create the basis for a common DR data exchange model. This standard, the Energy Interoperation (EI), was however not meant to be used by manufacturers to build interoperable products. It lacked (perhaps on purpose) specific implementation requirements, security, and transport protocol features.

In parallel to the EITC work on the Energy Interoperation, several stakeholders – both utilities and manufacturers – discussed the need for more specific branding and interoperability to make this standard successful. It was concluded that a new industry initiative was needed to put all of this in place and the OpenADR Alliance[2] was formed in late 2010.

The Alliance's goals were simple –

- Create implementable specification profiles based on the Energy Interoperation services descriptions.
- Create a testing and certification program to validate products and reduce the need for implementation testing.
- Create a broader ecosystem of companies that can provide interoperable solutions to utilities and customers alike (Figures 5.1 and 5.2).

Figure 5.1 OpenADR Alliance logo. Source: OpenADR Alliance.

Figure 5.2 The seal of compliance to OpenADR 2.0. Source: OpenADR Alliance.

2 www.openadr.org, https://www.openadr.org/overview

The Alliance made quick progress and derived two profile specifications from the OASIS Energy Interoperation – OpenADR 2.0A and OpenADR 2.0B Profile Specification – as well as a standardized test tool between the years of 2011 and 2013. Trials with the new OpenADR 2.0 standards started soon after and were generally very successful. The following defining features were identified for the standard.

- **Continuous, secure, and reliable**: Provides continuous, secure, and reliable two-way communications infrastructures where the endpoints at the end-use site receive and acknowledge the receipt of DR signals from the energy service providers.
- **Translation**: Translates DR event information to continuous Internet signals to facilitate DR automation. These signals are designed to interoperate with energy management and control systems, lighting, or other end-use controls.
- **Automation**: Receipt of the external signal is designed to initiate automation through the use of pre-programmed demand response strategies determined and controlled by the end-use participant.
- **Opt-out**: Provides opt-out or override function to any participants for a DR event if the event comes at a time when changes in end-use services are not desirable.
- **Complete data model**: Describes a rich data model and architecture to communicate price, reliability, and other DR activation signals.
- **Scalable architecture**: Provides scalable communications architecture to different forms of DR programs, end-use buildings, and dynamic pricing.
- **Open standards**: Open standards-based technology such as Internet Protocol (IP) and web services form the basis of the communications model.

The benefits of the framework were laid out by the stakeholders as well.

- **Open specification**: Provides standardized DR communications and signaling infrastructure using open, non-proprietary, industry-approved data models that can be implemented for both dynamic prices and DR emergency or reliability events.
- **Flexibility**: Provides open communications interfaces and protocols that are flexible, platform-independent, interoperable, and transparent to end-to-end technologies and software systems.
- **Innovation and interoperability**: Encourages open innovation and interoperability, and allows controls and communications within a facility or enterprise to build on existing strategies to reduce technology operation and maintenance costs, stranded assets, and obsolesce in technology.
- **Ease of integration**: Facilitates integration of common Energy Management and Control Systems (EMCS), centralized lighting, and other end-use devices that can receive Internet signals (such as the Extensible Markup Language – XML).

- **Supports a wide range of information complexity**: Can express the information in the DR signals in a variety of ways to allow for systems ranging from simple end devices (e.g. thermostats) to sophisticated intermediaries (e.g. aggregators) to receive the DR information that is best suited for its operations.
- **Remote access**: Facilitates opt-out or override functions for participants to manage standardized DR-related operation modes to DR strategies and control systems.

Today, the OpenADR Alliance has over 170 corporate members and over 250 certified solutions. It represents the largest specialized ecosystem in the demand response arena. Implementations now reach beyond California to many US states as well as countries in the Asia-Pacific Economic Cooperation (APEC) and European regions. Regulations in Japan and the UK now also prescribe OpenADR as mandatory.

Also around 2005, the ZigBee Alliance developed the Smart Energy Profile (SEP) 1.0 and later on 1.x. The initial rollouts of Smart Meters using ZigBee modules toward the home were thought to be extremely suitable to control many appliances and other consumers in the building. It was envisioned to create a ZigBee network within the house that would connect appliances to be controlled and also to make the customer aware of the current energy price. To accomplish the latter, the idea of in-home displays was developed. The displays worked either like a traffic light with different colors for different prices, or they would provide actual prices. Several of the early Smart Appliance manufacturers also looked at the possibility to add this networking capability but it was not clear how the system would be used. Furthermore, it became a concern that if appliances are controlled through the Smart Meter network, any issues with said appliances might be taken back to the utilities that operate the system. This combined with connectivity issues in bigger dwellings and potential delays in the communication from utility to meters caused this idea to not be realized.

In 2018, the OpenADR 2.0b Profile Specification was also published as IEC62746-10-1[3] and became an international standard.

5.3 How OpenADR Works

In essence, OpenADR is a web services-based message exchange protocol. It is fully bi-directional although not all functions are symmetrical – with other words, some signal elements are meant to travel primarily one way. Two actors are being used in an OpenADR exchange. The Virtual Top Note (VTN) is generally located at the utility side, and the Virtual End Node (VEN) is part of the demand side control system.

3 https://webstore.iec.ch/publication/26267

The VTN is the interface point of a larger system installed at the utility or an aggregator. There are several companies that build these Demand Response Management Servers (DRMS), Demand Side Management System (DSMS), or Distributed Energy Resources Management System (DERMS). Typically, these management servers integrate information elements from the utility control network to manage the downstream resources more effectively when and where they are needed. They can be integrated into the IT structure of the utility or cloud-based as software as a service (SAAS).

The VEN is the logical counterpart of the VTN. It represents a web services client and is in general the recipient of most on the information elements. There are however some services with information flowing from the VEN to the VTN. We will discuss these later. The VEN can appear in different forms. Early on these were almost always onsite, so in a gateway or even directly in a control system. However, lately, the trend of cloud-based control and Internet of Things (IoT) has shifted the OpenADR endpoint away from the resource into these cloud controllers. The VEN does not have to be the actual energy-consuming device or system. In fact, in most implementations, the VEN is rather a type of gateway between the internet and the devices (Figure 5.3).

In most cases, the VTN will communicate with the VEN using already existing internet connections. Broadband to building or generally through the cloud controller. In rarer cases, dedicated internet connections can be established using

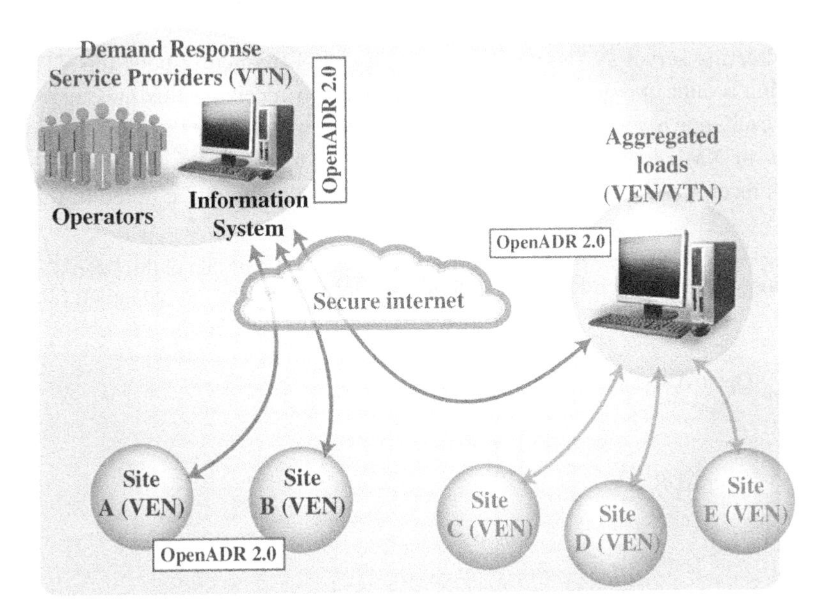

Figure 5.3 A relationship diagram between the parts of the system. Source: OpenADR Alliance.

cellular modems or other similar pathways. This can address connectivity concerns for high-value resources, or simply connect systems that are out of reach of typical internet connectivity. Examples of the latter would be irrigation pumps in fields or Electric Vehicle (EV) charging stations on parking lots.

OpenADR currently comes in two flavors, OpenADR 2.0a and 2.0b. The two profile specifications vary in feature content although the overall mechanisms are identical. OpenADR 2.0a could be considered a reduced function set. It was originally created to be used in resource-constraint devices like appliances, thermostats, and similar systems. The OpenADR Alliance working group assumed that these endpoints did not have extensive processing capabilities and could therefore not cope with large messages and extensive instructions. OpenADR 2.0a is limited to so-called simple signals, in which the signal content can only include discrete numbers from 0 to 3. These numbers could be associated with either prices or energy curtailment requests. Both associations needed to be pre-defined by a contract. An early program in California for instance defined the curtailment levels as follows: 0 – normal operation; 1 – medium shed; 2 – high shed; 3 – emergency.

OpenADR 2.0b in contrast includes all services, signals, and message contents, more about this later. Needless to say, with the increasing computing capabilities and cloud control trends, all recently certified OpenADR products use the 2.0b profile (Figure 5.4).

As already mentioned, OpenADR is a web services-based message exchange protocol. For the layman, it can be envisioned as a web browser (VEN) and website hosting server (VTN) combination. The VEN frequently polls the VTN by sending a data update request to the server Upon receiving this message, a security exchange is started and the VEN and VTN connect. Now the information elements in XML format will flow from the VTN to the VEN and the VEN confirms receipt and acts on the messages. Of course, it should be noted in this

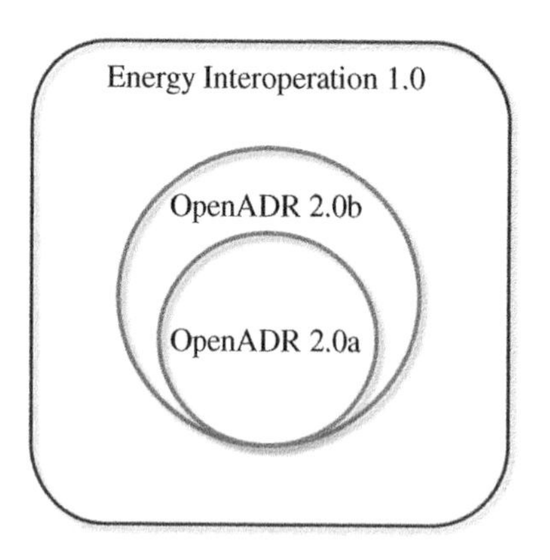

Figure 5.4 The relationship of the OpenADR profiles and OASIS Energy Interoperation 1.0. Source: OpenADR Alliance.

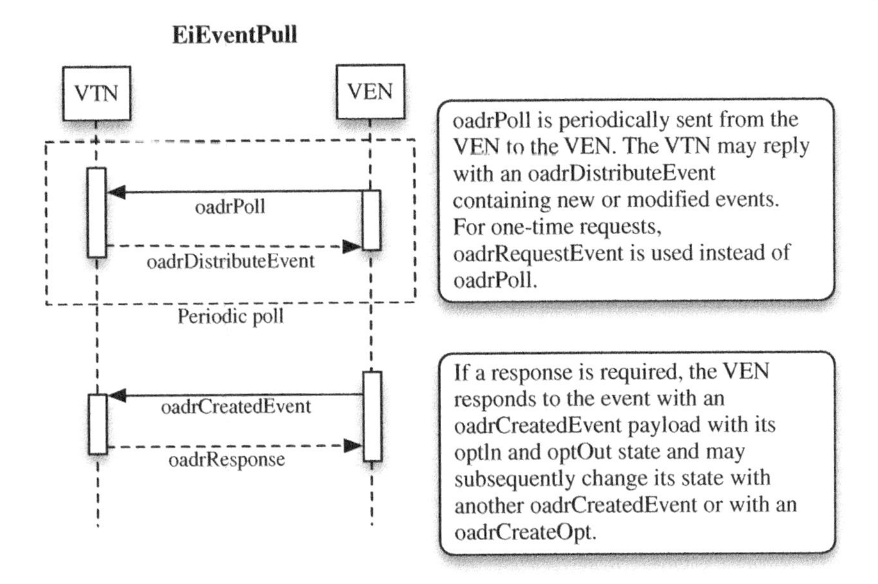

Figure 5.5 The Typical OpenADR Exchange Pattern. Source: OpenADR Alliance.

example that we are talking about machine-to-machine communication and not a graphical interface for the user (Figure 5.5).

The OpenADR services that are included in the 2.0b profile are the Event Service, the Opt Service, the Report Service, and the Registration Service.

5.3.1 Event Service (EiEvent)

So-called Events are the primary services in OpenADR. They can be envisioned to be very similar to a calendar invitation that you might receive through an email platform or other scheduling application. In fact, the XML schema for the OpenADR event service used the OASIS "WS-calender" structure in an effort to integrate existing internet functions instead of reinventing the mechanisms. An OpenADR event has a start time and an end time – or a duration. Within this timeframe, a large variety of signals and different time intervals can be incorporated into the information elements. It could be envisioned as a multi-track agenda within the event if one wanted to visualize this. Most common signal types are price and energy-related messages. It should be noted that from a technical perspective there are no limits as far as the event timing is concerned. It could start "immediately," essentially having a start time identical to the creation time, or it could be hours, days, or weeks ahead. Consequently, an event must have at least one-time interval with one signal. However, as already mentioned, it could have multiple intervals with changing values for signal types or different signals altogether. As an example, an event could be scheduled from 8:00 to 12:00. Within this timeframe could be for instance four one-hour time intervals with different energy prices for each of them. At the same time, there could be two intervals with different curtailment levels included. Appliances may also be faced with different signals. Most likely an appliance controller will simply ask for power cycling, however, if dynamic pricing becomes more prevalent then the consumer may need to be able to react to prices as well. In other words, one may set up a pool pump to run only up to a pre-selected price or a washing machine may delay the start until the energy price drops again (Figure 5.6).

The VEN can respond to an event by opting in or opting out of participation. At any time before or during the event the customer will still be able to reverse any decision made.

It is worth noting here that OpenADR is generally not considered to be a control and command protocol. The signals are not meant to directly switch off devices or change settings. Rather, the OpenADR signals are meant to inform and motivate customer systems to participate in a way that is either predefined or based on some processing logic in the controllers. Controlling the devices downstream from the controller is typically accomplished with other existing building control protocols. Anything from ZigBee-related standards (CSA – Connectivity Standards

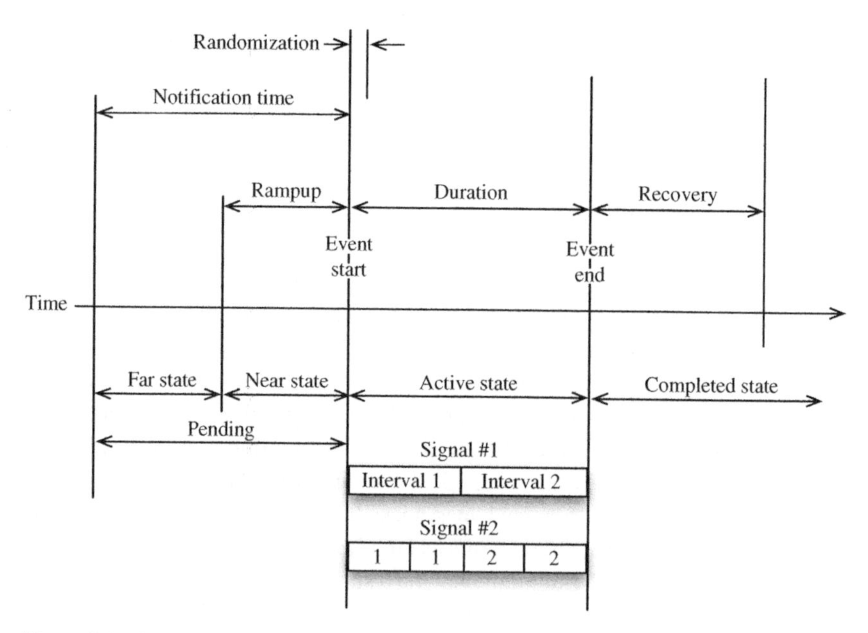

Figure 5.6 The structure of a typical OpenADR Event. Source: OpenADR Alliance.

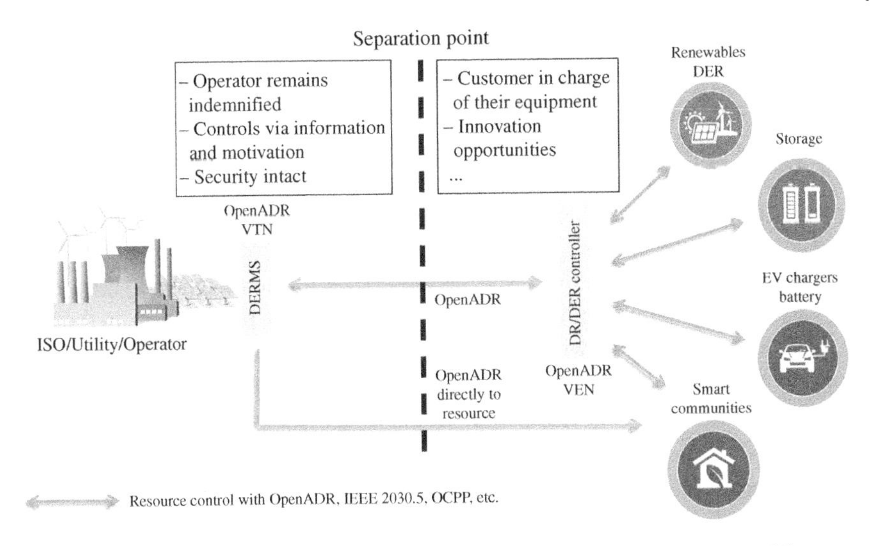

Figure 5.7 The typical information flow of the OpenADR signal. Source: OpenADR Alliance.

Alliance[4]), EEBUS,[5] etc., for residential applications over BACnet[6]/LonMark[7] in commercial buildings, to more specialized standards like OCPP[8] (Open Charge Point Protocol) for EV chargers and CTA-2045[9] (EcoPort[10]) for direct appliance connections.

The fact that OpenADR does not directly control devices also enables utilities to stay at arm's length from the customer equipment and therefore avoids the assumption of responsibility for failures at the customer equipment. The OpenADR link in this case provides a demarcation point between the two entities and also for cybersecurity (Figure 5.7).

5.3.2 Opt Service (EiOpt)

A less-used service in OpenADR, the Opt functionality enables resources to announce their availability ahead of time. We will likely see more usage for these services when more storage applications are available. In essence, the resource – through the VEN – can provide a schedule of available participation to the VTN. This will likely not affect appliances.

4 www.csa-iot.org
5 www.eebus.org
6 www.bacnet.org
7 www.lonmark.org
8 www.openchargealliance.org
9 https://shop.cta.tech/products/modular-communications-interface-for-energy-management
10 https://www.openadr.org/ecoport

5.3.3 Report Service (EiReport)

This service is a feedback mechanism from the VEN to the VTN. The OpenADR specification defines a set of mandatory reports. However, any custom assembly of data is possible within a program. Of course, not all resources can provide specific data. Imagine an older appliance being asked about its recent energy consumption history. However, the VEN is required to be able to form report messages even if no data is available.

In the initial exchange, the VTN sends a so-called metadata report to the VEN. This represents a template for what the VEN is supposed to provide back to the top node. After that, the VEN will provide the requested information at the specified time intervals back to the server. Often this is used to simply inquire about the status of resources, but it may also include more specific data.

5.3.4 Registration Service (EiRegister)

This is a low-level service that allows the VTN and the VEN to exchange capability and connection information when they first connect. It does not include any higher-level functions for demand response.

At the time of this writing, the OpenADR Alliance is reviewing future services and functions needed to address upcoming market needs. The proposed features can be categorized as follows.

- enhancements to existing OpenADR usage models
- extend OpenADR to support new use cases
- improve Infrastructure
- other.

OpenADR uses two transport protocols, Simple Hypertext Transfer Protocol (HTTP) (simplified REST interface) and Extensible Messaging and Presence Protocol (XMPP). The former is a typical web services mechanism used by websites and many other systems. It generally established a connection upon request, for instance, if the VEN is polling the VTN. The VTN can also push messages to the VEN but this may encounter problems with local firewalls that will block external commands from reaching the endpoints. Each time a connection is established, the VTN and VEN will also need to go through the cybersecurity authentication process, more details in the Cybersecurity session presented shortly.

XMPP is an open communication protocol designed for instant messaging and other applications that need constant connectivity. The advantage of this mechanism is that the connection between the VTN and the VEN is always open. You can compare this to any online chat tool you may use in daily life, or the so-called tickers on websites that keep updating automatically as long as the website is active, a communication heartbeat. This enables very fast response times to requests made by the VTN. The idea here is that customers and professionals in the space will eventually see DSM and flexibility programs that allow for almost instant demand response to help the grid balance fluctuations due to renewable energy resources as well as capacity-related variations. XMPP is not widely used at this time. Not only due to the fact that most programs do not require extremely fast response times but also because the VEN implementations are must faster and resource-rich than initially expected (cloud-based vs. gateway and so on).

Before discussing how OpenADR products can be tested and certified, it may be useful to talk about why the use of standards, testing, and certification makes sense for an industry and also for manufacturers. You may or may not be aware that we are surrounded by standardized products that make our daily life possible. We do not even have to reach for our mobile phones but simply look at a nut and bolt combination. If the nut has a different diameter or thread, it will simply not fit the bolt, no matter how similar it may appear to be. Therefore, the hardware industry has long ago standardized the sizing so that different manufacturers can build the individual components independently and still maintain interoperability. However, the manufacturing process is obviously not perfect. There are variations – tolerances – in the tools that make the components, the materials, and the finishes. These tolerances need to be defined in a way that the greatest positive deviation from the nominal specification will still fit the greatest negative deviation from the same nominal specification. Aside from the fit, the materials also need to have a defined strength in order to fulfill their intended purpose when used. As you can see, there is a variety of parameters that need to be maintained and tested before even a simple bolt can be put on shelfs in the hardware store. The same is true for any other system of components that are meant to interoperate.

This simple example can also be used to better understand the advantages of using standards in many areas. Here are the basic ideas:

- A consumer has a problem to solve and has a variety of products to choose from. It will not matter where one purchases the components, together or individually, because they will fit together.
- The consumer does not need to re-invent anything. The products are available, so it is much quicker and more affordable to purchase them instead of creating new components.
- Manufacturers in turn can specialize in specific components to maximize their expertise in that area. They know that these components will fit into the system later.
- If one vendor goes out of business, consumers can simply buy a replacement component from another vendor without having to change the overall system.

In OpenADR, for systems or components to be listed as certified products, devices need to go through a testing and certification process managed by the

OpenADR Alliance. The group partnered with a test tool company early in the process to create one reference test tool. This tool is a software implementation that simulates either the VTN or the VEN, exchanges and verifies messages, and eventually creates a verdict of pass or fail for each test case. The test cases in turn follow the specific conformance statements in the specification. The test tool can be obtained from the OpenADR Alliance,[11] and an identical tool is used at the approved test houses.

Once a manufacturer has implemented the protocol, they go through the testing at one of the test houses. The passing test report along with some other documentation is then sent to the alliance for review, certification, and listing on the designated certified products page.

5.4 Cybersecurity

Being a web services interface protocol, OpenADR is primarily interested in and required to secure the transport layer link from the VTN to the VEN system. Similar to the communication between an internet browser and for instance your bank, it needs to be ensured that both sides are in fact exchanging messages with the correct and real counterpart that they are intending to connect to. An additional layer of security included in the OpenADR specification lets the user encode the actual messages that are being transported. An XML wrapper is being used to ensure that the message that was sent is also the message that is being received. This mechanism is not widely used at this time but could be deployed to avoid any instances of repudiation, with other words any involved party claiming that the message did not contain what the other party claims.

OpenADR uses transport layer security – TLS 1.2[12] – as its main security mechanism. After the initial reviews by the US NIST (Smart Grid Interoperability Panel – SGIP), it was determined that the best approach is having server and client-side cybersecurity certificates. Additionally, the security reviewers required that the certificates have to be unique to OpenADR. To accomplish that, the OpenADR Alliance partnered with a third party to create a certificate policy and to establish an OpenADR specific Certificate Authority (CA). Using an existing provider, separate roots were created that correspond with the OpenADR requirements and are simultaneously unique to OpenADR products. In order to obtain these certificates, manufacturers need to be compliant with the OpenADR specification and demonstrate this during the product certification process. This ensures that the basic security functions have been implemented correctly and

11 https://www.openadr.org/openadr-test-tool
12 https://www.rfc-editor.org/rfc/rfc5246.html

will for instance reject incorrect security certificates among other tests. It should be noted that using the OpenADR-specific certificates – while strongly recommended for interoperability – is up to the final implementers of the technologies, usually utility companies. Some program operators may choose to use their own CA and certificates to have better control of the endpoints; others may even lower the requirements. This is outside the scope of the OpenADR Alliance.

Later reviews by the International Electrotechnical Commission (IEC) security working group also found this approach acceptable.

OpenADR allows for two types of certificates, Elliptic Curve Cryptography (ECC) and Rivest–Shamir–Adleman (RSA) types. The VTN (server) serves as the center of interoperability and needs to implement both certificate types. Please note that this is also true for all other optional features – the server must implement all of them and the client (VEN) can pick and choose between some available options. The used Cipher Suites are:

- ECC: TLS_ECDHE_ECDSA_WITH_AES_128_CBC_SHA256
- RSA: TLS_RSA_WITH_AES_128_CBC_SHA256.

When the connection between the VTN and the VEN is established, the systems exchange the required keys and create a secure tunnel.

5.5 Other Standards and Their Interaction with OpenADR and Energy Smart Appliances

As mentioned before, OpenADR generally stays away from direct load control. Therefore, it is ideally positioned to be combined with other in-building control systems. There are many standards that have gained market share including the ZigBee ecosystem (now CSA), Z-Wave,[13] EEBUS, IEEE 2030.5,[14] OCPP, and many more. In a typical architecture, OpenADR is used to convey informational and motivational messages from the utility to a gateway layer. This endpoint could be in a cloud-based control system or in a physical device on the demand side. Irrespective of where the endpoint is located, the OpenADR signal will have to be evaluated by the system logic so that the best possible actions can be applied to accommodate the requests. In most cases, these are pre-programmed scenarios and offer companies an opportunity to prove value-adding innovation to their products. Aside from these building automation protocols, there is one additional standard that provides appliance manufacturers a different avenue to connect with the Smart Grid. Therefore, we want to take a closer look at this in the context of the chapter – CTA2045 (EcoPort).

---

13 www.z-wave.com
14 https://standards.ieee.org/ieee/2030.5/5897

In the 2000s there was an influx of radio standards competing for the residential building automation space. The uncertainty about which of these standards would be prevalent in the future caused a lot of manufacturers to sit back and wait in order to avoid unnecessary expenditures and potentially failed targets. A thermostat company (RTA – Radio Thermostat Company of America) eventually brought up the idea of having plug-in modules that would provide different types of radio communications while using one very basic connection to the thermostat itself. The modules were approximately 6 cm by 6 cm by 1 cm big and could be slotted into the side of the thermostats. A blade of pins would connect to the internal functions and convey simple commands like changing the temperature or turning the system off and on. Different radio technologies were available in different modules and the homeowner could decide which is needed and exchange them if another technology became relevant (Figure 5.8).

The idea was soon extended to gateways which would include several slots for modules in order to be flexible and able to accommodate several technologies at the same time.

The concept initially gained a lot of interest and several companies banded together to form the USNAP Alliance, a non-profit organization "bringing together utilities, utility equipment manufacturers, and customer equipment manufacturers to develop and promote a modular communications interface (MCI) to enable customer equipment participation in energy management and demand response." However, with ever-improving technology, multiband radio chips, and cheaper implementations, the initial idea about different radio types in the modules alone did not make USNAP a relevant technology and the alliance was stagnant for a while.

In the late 2000s, the Electric Power Research Institute (EPRI) also worked on a related project with the intention to connect larger appliances. The same modular

Figure 5.8 USNAP add-on module. Source: Amazon.com, Inc., https://www.amazon.com/Radio-Thermostat-Z-Wave-Module-RTZW-01/dp/B00IA8ROSK.

concept was applied but the module was larger and could be securely mounted on washing machines, dryers, water heaters, pool pumps, and so on. The module also used line Alternating Current (AC) voltage as opposed to the Direct Current (DC) low voltage used by the original USNAP plug-ins. It was assumed that many of these larger appliances would not necessarily have easy access to DC circuit voltages. The stakeholders then decided that it would make sense to combine the two efforts into one standard. After some research and discussions, the two specifications were handed over to the Consumer Technology Association (CTA; formerly Consumer Electronics Association – CEA), and a project committee with working groups were formed to harmonize the standard. The new framework and standard were called CTA-2045 – The Modular Communications Interface for Energy Management.

This standard provided a solution to the problem of ever-changing connectivity and control technologies by defining the MCI enabling any product to connect to any type of demand response system (at the time summarized with "Advanced Meter Reading [AMI], Smart Energy Profile [SEP], OpenADR"), and/or home or building network. The concept was to be the same as previously in USNAP – to encourage manufacturers to build an MCI into their products that can accept a simple communications module. Consumers and program managers are then free to select whatever communication solution works best for their particular environment.

The technical concept was also kept simple. Utilizing the RS-485[15] and Serial Peripheral Interface (SPI) supported by most silicon chips at the time, the MCI protocol is capable of simply passing through standard protocols including Internet Protocol (IP), OpenADR, and SEP from the communications module to the end-device. Network security is supported through the selected transport protocol, such as Wi-Fi, ZigBee, HomePlug,[16] Z-Wave, LonWorks,[17] etc., in addition to network or application layer security. The passthrough concept however has not gained large-scale support to date.

The simplified communications messaging supported by the MCI standard became more popular and supports direct load control, Time of Use (ToU) rates, Critical Peak Pricing (CPP), Real time Pricing (RTP), peak time rebates, all kinds of block rates, and a range of ancillary services. The functionality of the removable modules can be tailored by utilities or other load-managing entities to provide support for the unique needs in a given region or service territory, without impacting the end devices.

15 https://en.wikipedia.org/wiki/RS-485
16 https://en.wikipedia.org/wiki/HomePlug
17 www.lonmark.org

The standard defines the two components as universal communication modules (UMC) and Smart Grid Devices (SGD – usual appliances of any kind) (Figures 5.9–5.11).

Upon completion of the standard in the mid to late 2010s, there was still not a lot of uptake for the technology. Manufacturers generally liked the idea and even joined a further development project with EPRI which created more implementation definitions and a test setup that could evaluate the UCM and SGD implementations. However, there were still no utility requirements that forced appliance vendors to enable any type of grid connectivity. Therefore, most companies chose to not invest in any technology. Around the same time the CTA-2045 standard was created, Portland General Electric, the utility company in Portland, Oregon, USA, observed an increase in heat pump water heater (HPWH) availability and sales. The water heaters – while efficient in general – represented an increased electricity usage. Discussions around how to manage this increase started and the MCI standard was assumed to be part of the solution. It still took a long time to define solutions and it was not until the 2020s that Washington State became the first area to prescribe CTA-2045 as a mandatory component of new heat pump water heaters. The requirements have since been refined and will enter into effect in the 2022/2023 time frame.

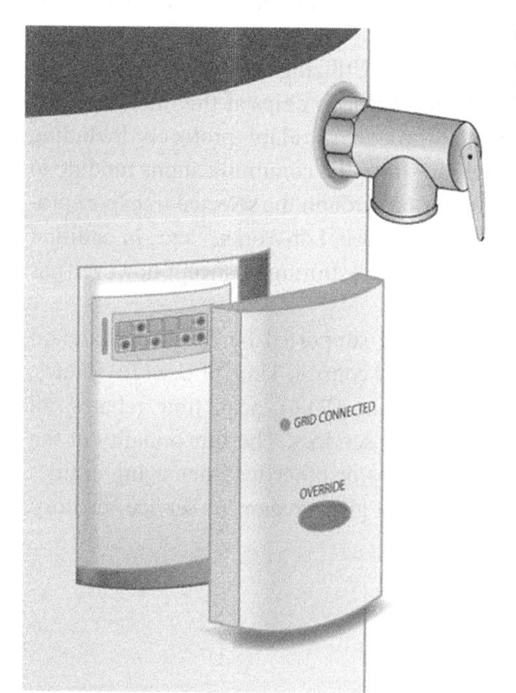

Figure 5.9 An AC Modular Communications Interface use case. Source: OpenADR Alliance.

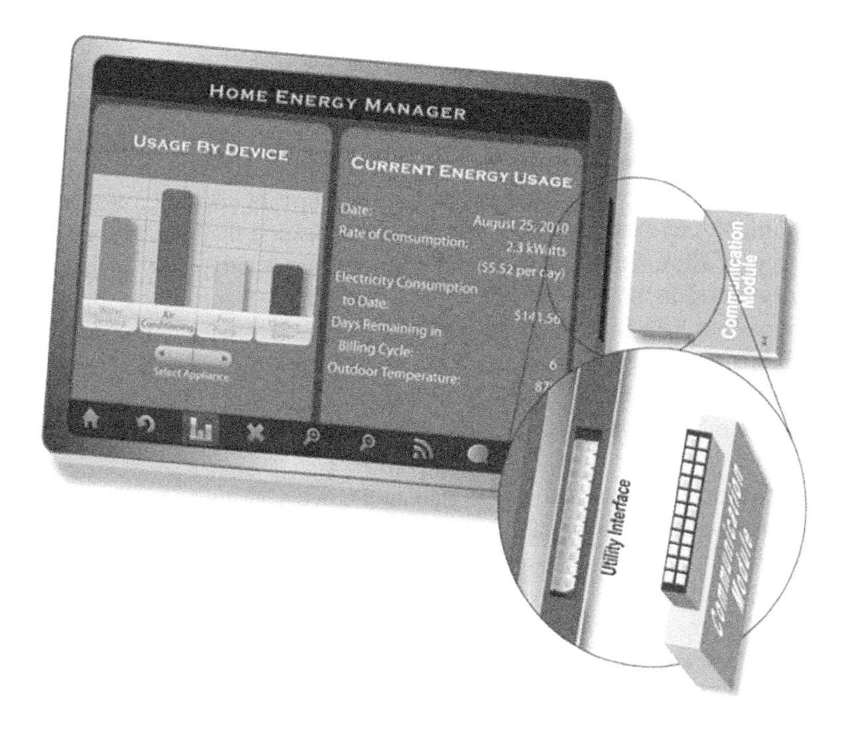

Figure 5.10 Example of a Home Energy Manager that supports DC Modular Communications Interface.

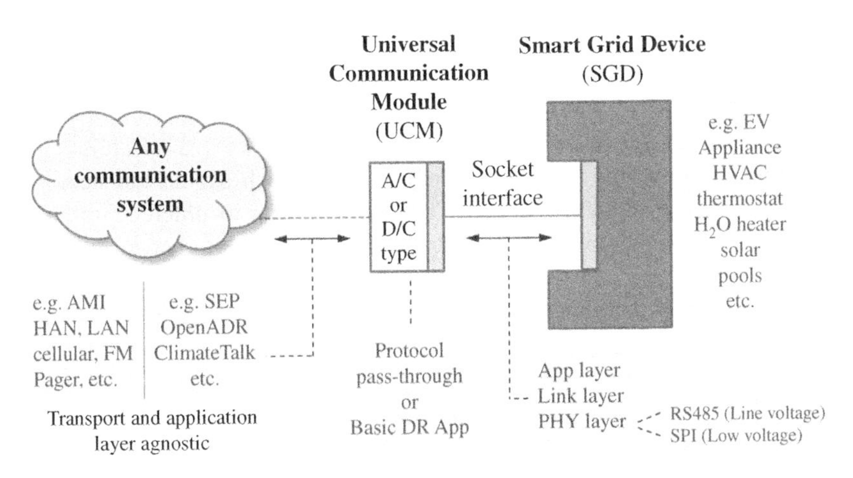

Figure 5.11 Block Diagram of the MCI (Socket Interface is defined in CTA-2045). Source: OpenADR Alliance.

During the discussions about these requirements, the contributors realized that there was no testing and certification program established for the CTA-2045 standard. The document itself included a variety of functions and models but this was not enough to build consistently interoperable products without defining the specifics of what is needed to effectively control HPWHs. NEEA, the Northwest Energy Efficiency Alliance, took on the task of creating a test tool and defining the test plan for the relevant communication modules and smart grid devices. NEEA, however, is not an organization that typically manages technology programs, so they approached the OpenADR Alliance to see if there is willingness to host the CTA-2045 (by now there is a version CTA-2045-B) testing and certification program. Since the MCI standards are a logical extension of OpenADR inside the building, the alliance agreed to add this program under their umbrella using the same certification process used for OpenADR products.

Aside from creating the test tool, test plan, and other relevant documentation, the task group also discussed additional branding for the CTA-2045 standard. While the term was reasonably well-known in the industry, it did not seem very intuitive and user-friendly. Many other successful standards have gone through this process successfully. For instance, many readers will not know what IEEE 802.11 is but will certainly have heard of Wi-Fi, the brand name for the profiled and tested IEEE 802.11 series of standards.

After a few months of discussions and the involvement of a branding consultant, the marketing name EcoPort was created and is shown in Figure 5.12.

EcoPort is designated for devices, both the modules and the appliances, that have passed the testing and conform to the requirements set forth by the regulations. At the time of this writing, the first group of manufacturers is working on the initial implementations. They are expected to be available in the third quarter of 2022.

EcoPort will enable water heater and other appliance manufacturers to create universally usable smart grid ready devices which should not need to be customized for different markets. The customization to different utilities, technologies, and environments will happen by selecting different modules to plug into the SGDs.

Another standard worth describing in more detail in the context of appliances is a lesser-known European control specification – EEBUS.

Similar to the OpenADR Alliance, the EEBUS initiative is a non-profit organization, based in Germany, with international members from the automotive,

Figure 5.12 CTA-2045 brand name. Source: OpenADR Alliance.

heating, ventilation, air conditioning, white goods, PV, energy storage, as well as energy management sector. On behalf of the industry, EEBUS describes the communication interface, meaning the application, data model, and transport protocol, to allow for the interconnection between energy management relevant devices as well as corresponding control systems. To arrive at a standardized communication interface, EEBUS is active in several national, European as well as international standardization bodies.

Effective communication between different stakeholders in the building requires more than just standardized data exchange. The communicating parties must also be aligned with respect to content and communication flow process via the same use case applications. By "communication interface" we therefore refer to the combination of the data model and transport protocol required for the data exchange as well as relevant use case descriptions. Such a communication interface to be successful and future-proof must fulfill several key requirements.

- **Technical standardization**: EEBUS standardizes the communication interface at and behind the grid connection point to allow for the development of energy-related and manufacturer-specific business models that are compliant with physical limits at the grid connection and thus ensure stable grid operation.
- **Easy installation**: Plug and play on the premises: EEBUS provides an automated plug-and-play mechanism without the need for specialists' costly configuration tools to integrate devices into an eco-system. This will lead to a broader acceptance for energy-optimized setups.
- **Holistic application**: EEBUS enables a holistic application by its use case specifications employing the domain unspecific data model SPINE (Smart Premises Interoperable Neutral-Message Exchange). Irrespective of device type or stakeholder role, it is the generic data model allowing for a broad range of use cases. Furthermore, information symmetry is implicitly guaranteed – there is no single point of information and control: any device can exchange energy-related data via its implemented use case(s), irrespective of the underlying system architecture.
- **Dynamic flexibility**: The EEBUS data model SPINE has been designed from the very beginning to allow for a maximum degree of flexibility. Its modular and flexible design is a prerequisite for consensus building among the various stakeholders.
- **Reliability**: Even the technologically most advanced and economically most feasible solutions must be reliable when it comes to grid-relevant implementations. EEBUS ensures reliability with respect to its SPINE data model in the following manner: use cases depending on a reliable connection implement a fail-safe/watchdog mechanism – if the communication behind the grid connection fails, the heartbeat mechanism of the EEBUS ensures that the

devices operate in a fail-safe mode and return to a normal operation as soon as the communication is re-established.

- **Security compliance for critical infrastructure**: In principle, any transport protocol can be used for transmission of the EEBUS SPINE data: EEBUS members have deliberately chosen an IP approach and developed the Smart Home Internet Protocol (SHIP) transport protocol[18] to be best-in-class with respect to cybersecurity and energy networking inside buildings.

  For finding and networking devices, EEBUS applies Multicast Domain Name System (mDNS)/Domain Name System – Service Discovery (DNS-SD) while secure communication is enabled through the Transport Layer Security communication standards Via SHIP EEBUS has implemented remote certificate update procedures that can be applied during ongoing device operation without requiring user interaction.

  Finally, a trust level mechanism allows to pre-configure which other devices shall be trustfully interconnected via public key acceptance procedure.
- **Broad support**: The data structure serves as a basis for the use cases that select and combine required elements in a modular principle: it is then these use cases that are agreed upon and standardized in domain-specific as well as cross-domain working groups of the EEBUS initiative.

  Figure 5.13 presents a diagram of the EEBUS application.

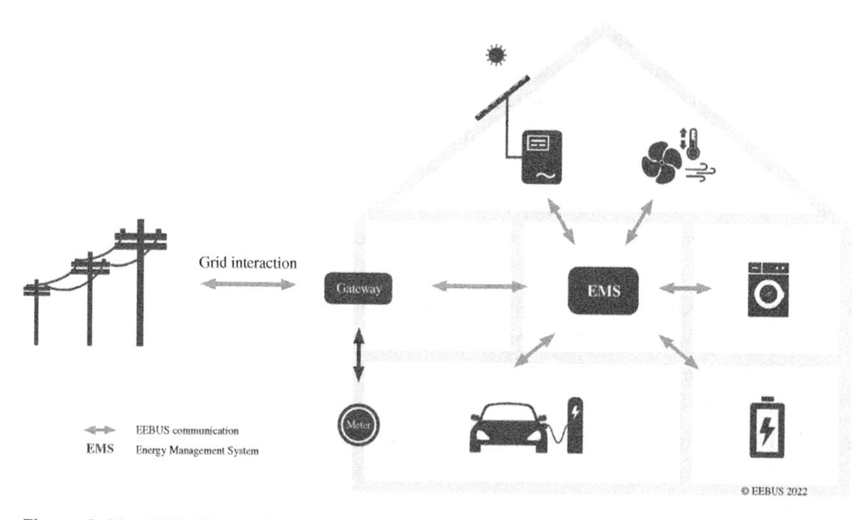

Figure 5.13 EEBUS overview. Source: EEBUS Initiative.

18 https://www.eebus.org/media-downloads/#specifications

5.6 Energy Market Aspects for Appliances

For the most part, appliances are small energy consumers when looking at the overall capacity of the energy grid. However, the great number of devices represents an interesting resource for utilities. Of course, we should not forget to mention non-traditional appliances like pool pumps, car chargers, and even the aforementioned electrical water heaters. These loads can add up quickly. Overall trends to electrification – exchanging traditional fossil fuel-powered systems with electric-powered integrations – are only adding to the need to better control the demand side. As discussed before, early utility ideas included direct control of many home appliances, but these ideas were discarded quickly for various reasons. Future programs for general appliances will more likely focus on two ideas. Price communication and energy requests. The United States sees a push for more flexible pricing structures. Dynamic prices that can vary during the day depending on transmission rates, renewable energy availability, and overall consumption could become reality sooner rather than later. Some utilities are also experimenting with so-called spot prices. Similar to online product retailers, the price will depend on the availability and location of the customer and could adjust up and down during the day. Preset gateway functions could then manage the customer devices without any input from the owner based on the current price. For instance, if the energy price is low, precool the refrigerator and freezer, heat the water in the water heater to a higher temperature, or run the pool pump at full speed to achieve the necessary filtering. When the price increases, dial back these appliances and delay washing machine or dishwasher cycles. The overall trend to achieve IoT could help with this automation process. However, to date, there are only a few vendors that provide solutions that integrate multi-vendor scenarios.

Many appliance manufacturers already have or are working on external control systems. Typically, a cloud-based controller will connect via a smartphone and add back to the appliance. Adding grid connectivity to the cloud would be the obvious next step. The OpenADR interface for instance can be easily integrated into existing web services environments. Appliance manufacturers can now shine by thinking outside the box and providing customers with a certain level of intelligence. However, of course, we are facing a chicken-or-the-egg scenario in which utilities may not know yet what appliances may be capable of participating and in turn, the manufacturers will not know yet what the utilities want. Ultimately it will come down to partnerships between some members of each group to define the applicable programs.

Aside from dynamic pricing, appliances may also see requests to simply reduce or increase power consumption. It will then be up to the control system to find the best approach to shifting energy usage. Also taking the opportunity to emphasize that we are not talking about energy efficiency here but rather the better timing of energy usage. Many common appliances will of course not be able to easily increase their usage. But similar to the price response, they can for instance pre-heat or pre-cool, charge more or less, and so on.

Another interesting aspect that is often cited when it comes to the smart grid is customer privacy. Consumers are concerned about the possibility that the utility can use smart meter data to extract behavioral patterns. The operators would then be able to tell when customers are usually home based on the energy use profile from the demand over time.

While it is certainly possible to analyze smart meter data to some extent, it would however be a large effort to derive any conclusions from this. A smartphone might be an easier choice for advertisers and other interested parties. Demand response systems in fact obscure the data even more by shifting the consumption to different times of the day. In addition, using a technology like OpenADR will also remove the appliance one step from the utility control, and the privacy is further increased.

5.7 Typical DR and DSM Use Cases

There are over 100 known DSM programs in place today globally.[19] Most of them are in the USA but other countries are starting to engage the customers now as well. Many of them are derived from the notion of peak consumption reduction (peak shaving). In general, most of these programs can be categorized as follows.

- **Critical Peak Pricing**: Price structure designed to encourage reduced consumption during periods of high wholesale market prices or system contingencies by imposing a pre-specified high rate or price for a limited number of days or hours.
- **Capacity Bidding Program**: A program which allows a demand resource in retail and wholesale markets to offer load reductions at a price, or to identify how much load it is willing to curtail at a specific price.
- **Thermostat Program/Direct Load Control**: A demand response activity by which the program sponsor remotely controls a customer's electrical equipment (e.g. air conditioner) on short notice. These programs are primarily offered to residential or small commercial customers.
- **Fast DR Dispatch/Ancillary Services Program**: A demand response program that provides incentive payments to customers for load response during an Emergency Demand Response Event. An abnormal system condition (e.g. system constraints and local capacity constraints) that requires automatic

19 https://www.dret-ca.com/wp-content/uploads/2022/05/OpenADR-Deployments-Survey-Report_Final-11-22-2022-v2.pdf

or immediate manual action to prevent or limit the failure of transmission facilities or generation supply that could adversely affect the reliability of the Bulk Electric System. These types of programs may sometimes be referred to as "Ancillary Services."

- **Electric Vehicle (EV) DR Program**: A demand response activity by which the cost of charging electric vehicles is modified to cause consumers to shift consumption patterns.
- **Distributed Energy Resources (DER) DR Program**: A demand response activity utilized to smooth the integration of distributed energy resources into the smart grid.

Few of these programs addressed appliances specifically. Rather, they focused on HVAC systems or aggregation of different energy-consuming devices in homes or small to medium offices. It should be emphasized again that residential consumers voiced out multiple times they do not want to constantly check their energy consumption, rates, grid status, and so on. There is a need for solutions that make this simple for the homeowner by providing a good collection of presets and decision-making help. Also, the ability of the consumer to override an event needs to be maintained. Just imagine you have a party starting and your dishwasher declines to run or you leave to work in the morning and your electric vehicle is only 10% charged.

There are by the way also large capacity demand response programs for commercial and industrial installations available. Often factories use a combination of automated responses and human-guided actions to fulfill the requests from the utilities or aggregators.

Here are some examples of typical programs.

The Sacramento Municipal Utility District (SMUD) in California has been running a CPP program for a long time. The installed equipment was pre-programmed to respond to price signals and the rate and/or price structure is designed to encourage reduced consumption. The California Public Utility Commission (PUC) was adopting CPP programs for residential and commercial customers. The program used the simple signals from OpenADR 2.0a with levels ranging from 1 to 3 and multiple prices in single events. The program is intended to support price-responsive demand for wholesale and retail prices.

Figure 5.14 shows an example of the flexible rate applied by the SMUD over the summer time in (year).

Another example is the Southern California Edison "Bring Your Own Thermostat" (BYOT) program. Customers were encouraged to choose from a variety of products to install in their homes. The program averaged a load reduction of 750 W per customer.

Figure 5.15 shows the SCE BYOT overall ecosystem.

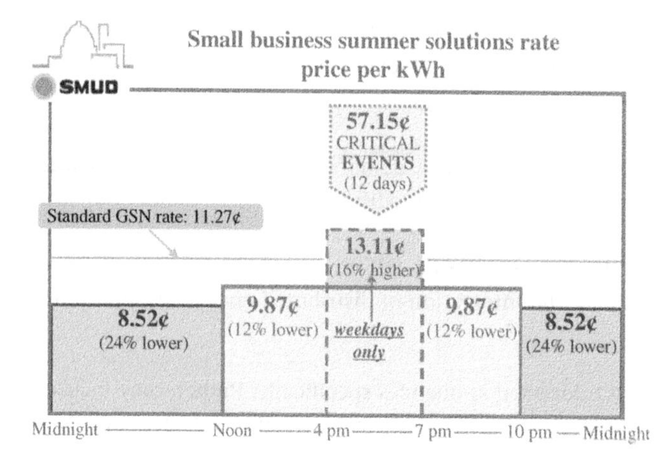

Figure 5.14 Sacramento Municipal Utility District summer rates. Source: SMUD.

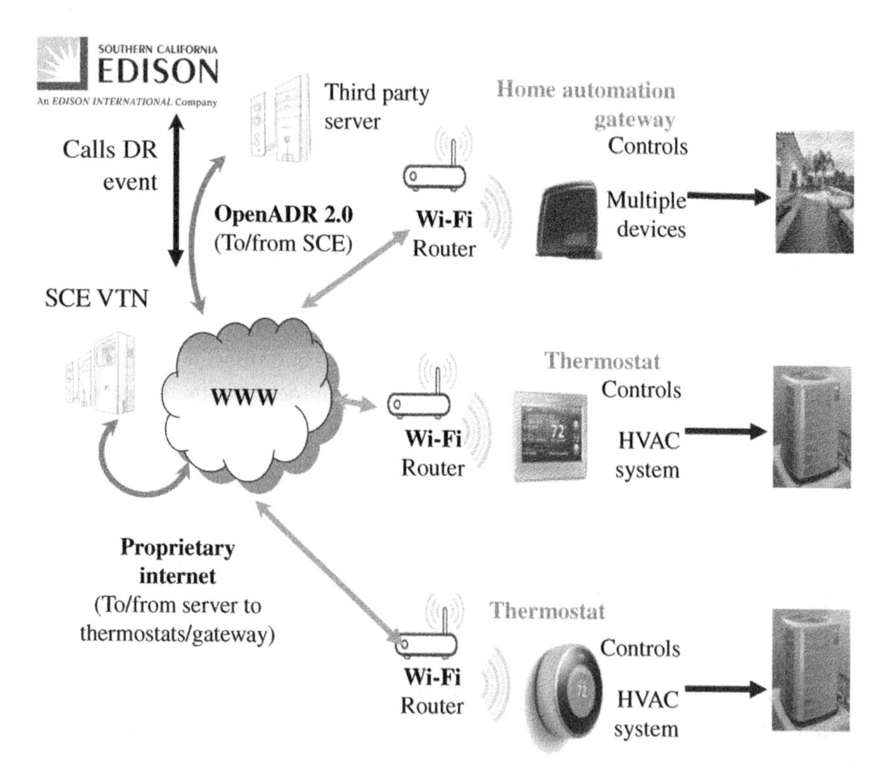

Figure 5.15 SCE Bring Your Own Thermostat ecosystem. Source: OpenADR Alliance.

Aside from these residential and small/medium business DSM programs the industry has seen a number of other areas that quickly became interesting. Some of these programs are worth mentioning here including specific industry use cases.

- **Hybrid Renewables and Storage**: Swell Energy is working to augment Hawaiian Electric's energy supply by absorbing excess wind energy when needed and providing 24/7 fast response to balance the grids. Once complete, the project will supply 25 MW of solar power and 80 MW of battery capacity to Hawaii electric's needs.
- **Electric Vehicle Charging – Chargepoint**: the EV-managed charging company, is partnering with German utilities including Vattenfall and Stromnetz Hamburg to deploy a network of EV charging locations that allow customers to participate in demand response programs. OpenADR protocol services employed in this process include Registration – EiRegisterParty is used to identify entities such as Charging Point Operators and other parties; this is necessary before an actor can interact with other parties – and the OpenADR Event/EiEvent are central event functions and information models that are used to reduce load. This service is used to activate a demand response. It should be noted that there are already many other managed EV charging programs up and running. However, this program is one of the first in Europe.
- **Battery Storage**: Consolidated Edison acted to make dispatch OpenADR-ready and OpenADR-friendly. In order to better integrate energy storage resources, Consolidated Edison was looking, in particular, to leverage the OpenADR Report Service for purposes of battery telemetry.

Symbols and Abbreviations

DER	Distributed Energy Resources
DR	Demand Response
DRRC	Demand Response Research Center
EI	Energy Interoperation
HTTP	Hypertext Transfer Protocol
IRC	ISO/RTO Council
LBNL	Lawrence Berkeley National Laboratory
OASIS	Organization of Structured Information Standards
OCPP	Open Charge Point Protocol
OpenADR	Open Automated Demand Response
PICS	Protocol Implementation Conformance Statement
TC	Technical Committee

VEN	virtual end node
VTN	virtual top node
XMPP	XML Messaging and Presence Protocol

Glossary

The OpenADR Alliance The OpenADR Alliance is comprised of industry stakeholders that are interested in fostering the deployment of low-cost price- and reliability-based demand response communication protocol by facilitating and accelerating the development and adoption of OpenADR standards and compliance with those standards. These include de facto standards based on specifications published by LBNL in April 2009, as well as Smart Grid-related standards emerging from OASIS, UCAIug, NAESB, and IRC.

OpenADR 2.0 Profile Specification The OpenADR 2.0a, or b Profile Specifications provide specific implementation-related information in order to build an OpenADR-enabled device or system. Developers shall use the Profile Specification in conjunction with the schemas, sample payloads, PICS, and test plans.

OASIS Energy Interoperation (EI) Energy Interoperation standard describes information and communication model to coordinate energy supply, transmission, distribution, and use, including power and ancillary services, between any two parties, such as energy suppliers and customers, markets and service providers, in any of the domains defined in the Smart Grid. The EI 1.0 standard was used as a basis for OpenADR 2.0 Profile Specification.

Demand Response A mechanism to manage customer load demand in response to supply conditions, such as prices or availability signals.

References

BACnet, *Information about the BACnet organization and the standard*, 2023, (online) - www.bacnet.org.

Connectivity Standards Alliance, *General information about CSA*, 2022, (online) - https://csa-iot.org/.

Consumer Technology Association, *Modular Communications Interface for Energy Management (ANSI/CTA-2045-B)*, 2021 - https://shop.cta.tech/collections/standards/products/https-cdn-cta-tech-cta-media-media-ansi-cta-2045-b-final-2022-pdf.

EEBUS, *Information about the EEBUS ecosystem*, 2022, (online) - www.eebus.org.

LonMark International, *Information about the LonMark organization and the standard*, 2021, (online) - www.lonmark.org.

OpenADR Alliance, *Overview of the EcoPort brand*, 2022, (online) - https://www.openadr.org/ecoport.

OpenADR Alliance, *Overview of the OpenADR Standard*, 2018, (online) - https://www.openadr.org/overview.

OpenADR Alliance, *The OpenADR Test Tool*, 2022, (online) - https://www.openadr.org/openadr-test-tool.

Open Charge Alliance, *Information about the OCPP organization and the standard*, 2021 - www.openchargealliance.org.

Southern California Edison, *OpenADR Deployments Survey*, 2021, (online) - https://www.dret-ca.com/wp-content/uploads/2022/05/OpenADR-Deployments-Survey-Report_Final-11-22-2022-v2.pdf.

UtilityDive, *Demand response failed California 20 years ago; the state's recent outages may have redeemed it*, 2020, (online) - https://www.utilitydive.com/news/demand-response-failed-california-20-years-ago-the-states-recent-outages/584878/.

6

Energy Smart Appliances

Neomar Giacomini

Senior Engineering Manager for Electronics Hardware Development at Whirlpool Corporation, Benton Harbor, Michigan, USA

6.1 Energy Smart Appliances

This chapter will present a debate on each appliance that is capable of energy collaboration and are present in current households, or available in the market for purchase. As the reader will notice many of these use cases are not available in current appliances, not for the lack of hardware and software technologies availability, but instead due to the lack of overall appliance market migration to a more collaborative energy ecosystem as depicted in Chapter 2.

"An appliance is a device or piece of equipment designed to perform a specific task, typically a domestic one" (LEXICO 2022), so traditionally speaking most consumers will relate specifically to Large Home Appliances (LHA) such as refrigerators, washer, and dryers. However, appliances go beyond that including HVAC (Heating, Ventilation, and Air Conditioning) and others.

Arguably, with the definition presented for appliances, the list can even include any other device that performs a specific task at home. Some houses use the garage door opener more than the water heating system during the summer, for instance, making that garage door opener an important task executer for the consumer. In another example through comparison, a washing machine cleans clothes and is a large home appliance, but at the same time, the vacuum and mop robots currently available in the market will clean the house daily. Going further, houses that are using lawncare robots are benefiting significantly from the reduction of labor these devices take from the homeowner.

This is an important debate not just from an energy perspective, but also in regard to how users will adapt, consider and reclassify these new devices in the incoming years.

Energy Smart Appliances: Applications, Methodologies, and Challenges,
First Edition. Edited by Antonio Moreno-Munoz and Neomar Giacomini.

That said, this chapter will include Large Appliances, Small Appliances and also other electronic/electrical devices capable of Energy Collaboration, even on a small scale.

6.2 Which Appliances?

As the title inquires, a major aspect of this session is to list and classify all devices capable of some sort of energy collaboration. It is important to look at a comprehensive list because some devices are easy to forget when looking just at house operation and maintenance.

Take a hair dryer as an example, a small portable device that in some homes depending on the size of the family is used multiple times a day and run between 800 and 1900 W usually used for about 10–15 minutes. It sounds insignificant to the Smart Energy Home scene, but that is nearly 2000 W of power added to the energy consumption that may already be at its peak at that given moment.

Examples like this show the importance of looking into all devices and discussing whether they can, join the collaborative Smart Energy Home scenario and support the efforts to flatten the demand.

To set the stage for the discussion, Table 6.1 presents a list of devices that are energy-dependent and can be connected. It also includes some devices that are currently not connected (e.g. hair dryers) but may play an important role in energy collaboration and demand peak avoidance.

This table is meant to provide a moment of reflection regarding most, if not everything, that consumers have in their houses consuming energy. Some devices like Large Home Appliances are obvious players in the energy consumption and often the target of energy consumption-related studies, but when the intention is not to dive on the absolute energy consumption, but instead on energy collaboration aspect, any device should be seen as a potential collaborator.

The remainder of the chapter will explore these devices looking at their use and how energy-dependent tasks can potentially be shifted in time to support flattening the demand. There is no intention to deep dive into algorithm development or even hardware design. But instead, the intention is to explore how critical loads within these devices can be controlled shifting priority across all that is taking place in the household to minimize peaks in the demand, sometimes directly or indirectly by suggesting to the consumer to opt for a delay.

It's important to mention the block diagram figures presented in this chapter intentionally omit minor elements like sensors, switches, and others, therefore focusing only on macro elements related to considerable Smart Energy Home use cases. These figures by no means intend to show functional block diagrams. Special attention as well to the presence of a connectivity module in the diagrams,

Table 6.1 List of appliances and other connected devices available in 2022.

Large home appliances	Health, comfort and care	Non-appliances
Dishwashers	Air purifiers	Electric vehicle (EV) cars and motorcycles
Dryers	Bed frames	Desktop computers
Grills and smokers	Cat litter robots	Modems and routers
HVAC	Hair dryers, brushes, and straighteners	Power banks
Microwaves	Health monitors	Power tools
Refrigerators and freezers	Humidifiers	Smartwatches
Stoves, ovens, and cooktops	Indoor exercise bike	Smartphones
Washing machines	Sleep sensors	Tablet computers
Water heating	Tracking devices	Uninterrupted power supplies
	Treadmills	Video games
Small appliances	Water filtration systems	
Coffee machines		**Entertainment**
Blenders	**House automation**	Aquariums
Faucets	Blinds and shades	Audio systems
Food processors	Garage door opener	Streaming receivers (cast feature)
Mixers	Gardening sensors	Televisions
Robotic lawn mowers	Light bulbs	Virtual assistants (multiple forms)
Toasters	Mop robots	Virtual reality goggles
	Pet feeders	
Monitoring	Power strips	**Security**
Energy monitors	Smart power switches	Alarms
Haptics sensors	Presence, proximity, and Movement sensors	Cameras
Leakage detection sensors	Thermostats and temperature sensors	Door locks
Smoke and gas sensors	Vacuum cleaners	Doorbell cameras
Water detection sensors	Vacuum robots	
Weather sensors		

Source: Neomar Giacomini (co-author).

current appliances in the market may or may not have connectivity, so that aspect is not to be taken as granted. Even those that are already connected, to this day, are not enabling comprehensive Smart Energy use cases, so the examples presented in this chapter are forward-looking.

The use cases described long this chapter may relate not just directly to devices, but to the user and actions taking place in the house, in any of such cases user consent to access activity trackers, motion sensors, or any other data input involving the user, should be requested.

6.3 Smart Energy Controller

The chapter presents potential Smart Energy use cases on the variety of devices shown in Table 6.1, that discussion will repeatedly refer to a Smart Energy controller. Figure 6.1 depicts the many features that may be involved in the

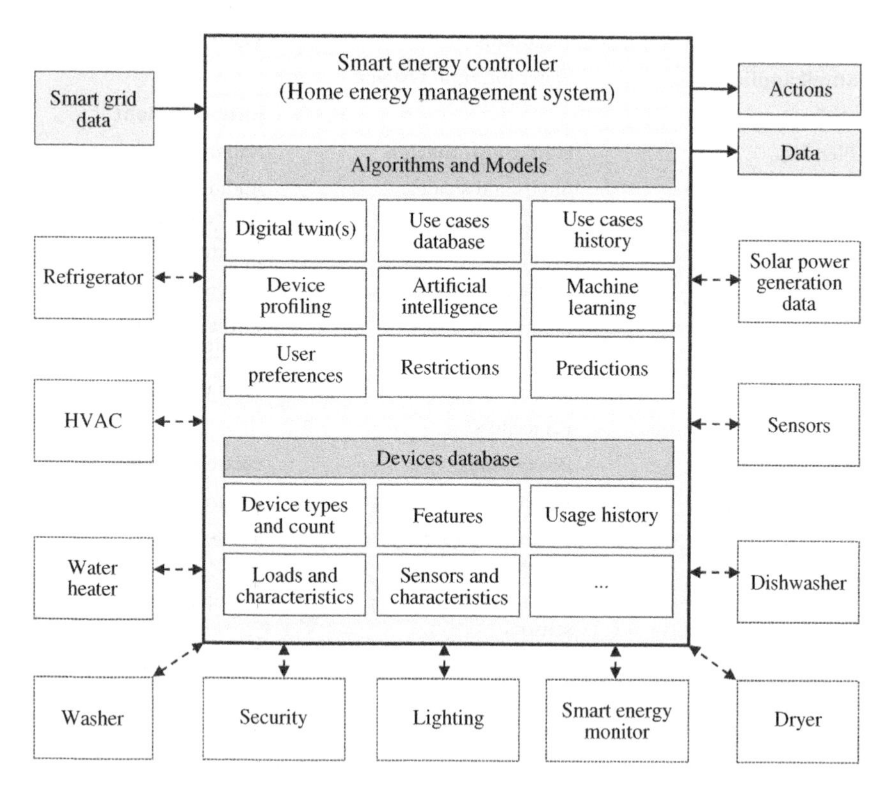

Figure 6.1 Smart Energy Controller block diagram. Source: Neomar Giacomini (co-author).

construction of a Smart Energy Controller. This device is also known in the literature as Home Energy Management System (HEMS).

Important to note this is not necessarily an independent device, it can also be a logical implementation within a Smart Assistant for instance. Either way, the fact is that a centralized control able to coordinate all energy aspects of the house and its devices is necessary due to the multitude of vendors, variety of devices and capabilities to generate Smart Energy use cases.

The use cases will refer to the Smart Energy Controller as a decision maker, the entity able to gather information from all possible sources across the house whether they are sensors or devices notifying energy consumption, prioritize the requests, generate energy demand optimal scheduling, and communicate back to all devices on the agreed plan of execution, doing so in a transparent manner not impacting the resident's lifestyle.

6.4 Large Home Appliances

6.4.1 Dishwashers

Present in about 64% of American homes (Statista 2021), dishwashers are key home appliances for many homeowners. The ownership rate of dishwashers in 2021 in selected countries is shown in Figure 6.2.

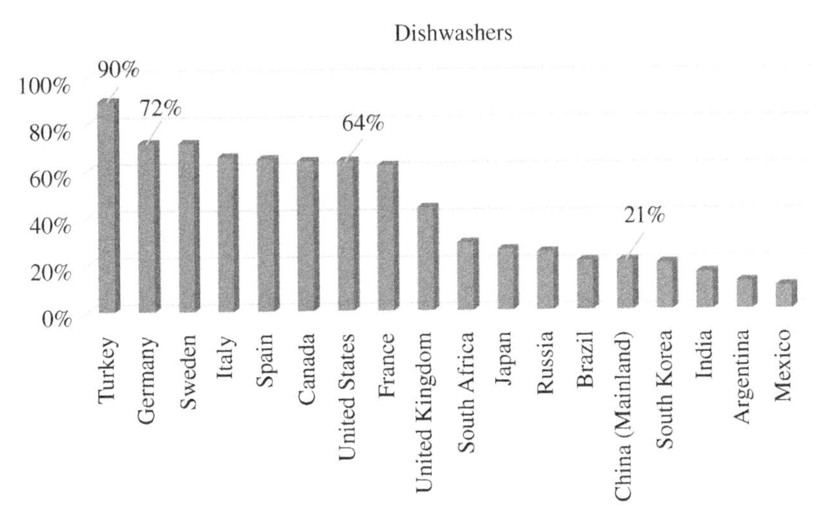

Figure 6.2 Ownership rate of major household appliances in selected countries 2021. Source: Adapted from Statista (2021).

These appliances are mainly composed of components to enable water intake and water disposal, water jets powered by a water pump, and a heating system to heat up the water. Additional elements, such as a human–machine interface (HMI), and sensors for temperature and water turbidity are also present in many models.

In terms of energy, dishwasher's key aspect is the water heating, and the energy consumption is directly impacted by the volume of water utilized in the cycle, temperature of the water received from the water intake, and the target temperature for the wash cycle. The heating process for the water is in general provided by a Calrod heater with power rating of 800, 1200, 1800, or even higher depending on the manufacturer and capacity.

Dishwashers are a critical appliance, however for the household, but they are not like an oven that the consumer must run at a specific given time to prepare a meal because it's about time for the family dinner. In some instances, the dishwasher can potentially have its cycle start delayed, given user approval for that of course.

For an energy perspective along with water heating, other main aspects of a dishwasher are the water valve for intake, water pump for disposal, and the water pump used to drive the water jets. These four elements comprise the subset that will build the macro energy profile of a dishwasher and are shown in Figure 6.3.

To support flattening the energy demand from a household there are two main aspects that involve dishwashers.

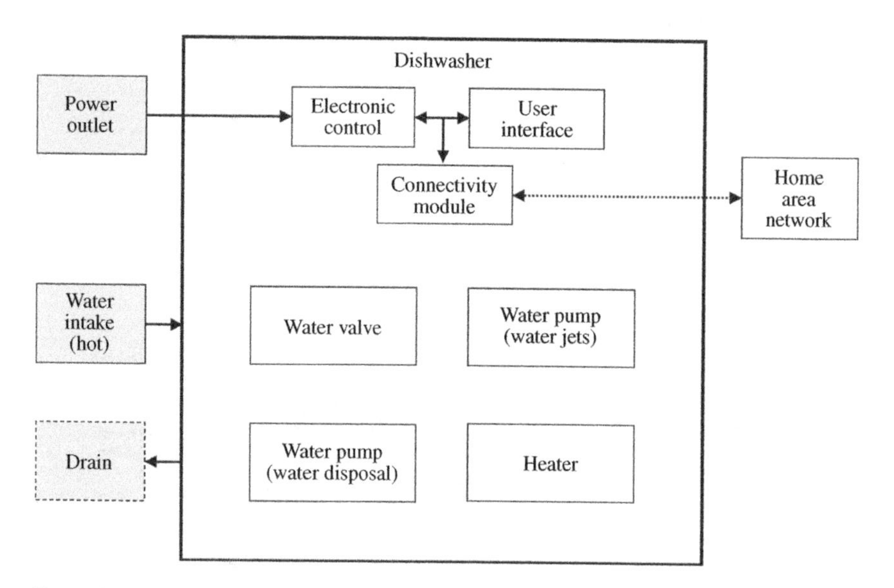

Figure 6.3 Dishwasher macro elements for smart energy applications. Source: Neomar Giacomini (co-author).

The first one and more obvious is to provide the ability to the consumer to enable cycle start delay. In this case, a dishwasher that is connected to the Smart Energy Home ecosystem would collaborate with the smart energy controller discussed in Chapter 2 to identify the most cost and demand-efficient time to run the cycle.

It is of course a critical feature that users should have easy access to enable/disable because certain dishwashing cycles may be critical to run on-time to enable dishes to be ready for the next meal. This is a feature likely to be used for dinner time for instance since the dishes have the entire nighttime to be washed.

And that does not mean a pre-wash cannot be executed right on-time of loading, but at least the major dishwashing and drying cycles can potentially be delayed.

Delaying the cycle would shift the energy consumed by the water valve, water pumps, and heater to a different, energy-optimal time.

The second aspect to consider is the intake water temperature. Depending on the house water blueprint, presence or not of a circulation water pump and other factors such as season and city water temperature, the initial water temperature at the start of the cycle may differ significantly from house-to-house, or day-to-day. This factor will impact the amount of energy delivered by the heater to reach the target dishwashing temperature.

Also, simple factors like the consumer using hot water in the kitchen faucet before starting a dishwashing cycle would enable the unit to intake hot water from the start, instead of loading a good amount of cold water before starting to receive warm water.

One solution to avoid intaking cold water is, in houses where the dishwasher is connected to a heater water line, is to have the dishwashing collaborating with the water system of the house and activating as needed a heated water circulation pump. That would make sure the water available in the inlet at the beginning of the cycle is already pre-heated, therefore saving energy at the appliance level.

This feature is meaningful because water heating systems may use solar heating, or other sources of energy such as natural gas that may be cheaper, or at least mitigate electrical energy consumption, therefore supporting the grid.

With these two aspects of demand shifting and intake water temperature management, the dishwasher can be integrated into the Smart Energy Home ecosystem and start providing benefits to reduce energy consumption and shifting demand.

6.4.2 Dryers

Another large appliance present in many houses that goes along the same lines as the dishwashers, in some instances the drying cycle is critical to the consumer and must be executed on time, in other moments it can be delayed for some time to accommodate demand management.

A consumer-accessible configuration to enable/disable such a feature must be available to accommodate that need.

In the market today consumers can opt to buy multiple types of dryers such as Air Vented which are based on heaters and spell out hot humid air usually ducted to the exterior of the house. Condensing dryers are also based on heaters, Heat Pump dryers that use a heat-pump cooling/heating system based on a compressor and refrigerant gas, and also Hybrid Heat Pump dryers that combine an extra heater to the regular Heat Pump models to speed up the cycle time (Whirlpool Canada 2017).

The approach to drying on all these variants is different, and although they may differ significantly in terms of absolute energy consumption, having the Heat Pump models a much lower use, for instance, in terms of electronics loads activation and the need to have them all running during a cycle is similar.

The vented and condensing models need the drum motor and heaters turned on, or cycling, during the operation process. On the heat pump models the drum motor and compressor also be turned, or cycling, during the operation process.

That said, dryers can have a significant impact on the total instant energy demand of the house by delaying their entire cycle, or potentially running a portion of it, stopping, and returning to operation after a given critical energy event in the house has passed. Drying cycles are different from cooking cycles, for instance, where a pause in the process would potentially ruin the food. The fact that the process is about removing water from clothes, it has more flexibility on how it can be delayed or paused.

The event of pausing a cycle can bring benefits to the energy demand management in case another critical load in the house needs to be in use. Picture a scenario where a house is already close to its energy demand limit, a limit due to grid overload, or a limit to avoid crossing the levels that would take the tariffs to a higher rate, pausing a drying cycle to run another appliance/load may avoid crossing such level.

Let us look at a vented dryer, Figure 6.4 shows the macro elements relevant to a connected electric vented dryer in a Smart Energy Home scenario. These models are available in the market with heaters in the range of kilowatts, 5400 W for instance.

One example of a situation with a beneficial pause in the operation would be the case the consumer decides to use a 1500 W microwave to make popcorn while the dryer is running, and the energy demand for the house is already high. The dryer could, without a noticeable difference in the performance, turn off the heating element and therefore releasing 5400 W of the current energy demand for the house, while even still maintaining the motor in operation. The popcorn cycle would run for a couple of minutes and after the microwave operation is ceased the dryer heater could return to its normal state. That action would avoid adding

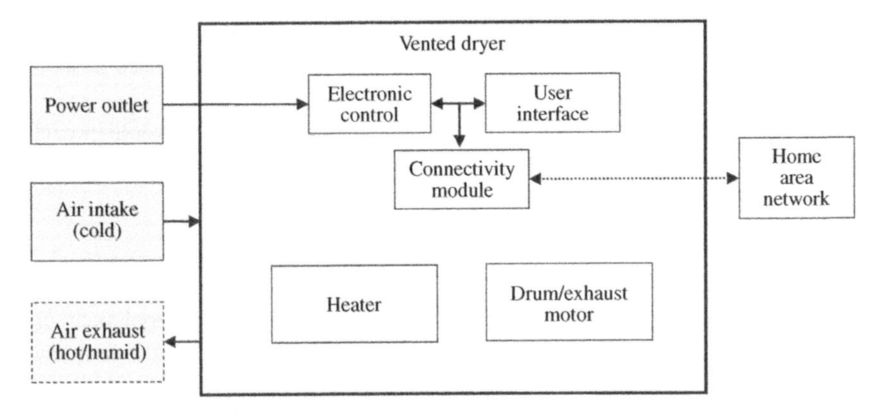

Figure 6.4 Typical Electric Vented Dryer macro elements for smart energy applications. Source: Neomar Giacomini.

up the power consumption of both appliances without a significant, and noticeable, impact to the consumer. The reason the pause would have a low impact to the consumer relies on the fact that a two minutes microwave operation may be negligible in time compared to a typical 60-minute drying cycle.

Maintaining or not the motor in operation is a design decision since models having the exhaust feature coupled to the same motor as the drum (via belt for instance) would keep cold air circulating during the pause, cooling down the clothes. Therefore, decoupling these features in two separate motors, or turning off a single motor completely would be a tradeoff decision for the engineering team.

Such energy cooperation across these and other appliances described in this chapter will require cooperation between the appliances, the Smart Energy Home ecosystem and the energy controller to enable and/or support the exchange of information across the appliances, therefore enabling proper decision-making.

A similar approach may be considered for the heat-pump models, or even for the extra heater in the hybrid ones.

This chapter is focused on the electric aspects of these appliances, but further consideration on how the hot air exhaust from a vented air dryer could be repurposed to support HVAC systems. These possibilities can be considered under thermal-focused studies.

6.4.3 Grills and Smokers

Grills and smokers same as dryers can be found based on many different fuel sources. Electric, gas, charcoal, hardwood, and so on.

For the context at hand, the electric ones are commonly found from 1500 to 8500 W. Although this is a considerable power this is an appliance used in food preparation and at specific times critical to the user.

That being the case reducing power, delaying or pausing operation would impact the expected performance and cause customer dissatisfaction. For that reason, no viable use case was identified.

6.4.4 HVAC

HVAC have a major presence across the globe. The need for such appliance, or appliances, depends heavily on the climate of the country. HVAC comes in many formats. In the U.S.A. many installations are comprised of a furnace that integrates heating and ventilation plus an external unit for cooling. In other countries or areas without significative winter temperature drop, households are equipped just with an air conditioner that may be a wall unit installed through a unit installation window, or a split unit with an internal and an external unit.

On the opposite side, houses in cold climate tend to have only a furnace and ventilation, and on top of that many houses across the globe are equipped with ventilation systems, whether in the form of ceiling fans, stand-up fans, and other variants.

Vendors also offer units with all HVAC features all-in-one such as the packaged HVACs.

Cooling units these days are quite straightforward on the use of compressors and refrigerant gas-based units. Heating on the other hand comes in many forms, electric, gas, and hydronic based on steam or water in the modern implementations.

In short words, the HVAC space is crowded with different options that consumers may pick from.

Before debating on possible solutions, it is worth pointing back to Chapter 2 where the need to have a Smart Energy Controller and Smart Energy Home ecosystem capable of collaboration across all these devices to map what type of HVAC was mentioned, features, and existing loads for advanced energy use cases should be identified to build an energy model for the house including all its devices, a digital twin.

Thermal feeling regarding environment comfort is a must for the users, typical settings for the temperature while using a Smart Thermostat is about 3–5°F. It is a small range but allows the opportunity to use it for Smart Energy use cases.

Figure 6.5 shows a standard HVAC split system macro elements for smart energy applications.

To provide a baseline in terms of power in use on such systems, a 240 m^2 home in Michigan, USA, typically requires an air conditioner with a 3 ton[1] nominal

---

1 One ton equals 12,000 BtuH cooling capacity.

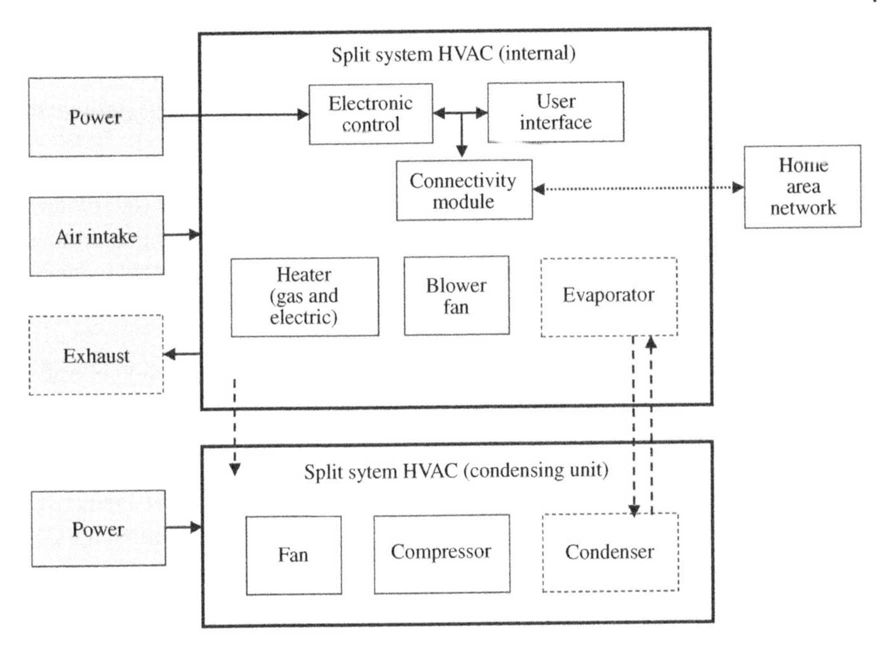

Figure 6.5 Standard HVAC Split System macro elements for smart energy applications. Source: Neomar Giacomini (co-author).

capacity. Such units contain a compressor and fan that operate at the same time, including also the ventilation fan, combined this group may surpass 2500 W of instantaneous power consumption.

The furnace for a similar application will have a heating capacity of 60,000 BUTH, aside from the thermal aspect that will not be covered, electrically speaking these units run at about 1100 W due to the circulation fan, the major electrical load present in the unit.

These loads will by the nature of HVAC systems already cycle according to the thermostat needs.

Based on the characteristics described, the options on HVACs rely on pausing a cycle, cutting it short, or delaying its activation. Pausing for an acceptable period or cutting the cycle short may be done without impacting the expected temperature range. Delaying the start of a cycle may result in the temperature of the house going slightly lower or higher than the settings, so in that case user approval through settings may be required. The ECO mode available in certain smart thermostats benefit from additional temperature range, especially while the consumer is away from home.

Pausing or cutting a cycle short must be considered cautiously, depending on how the algorithm uses this feature it may result in an increase in the number of times motors (e.g. compressor, fan, etc.) start up, and this process is known to use additional energy due to system inertia. The trade-off between supporting peak demand and accepting potential energy consumption due to additional motor starts should be evaluated in detail based on components characteristics and system design.

Due to these multiple factors involved it is not recommended to consider interfering with HVACs cycling unless there is a considerable need. Having in mind a scenario where the house is already at peak demand and the user is about to make popcorn using a microwave, interfering with an HVAC to accommodate for such a short (e.g. two minutes) cycle will be unreasonable.

What could be done, however, is to shift fuel sources in case that option is available. A furnace able to run both gas and electric may shift to gas to free up electrical energy for another appliance.

Considering the options described above it is clear that delaying the HVAC start is the simplest option and required less coordination across appliances. Predictive algorithms, however, may help expand to additional use cases, including anticipation of the cycle. Appliances such as microwaves, electric ovens, and induction cooktops are likely to be used around mealtimes like lunch and dinner.

A predictive algorithm capable of identifying user patterns and feeding this data into the Smart Energy Home model would enable HVAC and other systems to anticipate energy needs to free up demand for a given incoming predicted energy consumption. A sample case would be the predictive algorithm having identified the user usually starts preparing a meal at 11:30 a.m. and crossing that information with the house temperature cycling behavior could anticipate a cooling/heating cycle.

6.4.5 Microwaves

In 2022 in the United States 84% of households had a microwave (Statista 2022a), refrigerator came in second with 83% and 74% for washing machines. This food preparation appliance uses 2.4 GHz electromagnetic waves to heat food items and became popular due to the easiness of use and time required to heat up or prepare a meal.

Different models are rated at a cooking power of 750, 1000, and 1500 W, with certain units reaching even 2200 W. Microwaves in general are comprised of a Magnetron which is typically the main load in this appliance, a turntable motor, lamp, door switch, user interface, control board, and in combo models also Broil and Convect heaters and a convection fan. The power in combo units can of course reach much higher levels.

Microwaves are available as countertop appliances, built-in as single and combo units paired with wall ovens, and as microwave, hood combos to be installed over the cooktop area. Any of these variants are currently available in the market with the aforementioned features/loads.

A typical high-end unit with all the described cooking features as of 2022 utilizes a convection heater of 1400 W, microwave cooking power of 1000 W, and a grill heater of 1300 W. Combining these with the turntable motor and convection fan units may reach considerable power depending on the selected cooking cycle. It's also worth noting modern driving technology also offers Magnetron activation through inverter technology instead of the traditional transformer-based approach, enabling further power control.

As mentioned in Section 6.4, cooking appliances are not simple to interfere with their operation, the cooking item being prepared requires a set step of actions including given temperatures and timing to reach the desired outcome. The appliance, or even the user, needs to be cautious while pausing, delaying, or cutting short the appliance operation risking not achieving the desired look, flavor, or texture expected from the cooking process.

Figure 6.6 shows a microwave with broil and convect features macro elements for smart energy applications.

The use of this type of appliance, similar to stoves, cooktops, and ovens that will be discussed in a further session is based on the user's on-time needs. Meaning that a consumer would not set a food item for preparation and be able to wait for that process to start at a later moment. That is not an expected user behavior.

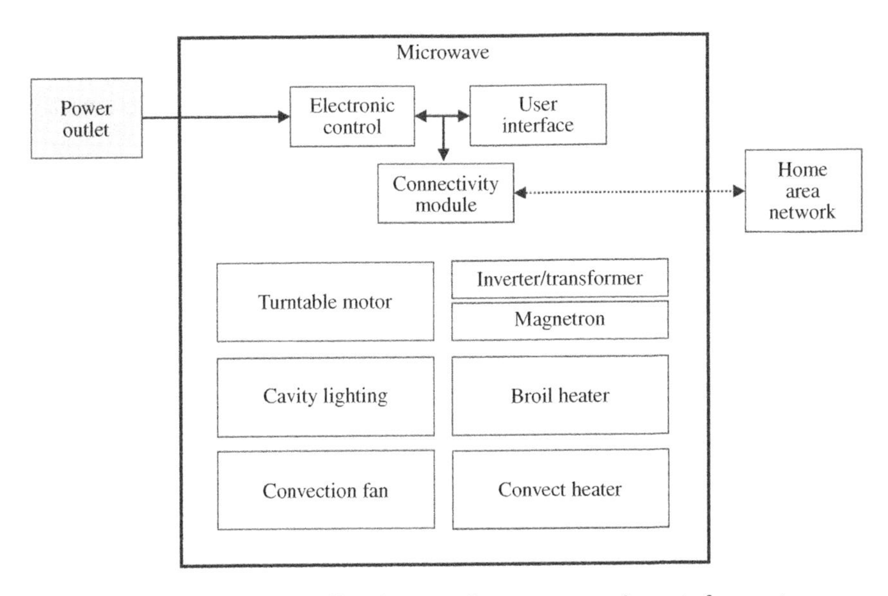

Figure 6.6 Microwave with broil and convect features macro elements for smart energy applications. Source: Neomar Giacomini (co-author).

That said, the energy management of cooking devices is in essence an in-product aspect to be dealt with during design. Smart Energy Home use cases may use cooking appliances to trigger load shifting in other devices, but not the other way around, or at least in a minimalistic way.

For cooking appliances Smart Energy use cases, users' approval and participation may be required. Alerts showing the user that using a given cooking appliance at that moment will result in peak or over demand may be an interesting approach to shift behavior and drive the adoption of another food preparation method at that time. Like using a gas stove for instance.

Cooking appliances therefore present more opportunities for triggering Smart Energy use cases in other appliances than the other way around. Smaller loads like cavity lighting may present an easy opportunity for control; however, these are not major loads in the system.

6.4.6 Refrigerators and Freezers

Food preservation appliances such as refrigerators and freezers are basically a must in modern society. In 2022, a unit sale of refrigerators is surpassing 200 million units and freezers 45 million units, a year globally (Statista 2022b). These appliances vary in capacity and complexity, and the connected models still represent a small portion of the installed base.

Refrigerators and Freezers are very similar in terms of operation, mostly based on the coolant gas' compressing and expansion thermal characteristics. The principle of operation involves a closed loop of coolant gas, a compressor, evaporator, condenser, and expansion valve. Multiple auxiliary components such as fans, air dampers, sensors, electronic control, and user interface are also part of the system. The major difference between refrigerators and freezers is in the insulation material, control scheme, and esthetics.

To avoid external condensation, characteristic of surfaces that get too cold and reach the dew point bringing air humidity to condensate on it, it is very common to find these appliances also using heaters underneath the external surfaces accessible to the user. Additional features such as icemakers, water dispensers, and others also contribute to power consumption.

Figure 6.7 shows the refrigerator macro elements for smart energy applications. It's important to emphasize these sub-components are not present in all types of refrigerators. Cold plate, frost-free, dual, or even triple evaporator units will differ on the types and number of subcomponents.

Refrigerators and freezers are food preservation appliances, therefore the temperature ranges are critical for the proper function of the appliances. That means Smart Energy use cases must never influence the operation taking them out of regular operation conditions pre-established by the vendor.

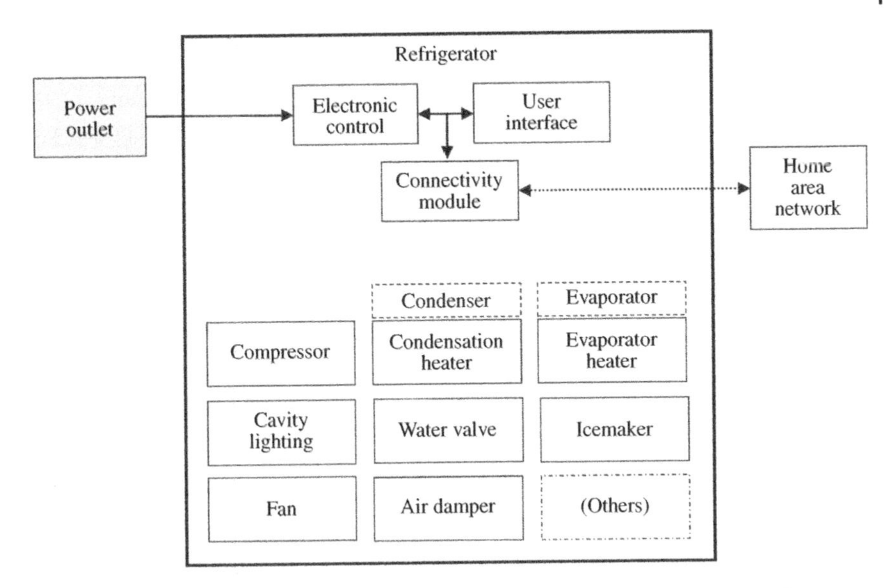

Figure 6.7 Refrigerator macro elements for smart energy applications. Source: Neomar Giacomini (co-author).

Additional features such as cavity lighting or water dispensing are on-time functionalities that are activated with the user's presence and therefore cannot be anticipated, delayed, or even paused.

That brings us to extra cooling and thermal storage use cases. With proper collaboration with the Smart Energy Ecosystem, these units may increase their cooling cycles to store extra cold before higher energy demand takes place across the house. This is an option, especially for freezers. Whether the appliance is a freezer only, or a combo unit, the freezer compartment can be taken at even lower temperatures, demand requiring, and last longer before the next cycle starts. The refrigerator compartment has a limitation as temperatures too close to 0 C will cause frosting and compromise food storage performance.

Future applications may even consider thermal storage to take place overnight using high latent heat chemicals. A low-temperature thermal bank would be able to store energy during low energy demand periods of the day and deliver it back to the system as needed.

Other functions may also benefit from a predictive algorithm capable of identifying user patterns. Feeding this data into the Smart Energy Home model would enable the refrigerator to identify whether icemaking for instance could be done only during nighttime, minimizing energy consumption during higher demand times of the day.

6.4.7 Stoves, Ovens, and Cooktops

Another group of core appliances broadly used consists of stoves, ovens, and cooktops. The main function of these is to support food preparation through processes such as boiling, frying, baking, broiling, and so on.

Like the microwave, these food preparation large appliances are mostly used shortly before meals takes place. The cooking process time for regular meals is not extensive, but for special preparations such as a large turkey, it may take hours.

Due to the food preparation-related nature of these appliances', delaying the cooking process is not an option, and so it is not to pause or cut the process short before completing the intended function.

Interfering with the cooking process, same as mentioned for microwaves, would risk not achieving the desired look, flavor, or texture expected from the cooking process.

Figure 6.8 shows oven macro elements for smart energy applications. Stoves and cooktops differ from this block diagram, but since no use cases specific to the internal components of those appliances will be discussed, the block diagrams will be omitted.

In regards to Smart Energy use cases there is not much the Smart Energy Ecosystem can do to interfere with cooking appliances, so the option in this case is to understand the loads and their power to make decisions on how to trigger demand shift in other devices in the house.

Typical bake and grill heaters can be found in these appliances ranging from 3 to 4 kW, convect heaters usually lower than that around 1 kW, and convection fans around 150 W. Other loads present in such ovens are not that significant to smart energy use cases.

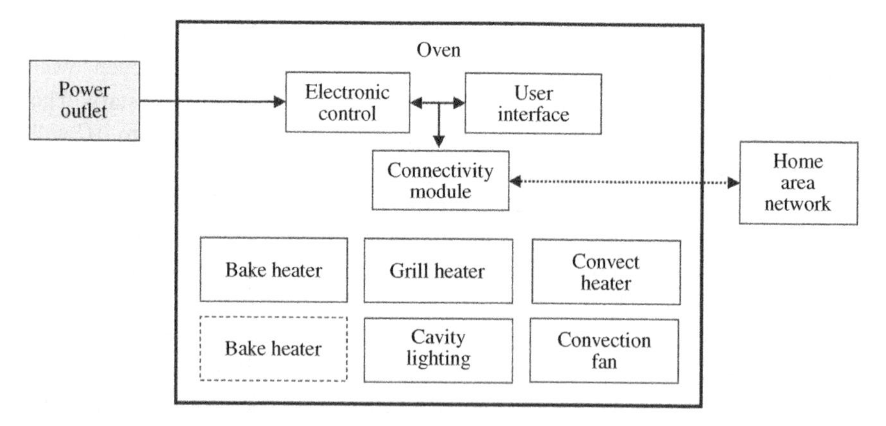

Figure 6.8 Oven macro elements for smart energy applications. Source: Neomar Giacomini (co-author).

The cooking table on electric stoves and cooktops also have considerable loads ranging from 100 W warm zones up to over 3 kW on large. Induction cooktops have a dedicated session in Chapter 11.

With the understanding of these loads and properly designing the appliance control algorithms to interface with the Smart Energy Home and Smart Energy Controller, provide quantitative data on the power each feature will use and events that are starting, other appliances that allow flexibility to free up demand, if necessary, will be able to take action.

6.4.8 Washing Machines

Another laundry appliance is present in many homes, after 2020 the annual global volume sold surpassed 160 million units and continues to grow (Statista 2022b).

Washing machines come in many formats, top-loader vertical axis, top-loader horizontal axis, front-loader, combos with drying functionality integrated, and so on. Although they differ in design, the loads existing in the system are quite similar.

The simplest electromechanical models, due to the lack of advanced electronics, will not be discussed as they are less likely to become connected in the future. The mid and high-end models will be the target of the analysis.

These models typically have a combination of the following components, or a subset of, an intake water valves, sensors, a motor, water pumps for circulation and drain functions, heater(s), door switch, door lock, a control unit, user interface, and other supporting components. Once again worth noting, not all these elements are crucial for Smart Energy use cases.

Figure 6.9 shows the washing machine macro elements for smart energy applications.

In terms of Smart Energy use cases, anticipating, delaying or pausing the washing machine operation are actions to be considered, this is possible depending on the stage of the washing cycle and also for not being related to food preparation, which was shown as a limiting factor for Smart Energy use cases in cooking appliances.

The main loads in this system are heaters and the motor itself. Typical water heaters are in the range of 900–2000 W, and the motor varies significantly due to motor type with models in the USA driving up to 1000 W on inverter-based models during the ramp-up of the spin cycle. Washing machine capacity, washing cycle stage (e.g. agitation compared to water removal spin), and so on also impact the power consumption. A drain pump for instance would be in the range of 50–70 W.

Important to note that all these power values are a reference to the instantaneous individual power measured while loads are activated. No considerations are being made on energy consumption itself in kilowatts-hour (kWh). Reason being that for instant demand it's more important the momentary state of the

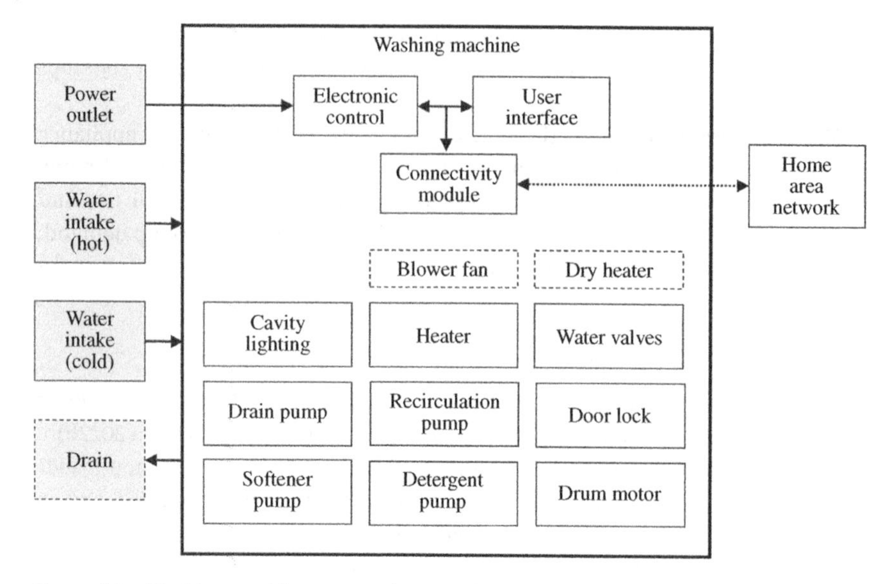

Figure 6.9 Washing machine macro elements for smart energy applications. Source: Neomar Giacomini (co-author).

loads and their power, than the average energy consumption over time. Energy consumption is of course very important, but the focus of this chapter is to discuss demand-shifting opportunities, not to judge the energy consumption required to perform the intended function of any of these appliances.

Along the many times, the user will utilize this appliance, there will be instances where the operation will be on-time, for example when a quick wash is required. In those cases, the process will be straightforward starting at moment zero and running the cycle as usual.

In other circumstances, with user permission, the washing machine can be set in Smart Energy mode and follow recommendations from the Smart Energy Home ecosystem and Smart Energy Controller.

Simply delaying a wash cycle is a straightforward implementation, clothing items will still be usually dry, and assuming no major contaminants are present the cycle of operation could be delayed for hours. Conditions where this approach would not be optimal would be the presence of clothing items with extreme cleaning needs, such as soiled linen. Soiled linen means fabric that is dirtied from blood, saliva, and vomit for instance.

Anticipating is a bit odd at first but washing machines with a delay function set by the user to perform a wash overnight, let us say 3:00 a.m. for the sake of this discussion could, based on guidance from the Smart Energy Controller anticipate the cycle to 1:00 a.m. to support demand needs and pricing.

Washing machines can also afford pausing the cycle with minimal energy waste depending on the step of the process. Pausing during water heating is possible, but the water would lose thermal energy, impacting efficiency. Pausing during the spin process would drive the need for a new acceleration from zero speed, also impacting efficiency. However, pausing between phases for a reasonable amount of time is an approach that can be taken without energy impact. Pausing between pre-wash and wash cycles, wash and rinse, rinse and spin, these wash process phase transitions will still have the clothing items wet, they will lose thermal energy which in case of a hot wash is not ideal, but at least the macro water volume would have been drained. So reasonable accommodation of a pause can be evaluated during the design and efficiency portions of the project and made available to the Smart Energy Home Smart Energy Controller for collaboration.

6.4.9 Water Heaters

Very typical in cold locations, water heaters also come in many forms. Solar, gas, electric in-line, and electric are the most common.

The opportunities for different water heaters systems in terms of Smart Energy use cases vary especially whether they have hot water storage capability or not. An inline water heater for instance can be turned on and off, but if the consumer decides to take a shower the Energy Controller cannot take any action to just turn off the heater, that would be unreasonable. That said, for this session, an electric tank-based water heater will be taken as an example.

These water heaters, due to tank capacity, will vary in heating power needs. A typical 190 l tank comes with two 4500 W heaters, an upper and a lower in regard to positioning within the tank.

These heaters typically operate in an alternating state. As water is used the upper heater acts first and based on the control operation will stop and activate the lower heater, cycling as needed one at a time. Note that this example shows an electronic water heater, not necessarily a regular thermostat-based unit. Reason is that a connected water heater able to participate in Smart Energy use cases would require connectivity, therefore a higher-end unit with additional electronics is applicable.

Figure 6.10 presents an electric tank-based water heater macro elements for smart energy applications.

The opportunities for water heaters are like those discussed in other appliances, delay, anticipate, and pause. A Smart Home Ecosystem and Smart Energy Controller capable of mapping the Water Heater loads and the user's behavior is key to enable all three cases.

Knowing the user's behavior and knowing hot water is never consumed during certain periods of time (e.g. from 12:00 p.m. to 6:00 a.m.), the water heater could

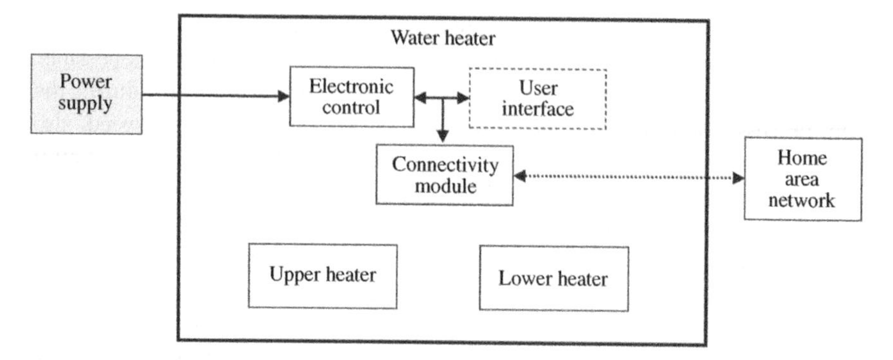

Figure 6.10 Electric tank-based water heater macro elements for smart energy applications. Source: Neomar Giacomini.

avoid keeping the target temperature during this period, therefore recovering just near the time of use.

Delaying is also an interesting option for a couple of reasons, if the usual thermal load spent at a certain time of the day is known (e.g. equivalent to one shower), the Water Heater may delay the temperature recovery due to the fact the user will be out of home all day for instance.

Triggering a pause in the Water Heater operation is also possible. If the temperature in the tank is within an acceptable range but another appliance calls for extra demand, the heating can be paused and restarted later on.

An interesting aspect of the delay case is that there are no additional/unexpected thermal losses other than the regular system losses. Pausing a washing machine or dishwasher during operation would cause the water to cool down and lose performance, but on a heater, the losses will be the same as during regular operation.

That said, electric water heaters allow great flexibility for Smart Energy use cases, and considering user acceptance to stretch the settings for a wider range can bring even further benefits.

6.5 Small Appliances

6.5.1 Coffee Machines, Blenders, Faucets, Food Processors, Mixers, and Toasters

Small appliances in general have very random applications. Identifying their specific time of use may be a challenge for the Smart Energy Controller. These devices also come with considerable power.

Espresso coffee machines with a heater over 100 W, toasters also at or over 1500 W, portable heaters reaching 2500 W, and so on.

The demand-shifting-related use cases end up being very restricted, especially due to the fact the use of Small Appliances is usually at the time of need. The user will start the expresso maker because the preparation of a coffee will follow right after, thc toaster will follow the same on-time usage needs, and so on.

6.5.2 Robotic Lawn Mowers and Electric Tools

A revolution in lawn care, these robots are still entering the market at a considerably high price but are gaining adoption among technology aficionados. These robots are battery based and are currently capable of mowing from 500 m^2 up to over 2000 m^2.

Larger models use for example a 20 V, 6.0 Ah battery, which is charged as the unit returns to it is base and for that reason, the energy use case both for lawncare robots, vacuum robots, mop robots, and others must be specific to the charging phase.

Even if the charging takes place over a period of time with a reasonably low instantaneous power usage, use cases related to delaying the charge for late night while energy demand may be at minimum, will contribute to the overall grid demand. A single smaller load may look insignificant to the grid but combining millions of robots recharging outside high-demand hours will bring reasonable benefits.

With this benefit in mind, charging stations for these autonomous service robots must consider a future where re-charging is based on low-demand hours.

6.6 Monitoring

6.6.1 Energy Monitors, Haptics Sensors, Weather Sensors, and Others

Monitoring in general is not an energy-consuming task. Sensors and minor processing may even be done using batteries, even in connected applications. In the market today devices that are even WiFi™ but do not need bidirectional communication can be found in battery-powered versions.

Devices targeted for monitoring, which are basically executing data collection, do not have any meaningful loads to be on the receiving end of energy use case actions, but they do provide valuable data and insight on how the overall Smart Energy Home can adapt, predict and act on other loads.

The energy monitors added by the consumer in addition to the Smart Energy Meter can provide valuable information per circuit, or even per load. Advanced energy monitors based on Machine Learning allow device detection, historical usage and trends, bill tracking, and goal-setting (Sense 2022). Device detection in this case is a key feature as it allows the energy monitor to support the creation of a digital twin of the Smart Energy Home.

Haptic, movement, presence, and proximity sensors can provide valuable data related to the number of people in the house, which dependencies are in use, whether people are active or not, and so on. These are valuable information to enable advanced HVAC control, lighting, blinds, and other features that will enable energy gains. Simply knowing the users are not home and, with user consent, close the window blinds to minimize heat intake through the windows will in the long run make a difference.

Along the same line, connecting to weather services or having an external weather sensor allows HVAC and other systems to consider data input such as humidity of the intake air and its temperature on the expected performance of the air conditioning (e.g. how long will it take to reduce the temperature by *x* degrees.)

6.7 Health, Comfort, and Care

6.7.1 Air Purifiers, Humidifiers, Health Monitors, Sleep Sensors, and Tracking Devices

These devices may play a crucial role for the user, especially if in use due to any existing health condition. That said, no Smart Energy use case should act on them, but instead, learn from them to act on other devices.

Air purifiers may provide data on air quality, temperature, humidity, CO_2, and other factors, which can then be fed to the HVAC control algorithm, cleaning robots, smart vents, etc. The use case may not be energy optimal but the focus on the comfort and health must be a priority.

Humidifiers could anticipate the target humidity to the HVAC control algorithm and therefore work together, and more efficiently, to balance both temperature and humidity targets. Knowing about the presence of a humidifier and how much water will be injected into the house can even help the HVAC estimate ice formation in the evaporator coils, estimate efficiency, etc.

Tracking devices, sleep sensors and also smart bed frames, are a source of data related to user presence and activity. The Smart Energy Home, with consent of the user, may use information from these devices to identify location, activity, rest time, and profile the behavior and daily routine in order to optimize energy use. Knowing that a user wakes up consistently at 4:00 a.m. can drive different energy use cases from another user that wakes up at 8:00 a.m. An example of a different adaptation is related to the tank-based electric water heater. For the user waking up at 4:00 a.m., the heating may start to recover at 3:00 a.m., one hour before. For the user waking up at 8:00 a.m., the heating may start at 6:00 a.m., two hours before as 7:00 a.m. may be a known peak demand time previously identified and that the system should avoid.

6.7.2 Cat Litter Robots, Pet Feeders, and Other Pet-Related Connected Devices

Most of these small devices focused on pets, besides being low energy, are intended to increase the well-being of the furry friends.

Similar to health and comfort for humans, these devices should be considered as data sources only. Simply detecting a connected pet-related element in the Smart Home Ecosystems is enough to drive certain use cases. Additional data related to the pet activity also carries value to the algorithms.

Detecting a pet-related device, or pet activity, for instance, will inform the HVAC to avoid getting out of the preset temperature range. Detecting acclivity may also drive actions on smart vents to assure all rooms in use are properly ventilated and so on.

6.7.3 Hair Dryers, Brushes, and Straighteners

This group is another set of devices that present a meaningful load to the system. Hair dryers range from 500 W to as high as 3600 W, and hair dryer brushes use between 800 and 1300 W, with 1000 W being the most common.

Although most of these devices are not connected and are unlikely to be in the future at least in scale, a Smart Energy monitor capable of identifying random loads in the system can learn and support system-level adaptations. One sample use case around these devices relates to a user having a shower and overall user preparation to get ready for the day.

Picture a scenario where the user just finished a shower and will use these hair-related devices such as a hair dryer and straightener. The water heater tank may have dropped in temperature and is ready to start the temperature recovery process. As discussed previously a typical water heater will activate a 4500 W heater, and the user is about to add another 2500 W on top of that due to the use of a 2500 W hair dryer, and after that process continue for some time with around 300 W for a hair straightener.

Even not having direct connectivity to the hair-related devices, a Smart Energy monitor that already identified the use of such loads (through electrical current usage patterns measured at the electrical distribution box) as occurring right after a shower (identified by the temperature drop in the water heater tank), would enable the Smart Energy Home Smart Energy Controller to signal to the heater to delay the start of the heater to initiate just after the use of the hair devices has ceased.

It may sound like a stretch use case, but the technology involved already exists, it is just not ready in terms of logical collaboration via a Smart Energy Home ecosystem. The benefit of the case described would result in avoiding having a peak of around 7 kW by spreading it as 2.5 kW for the hair dryer, followed by 300 W for the straightener and then 4.5 kW for the water heater.

6.7.4 Treadmills, Indoor Exercise Bike, and Other Fitness Equipment

Fitness equipment is a class of devices that many consumers found crucial for their exercising habits during the pandemic. Multiple models with connectivity reached the market with features such as connected exercising statistics, ability to present streamed audio-visual content to the user, link to health tracking apps, and so on.

Treadmill motors for instance range from 1 to 3.5 HP, therefore resulting in around 700–2400 W. Other fitness devices present smaller loads, but the use case presented in this session would apply, in general, to most fitness equipment.

A treadmill itself has information on the exercise routine based on a user-selected preset or manual configuration, and with the connectivity feature available it enables the treadmill to cooperate with the Smart Energy Controller to notify of the power and duration that will be in effect. The Smart Energy Controller, same as described in previous use cases, can delay, pause or take other actions through devices that are part of the Smart Energy Home ecosystem to free up demand during the exercise routine.

This example would require the treadmill, or any other fitness equipment, to be designed enabling energy cooperation.

6.7.5 Water Filtration Systems

Water filtration systems are evolving significantly in the recent years in terms of electronics. With the concerns and awareness raised during the COVID-19 pandemic, and the option to add sterilization and sensing, consumers started to opt to bring these additional features to their homes. Although connectivity is not a broadly deployed feature in this type of device, consumers can already find products in this class with interesting features.

Constant water flow alerts to detect leaks, remote shut-off, peak flow rate, and daily water usage are a few of the features available in the market.

Smart water filtration systems may be powered with electricity, but the bulk of the mechanical actuation is performed directly by the water pressure present in the city supply. This aspect therefore prevents the Smart Energy Controller from having loads to actuate on and implement demand-shifting actions on the unit.

However, the data from peak flow rate and daily water usage are data points that can serve as input to the water heating system and other appliances, but mainly to be fed into the Smart Energy Home digital twin. The total daily consumption was either used in showers which likely went through heating, or in the kitchen for cooking purposes, the dishwasher, washing machine, and so on. In other words, crossing the daily water usage with the history of appliances activated each day will, in a reasonable timeframe, result in a comprehensive water use profile for each appliance/application, whether it requires heating, cooling, or any other action.

Although this is a water use profile and not an energy use profile, the data can be used in predictions for each day of the week in regards to anticipating water heating needs.

6.8 House Automation

6.8.1 Blinds & Shades and Light Bulbs

Window covers and decorations are available in many forms and shapes. Smart window blinds and shades, however, are commonly found in the form of motorized rollers with fabric or corrugated materials that be moved up and down by the use of integrated and motorized string actuators.

Versions that are both batteries based or directly connected to a power supply are available in the market. Independently of the type, the energy consumption is significantly low.

The benefits of Smart Blinds & Shades, besides privacy, comfort and decoration, relate to environmental effects of allowing sunlight to get through the windows.

Receiving sunlight through the windows impacts both lighting and temperature, therefore making these devices suitable for Smart Energy Home applications.

As described in this chapter, a Smart Energy Home ecosystem that has a comprehensive Digital Twin would allow the Smart Energy Controller to gather sensor data from all available sources, understand user behavior and current actions, evaluate the loads available for actuation and their effect on the system and its energy consumption, and putting all that together make educated decisions to optimize demand and energy use.

This quick recap was provided as a base for comparison to the following use cases.

The Smart Energy Controller, knowing the day of the week, knowing the traditional user behavior for that day, knowing the local weather through online services, and correlates that data to on-time sensors data such as movement, presence, a wake-up signal from a connected bed, and so on, is able to coordinate lighting where needed and if needed.

Lights may be set at 100% brightness, or 50%, and the windows blinds may be activated to open, or if it's a sunny day identified through the online weather services data the lights may be set to be completely off with window blinds open and potentially providing enough light to provide the expected user comfort.

Additional cases where window blinds are automatically closed when the user leaves the house, reopen upon arrival, and others to support, even if in a minimalistic way, Heating or Air Conditioning are possible.

In very elaborate scenarios the Smart Energy Controller could even consider the geographical orientation of the house and actuate on window blinds as the sun moves along the day.

Although some of these use cases sound extreme and not necessarily providing large gains on instant power and overall energy efficiency, they are being provided to instigate further reflection on the power of having comprehensive data related to devices, sensors, the house structure, location and orientation, online services, and others.

6.8.2 Garage Door Opener

Besides being an important safety-related aspect of the house, the garage door allows the intake of external air, which in houses with a garage temperature control will play an important role in how much hot or cold air will be taken in.

In locations with extreme weather conditions, even a short amount of time may result in taking/losing thermal load that will impact HVAC operation.

Keeping the door open is not an option without user consent, but assuring the door closes as soon as possible for instance will make a difference in thermal load exchange.

6.8.3 Sprinklers, Gardening Sensors, and Accent Lighting

Gardening in terms of sprinklers, humidity sensors, and accent lighting is very common in some countries. Sprinklers water flow is traditionally provided by the city water supply pressure, therefore using electricity only for water valve solenoids and the controller, which are not a considerable load to the system. Other loads used in landscaping such as accent lighting will vary per project, however, and those being related to user comfort and visual appeal may not directly participate in Smart Energy use cases.

It is possible, however, that landscaping accent lighting presents a significant load, but the perceived impact from a user's perspective in case the Smart Energy Controller acts on them may result in user dissatisfaction. Features in the house that are intendedly added by the user to increase visual comfort need to be addressed carefully.

Gardening sensors, however, and information related to the state of the sprinklers may be useful to certain use cases. The HVAC will be informed the intake air will be cooler than the moment before, that the humidity of such intake will be higher, and so on. Even the overall temperature outside the house may be impacted while sprinklers are active.

Another interesting aspect of sprinklers is the pressure drop they may apply to the house water supply. The usual recovery time of a water heater and its behavior may change. An interesting aspect related to this combination may be in case of an inline water heater that usually operates for let's say five minutes for a quick heated shower, and in case of a concomitant operation along with the sprinklers will provide the user a lower pressure shower that may result in an increase of the shower time, let us say to seven minutes or so. Or the time it may take to fill up a bathtub the same.

On an inline water heater with proper temperature control and modulation the total energy spent may be similar in both cases, but the time the heating unit will be active will increase, therefore potentially delaying other loads from getting activated in other use cases. These other loads may be in the queue of the Smart Energy Controller waiting for their authorization to be activated, avoiding peak demand.

6.8.4 Smart Power Strips and Smart Power Switches

Along with light bulbs, smart power strip and smart power switches were part of the initial phase of IoT deployment consumers had easy access to. Commonly used to drive and automate lighting, heating, Christmas trees, bedside lamps, fans, and other devices. Some of these smart power strips and power switches also include the capability to measure energy consumption.

A Smart Energy Meter capable of device characterization, through coordination with the Smart Energy Home ecosystem, may even be able to identify the loads in use in these IoT energy delivery devices.

Identifying that a smart power switch is connected to a bedside lamp enables the Smart Energy Controller to automatically turn it off once the user has slept for instance. Identifying the load as a fan on a room the user has left may also enable the Smart Energy controller to turn it off.

Given user acceptance for such use cases, the Smart Energy controller would be able to reduce, even if on a low scale, the loads active in the house.

6.8.5 Presence, Proximity, and Movement Sensors

Sensors in general that provide input regarding user behavior, number of users in the house, pets, and any other moving device are of great importance to a Smart Energy Controller. With user consent to use such data, the Smart Energy Controller is enabled to build a comprehensive Smart Energy Home digital twin to build a broader set of smart energy use cases.

By identifying whether the house is having a party in progress, or that it is unattended during work hours, for instance, these inputs may drive actions on HVAC, cleaning robots, window blinds, and others even with the benefit of not impacting the user comfort. User preference settings may request the Smart Energy Home to show minimal automation when users are in place. Not everyone will enjoy the feeling of living in the future when a house takes action automatically at any time.

6.8.6 Thermostats and Temperature Sensors

Present in most homes in the United States and many other countries, thermostats are key to assume thermal comfort in the house. These devices along with temperature sensors distributed in the house provide valuable information for the HVAC control, but also provide information for Smart Energy use cases related to window blinds, fans, smart vents, and other environment-related devices.

One important aspect of smart thermostats is that most, to not say all of them, in the market today take actions based on their own control algorithms, therefore not relying on data and suggestions provided by a Smart Energy Controller.

Like most devices described in this chapter, Smart Thermostats will have to evolve to be part of these advanced use cases.

6.8.7 Vacuum Cleaners, Vacuum Robots, Mop Robots, and Power Tools

House tasks-related robots such as vacuum, mopping, and even handheld devices like vacuum cleaners and power tools rely mostly on batteries. These devices whether their use occurs automatically or with user intervention, once the task is completed, they are set back to the charging station.

As described in the Robotic lawn mower session, with proper coordination along with the Smart Energy Controller a decision can be made to charge on-time or to delay for a moment later in the day. This coordination needs to consider usage habits and user consent as a device such as a power tool may be required for use sooner than usual.

6.9 Non-appliances

6.9.1 Electric Cars and Motorcycles

Electric cars and motorcycles, considering user transportation needs, may be a viable options for Smart Energy use cases.

Consider a scenario where a user sets the minimum required to charge for the electric vehicle to be 50% at all times.

The car coming back to the house at 95% battery during peak demand could enable a use case where energy flows back into the house, and the electric card is recharged later that day during lower demand hours, still respecting the minimum 50% charge of course. Newer electric vehicles launched in 2022 are starting to offer the bidirectional capability (Woody 2022).

Or another situation where the car returns at 10% battery still during peak demand hours, and the Smart Energy Controller is able to delay other loads to focus on charging the vehicle back to the minimum expected 50%, then stop the charge and restart later on when the peak demand has ceased.

These and many other use cases are possible to consider the fact these vehicles will be capable of acting as energy storage and energy transportation units.

Considering the vehicle as an energy transportation unit applies to situations where the vehicle was charged outside the house and return energy to the house or the grid when back to the garage. Or vice versa when the car leaves the house at 100% and is re-plugged somewhere else and can support peak demand during the day, in an industrial zone for instance while sitting still in the parking lot.

Given the importance of Electric Vehicles in this scenario, the book has two Chapters 9 and 10 dedicated to the topic.

6.9.2 Desktop Computers

Computers in general are not major loads in a home scenario, even advanced gaming computers run much below 500 W. At rest, these machines will indeed have stand-by consumption but that is a topic not being addressed in this chapter.

Acting on a computer as a load is not a straightforward operation as there may be complex tasks running and it would be unviable to the Smart Energy Controller to deep dive into that aspect, and the possibility to shut off the computer or set it for hibernation is already a common automatic feature current Operating Systems support.

That said, with proper software tools interacting with the Smart Energy Controller, computers can become a source of user presence and activity. A single flag, given user acceptance to do so, provided by the computer to the Smart Energy controller to state a given room is in use and the user is present and can drive advanced HVAC, lighting, and other use cases.

Picture the scenario of a house with a single resident and a mapped behavior or spending working hours mostly in a room designated as an office. The Smart Energy Controller would be able to drive Smart Energy use cases across the rest of the house most of the time.

6.9.3 Modems and Routers

Smart home networking devices maintain the infrastructure needed for all other devices to collaborate in the use cases under discussion. They are definitely not to be considered as a load in any way, and also do not provide any sensing data.

However, these devices will have important information related to devices that may not be part of the Smart Energy ecosystem.

A smart television that is connected to WiFi but is not necessarily added to the Smart Home ecosystem will be listed in the router's active device list, and so any other device in similar setup condition.

The reason this may be an important data point is that the simple increase or decrease in bandwidth use from that device may provide important statistics on what is happening in the house. A significant drop in bandwidth usage will be characteristic of nighttime or the absence of residents.

The list of active devices in the router could also be provided, with user consent, to the Smart Energy Controller to enable the identification and presence of the resident's phone in the house, therefore an indication of presence that can drive the importance of HVAC, water heating, window blinds, and other use cases.

6.9.4 Power Banks, Uninterrupted Power Supplies

A whole house power bank, a feature that has become available in recent years due to the popularization of solar panels and electric vehicles, adds interesting use cases to the smart energy home scenario. The availability of power storage whether at a macro house scale through a power bank, or minor through distributed Uninterrupted Power Supplies (UPSs) offers flexibility to prevent peak demand from reaching the grid.

Currently, UPSs that are for home use on single devices like computers are usually specified from 750 VA to 2 kVA and are not bidirectional in regard to the Alternate Current (AC) input, but future products could add this feature.

In a scenario where the house is reaching peak demand, these power banks and UPSs which are usually backups for power outage events may be able to act as short-term energy sources. A 2 kVA unit could supply half of its capacity for a short time as a contribution to reduce overall demand for instance.

Based on user settings, such as a minimum 50% capacity while in normal operation, the extra 50% flexible capacity can add interesting use cases to the portfolio.

6.9.5 Smartphones, Tablet Computers, Smartwatches, and Video Games

Another set of consumer electronics that besides the charging time had little to offer in terms of actions as loads to the Smart Energy System. Video games are in this group but of course, are available in both battery and direct AC-powered versions.

These battery-operated devices have already evolved in terms of charging methods, recent launches are already coming with features for a quick charge during the day, or slow charging while sitting in a charging station overnight.

That said, and due to the low power associated with these devices, their contribution can be focused on the data input to the Smart Energy Controller directly, or via other devices.

The simple presence of a Smartphone in the house may indicate the resident is at home, multiple Smartphones detected by the router may indicate the entire family is at home, and so on. With user consent even the calendar maintained on a smartphone could allow the house to anticipate or delay energy use cases.

In case of a party event at the house, the refrigerator could trigger extra icemaking ahead of time, the HVAC could anticipate or get ready to provide extra cooling, the water heater could potentially increase water temperature a couple of degrees, and so on. Although all these examples sound just like extra energy use, properly coordinating them on time can reduce significantly the occurrence of a high-demand situation.

With a smartphone being such a personal item that rarely leaves the user's possession, future use cases considering it as a central source of data will be of great benefit to the Smart Energy ecosystem.

6.10 Entertainment

6.10.1 Aquariums

Aquariums, terrariums, and other ecosystems that are pet-related should be excluded from the list of loads available for energy use cases. Many of these pets are sensitive to any changes in the environment making such use cases a risk. However, knowing these pet ecosystems are in the house, or even in a specific room, enables the Smart Energy Controller to Live, don't do anything.

6.10.2 Audio Systems

A piece of equipment audio enthusiasts may put considerable investment on. Depending on the system may range in the thousands of watts at maximum power. Such high-power conditions are unlikely to apply to day-to-day use and be focused on special occasions.

The sound volume implies in more or less power in the system, but if Smart Energy use cases act on it the user experience will be compromised. So it is not a viable option to act on.

That said, audio system may instead be a source of information related to user presence depending on the room in the sound is active, or the content it is being used to play, whether pure audio or audio from a movie which would imply even more the user is at that room and so on.

Although acting in the volume/power is not a reasonable approach, potentially turning it off completely in case the resident leaves the house can be an option.

6.10.3 Televisions and Streaming Receivers (Cast Feature)

Televisions are available all around the globe, even multiples per home on many occasions. Being an entertainment device similar to audio systems, acting on the image and sound may compromise the user experience.

Decisions on Smart Energy use cases related to televisions should ideally be based on presence sensing. There would be no impact to customer experience for instance if the display brightness were taken down even to zero while the user leaves to room for whatever reason. Keeping the sound on would keep the user experience as if the television was still on.

Other energy use cases are also already native to many of these devices and can be triggered by the user such as the sleep function. However, today's implementation is based on user activation, what about the smart bed sensors triggering a television shut off when sleep is detected?

6.10.4 Virtual Assistants (Multiple Forms)

Virtual Assistants are from a user's perspective the center of a Smart Home. Throughout this chapter, a lot was discussed about the Smart Energy Controller, which may or may not be integrated into the same device along with the Virtual Assistant features.

One way or another, even if the Smart Energy controller which is the center of all Smart Energy use cases described in this chapter is on a separate device, the user will likely always keep referencing to the Virtual Assistant as the central device of the Smart Energy Home. This view is important to consider as the adoption of Virtual Assistants is already taking place, but to convince consumers a new separate device will be required may not be appealing, so integrated versions may be the solution, especially as it refers mostly to software implementation and algorithms.

Virtual assistants are available in many forms and shapes, also from different vendors for their respective Smart Home Ecosystems. Figure 6.11 shows typical blocks that comprise an relatively advanced Virtual Assistant.

6.10.5 Virtual Reality Goggles and Other Gadgets

Virtual reality (VR) is a growing technology boosted by the video gaming industry and metaverse[2] implementations. These devices are wearables that do not have

---

2 An immersive virtual world enabled by the use of virtual reality (VR) and augmented reality (AR) headsets.

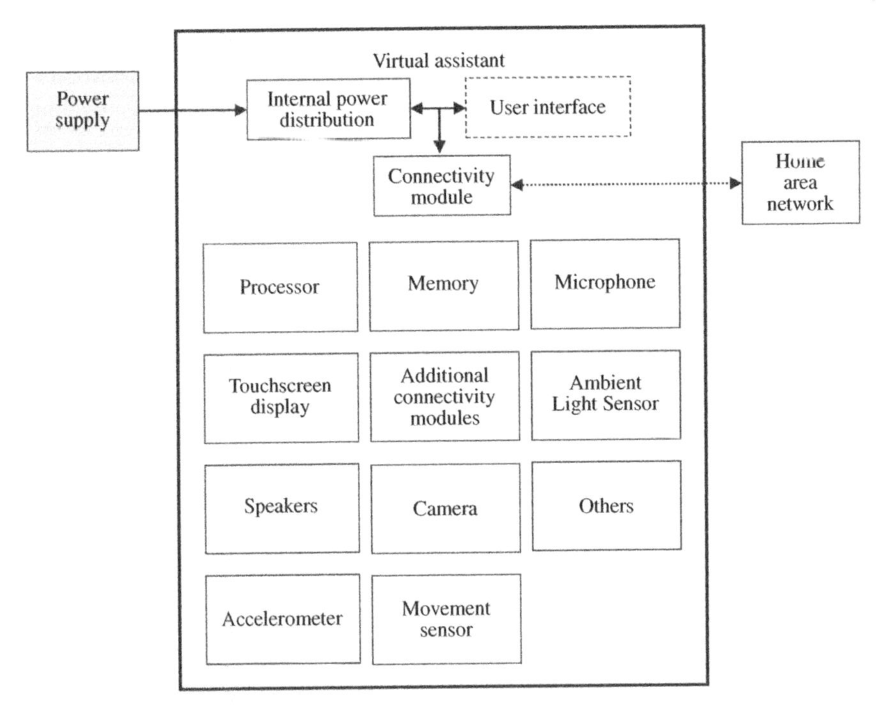

Figure 6.11 Virtual Assistant block diagram. Source: Neomar Giacomini (co-author).

any significative loads to be driven on Smart Energy uses cases. However, they do have sensors and their mere use can be used as data for Smart Energy use cases.

A user profile linked to the Smart Energy Controller by itself would provide valuable information on the frequency and duration of VR goggles usage. During those times lighting in unused rooms can be turned off, the lighting in the occupied room during the VR session may have the brightness toned down, window blinds can be controlled accordingly, and so on. In summary, the user immersion into the virtual environment may allow considerable time the Smart Energy House can use to drive feasible use cases.

6.11 Security

6.11.1 Alarms, Cameras, Door Locks, and Doorbell Cameras

Like all other devices that include sensors, security systems are a valuable source of data. Cameras, door locks, and others enable the Smart Energy Controller to identify events across the house.

The resident's arrival at home through its car being identified by a driveway or garage camera, a window being opened and potentially left open, a backyard camera showing that users left the house but not the property, a door camera providing a count on the number of individuals at home and so on so forth.

These inputs may drive Smart Energy uses cases related to HVAC, smart vents, lighting, water heating, and others.

6.12 Conclusion

A holistic discussion on appliances in general was presented in this chapter, the traditional understanding of an appliance was also extended considering all other connected devices that are able to participate in Smart Energy use cases along with Large Appliances such as kitchen appliances, laundry, HVAC, and others.

The possibilities for use cases presented showed that deep diving into an input/output analysis across all devices in a one-to-many and many-to-one can result in a significantly large amount of possibilities in terms of energy management.

A central decision-making entity described as the Smart Energy Controller was introduced and linked through all the use cases, showing its importance and need to fully implement this scenario.

The technologies required for the implementations described are known, but not yet put together in such a comprehensive manner, it will take time and collaborative effort across impacted industries to reach the fully automated state described in this chapter. As mentioned in Chapter 2, to enable this level of interoperability a shared model of consensus that facilitates the matching of existing assets in the smart applications domain is needed, and Chapter 7 will discuss the Smart Applications REFerence Ontology (SAREF) and its relationship with existing IoT ontologies focusing on the Smart Home and Smart Appliances domains.

Symbols and Abbreviations

AC	alternating current
AR	augmented reality
HEMS	HOME ENERGY MANAGEMENT SYSTEM
HMI	human machine interface
HVAC	heating, ventilation and air conditioning
SAREF	smart applications reference ontology
VR	virtual reality

Glossary

Augmented Reality A technology that superimposes a computer-generated image over the real world creating a composition with added value either cosmetic or with additional data.

Appliance A device designed to perform a domestic task.

Device An apparatus for a particular purpose, usually mechanical or electronic equipment.

Load A device or element used to execute a particular action, in this context related to the use of electrical energy. Any load is meant to be an element that consumes considerable electrical energy.

Sensor A device which detects or measures a physical property.

Smart Energy Related to the use of energy in a coordinated way across devices and providers usually by means of additional data sharing/connectivity means.

Virtual Reality A computer-generated three-dimensional environment that can be interacted with in a perceived physical way by a person using virtual reality goggles and controllers.

References

LEXICO (2022). APPLIANCE English Definition and Meaning | Lexico.com. https://www.lexico.com/en/definition/appliance (accessed 6 January 2023).

Sense (2022). Welcome to Sense. sense.com/ (accessed 6 January 2023).

Statista (2021). Household ownership rate of major appliances in selected countries worldwide in 2021. https://www.statista.com/statistics/1117972/major-appliances-ownership-selected-countries/ (accessed 6 January 2023).

Statista (2022a). Household appliances ownership in the U.S. in 2022. https://www.statista.com/forecasts/997158/household-appliances-ownership-in-the-us (accessed 6 January 2003).

Statista (2022b). Major Appliances. https://www.statista.com/outlook/cmo/household-appliances/major-appliances/worldwide (accessed 6 January 2023).

Whirlpool Canada (2017). Learn about Whirlpool® Heat Pump Dryer Technology. https://www.youtube.com/watch?v=5wmFoWOujZ8 (accessed 6 January 2023).

Woody, T. (2022). How Ford's Electric Pickup Can Power Your House for 10 Days. https://www.bloomberg.com/news/articles/2022-05-31/how-the-ford-f-150-can-be-a-backup-home-generator?cmpid=BBD060122_TECH&utm_medium=email&utm_source=newsletter&utm_term=220601&utm_campaign=tech (accessed 6 January 2023).

7

The ETSI SAREF Ontology for Smart Applications: A Long Path of Development and Evolution

Raúl García-Castro[1], Maxime Lefrançois[2], María Poveda-Villalón[1], and Laura Daniele[3]

[1] *Ontology Engineering Group, Universidad Politécnica de Madrid, Madrid, Spain*
[2] *Mines Saint-Étienne, Univ. Clermont Auvergne, INP Clermont Auvergne, CNRS, Saint-Étienne, France*
[3] *TNO, Netherlands Organization for Applied Scientific Research, The Hague, The Netherlands*

7.1 Introduction

Building smart applications in the Internet of Things (IoT) field requires interchanging and using information from others whether people or machines but also being able to understand such information unambiguously. To address this issue, the European Telecommunications Standards Institute (ETSI) Technical Committee (TC) on Machine-to-Machine, SmartM2M, leads the Smart Applications REFerence ontology (SAREF) initiative with the goal of bringing a common understanding among cross-domain heterogeneous systems.

IoT fragmentation is one of the main threats to the adoption of IoT technologies on a large scale. To overcome this, the current fragmented landscape of IoT technologies requires standardized interfaces and data models to ensure interoperability. In this scenario, one of the main challenges to ensure interoperability is having a set of standard data models that enable interchanging not only information but also the meaning of such information to avoid misinterpretations between senders and receivers.

To cope with this, the European Commission has promoted the SAREF ontology in collaboration with ETSI SmartM2M TC since 2014 with the goal of providing and maintaining a common data model over time to ensure interoperability. The SAREF ontology is a reference data model enriched with formal semantics, intended to enable interoperability between solutions from different providers and among various activity sectors in the IoT, thus contributing to the development of

Energy Smart Appliances: Applications, Methodologies, and Challenges,
First Edition. Edited by Antonio Moreno-Munoz and Neomar Giacomini.

the global digital market. Over time, SAREF has evolved into a suite of ontologies that includes a general-purpose ontology (i.e. SAREF core, which is currently in its third release), 11 extensions, and an ontology pattern. It further defines a clear workflow for development and versioning and provides a portal to its user community for documentation and collaborative development. Several experts have made great efforts in the past years (2014–2022) to document the extensive work of SAREF in technical reports and specifications, scientific papers, project deliverables, websites, etc., but this information is often perceived by stakeholders as scattered and not easy to find. This book chapter provides stakeholders with the opportunity to find this information in one single place, including a historical overview of the SAREF activities since its first release, a description of the main SAREF concepts that are relevant for the smart home environment (from SAREF, SAREF4ENER, SAREF4BLDG, SAREF4WATR, SAREF4CITY, and SAREF4SYST), a clarification of the mechanism for version control and the automatic tool support for the ontology developers. Everything according to the best practises for ontology standardization adopted in ETSI SmartM2M TC.

The rest of this chapter is organized as follows. Section 7.2 presents the need for a standard ontology that can serve as an umbrella to represent contextual and multisectorial IoT-related data and will review the most relevant existing IoT ontologies. Section 7.3 introduces the SAREF initiative with a brief history of the SAREF ontology, promoted by the European Commission and the European Telecommunications Standards Institute (ETSI), from its initial conception to its current third version and 11 extensions. Section 7.4 describes the main ontology requirements followed when developing the SAREF ontology, as well as the design patterns implemented in the ontology and the main design decisions and lessons learnt during development. Section 7.5 provides an overview of the SAREF ontology and its main classes and properties. Section 7.6 describes how the SAREF ontology and its extensions can be applied to smart appliances and the smart home environment. To do so, it will describe the main extensions that are relevant for this environment: those for the energy, water, building, and smart city domains and the extension to represent systems. Section 7.7 presents examples of the use of the SAREF ontology focussing on the environment of smart homes. Section 7.8 discusses some lessons learnt from the development and evangelization of SAREF to the industry. Section 7.9 concludes the chapter and discusses future work.

7.2 IoT Ontologies for Semantic Interoperability

One of the cornerstones to making the IoT a reality is the interoperability between heterogeneous services and actors. Such heterogeneity appears at different levels, for example at the protocol level for connectivity, temporal validity or update of data, data storage, language, etc. Focussing on data interoperability, different levels of interoperability should be addressed: transport (how data are accessed), syntactic (how data are expressed), and semantic (how data are modeled). Ontologies play a key role in addressing and allowing semantic interoperability in many heterogeneous data scenarios, including IoT. For this reason, a number of ontologies have been developed.

Hundreds of ontologies for IoT and related domains are listed in ontology repositories such as the LOV4IoT[1] catalogue (Gyrard et al., 2016), which registers and classifies IoT ontologies considering a wide number of domains such as transport, healthcare, energy agriculture, city, weather, etc. Another ontology registry in which IoT ontologies can be found is Linked Open Vocabularies[2] (LOV) (Vandenbussche et al., 2017), a general-purpose ontology index. The ontology landscape maintained by the Alliance for the Internet of Things Innovation (AIOTI)[3] provides a more focused overview of the main IoT ontologies structured by their domain of interest, taking into account aspects such as sustainability (i.e. who is maintaining it) and technology readiness level (i.e. how mature it is).

One of the main ontologies developed in the context of IoT is the Semantic Sensor Network (SSN) ontology developed by the first joint working group of the Open Geospatial Consortium (OGC) and the World Wide Web Consortium (W3C) on *Spatial Data on the Web*. SSN[4] was released in 2017 and is composed of the SSN module[5] and a module called SOSA[6] (Sensor, Observation, Sampler, and Actuator), among others. The SOSA module replaces the SSO pattern (Stimulus Sensor Observation Pattern) and could be used in isolation in a lightweight fashion. The main components of the SOSA module are "sensors and observations," "samplings and samples" and "actuators and actuations." SSN augments SOSA with additional terms and a stronger axiomatization.

Another relevant ontology in the IoT domain is the oneM2M base ontology (ETSI, 2016; oneM2M, 2018), created to provide the definition of the concepts, relations, and restrictions needed to allow the semantic discovery of entities in oneM2M systems. The objective of oneM2M for this ontology is to provide syntactic and semantic interoperability between oneM2M systems and external systems that could derive new ontologies based on the oneM2M base model. The oneM2M ontology focusses on the description of oneM2M:Thing which is specialized as oneM2M:Device. For devices, their

1 http://lov4iot.appspot.com
2 https://lov.linkeddata.es
3 https://aioti.eu/aioti-ontology-landscape-report
4 https://www.w3.org/TR/vocab-ssn/
5 http://www.w3.org/ns/ssn
6 http://www.w3.org/ns/sosa

functions (oneM2M:Function) and services (oneM2M:Service) are modeled. Services are linked to the operations (oneM2M:Operation) they provide, which can expose commands (oneM2M:Command).

Finally, the Web of Things ontology could be mentioned in relation to IoT ontologies based on the Web of Things (WoT) Thing Description[7], developed under the W3C WoT working group activities. The core vocabulary of the Thing Description includes the representation of interaction affordances in addition to the metadata needed to represent the abstraction of physical or virtual entities, called Things. The interaction affordances are classified into Properties (used for sensing and controlling parameters by exposing internal state of the thing), Actions (used to invoke physical processes and set the internal state of the thing not exposed as properties), and Events (used to describe the event sources that asynchronously push messages).

7.3 The SAREF Initiative

SAREF originated from a standardization initiative launched by the European Commission in collaboration with ETSI in 2013, when more than forty percent of the total energy consumption in the European Union used to be produced by the residential and tertiary sector, of which a large part were residential houses. Appliances, which are inherent in the building ecosystem, were considered among the culprits of this high energy consumption. Therefore, the European Commission (EC) identified an immediate need for the market to optimize energy use by managing and controlling appliances at the system level. In particular, industry and the EC raised the need for standardized interfaces and a common data model to ensure interoperability and overcome market fragmentation. The development of a reference ontology was targeted as the main interoperability enabler for appliances relevant for energy efficiency, and ETSI was accepted to cover the communication aspect and provide the necessary standardization process support. As a result, after a comprehensive consultation with stakeholders to address clear market needs, the EC financially supported a study (SMART 2013/01077) to create a reference ontology for smart appliances (Daniele et al., 2015b). The study was carried out from January 2014 to April 2015 in close collaboration with smart appliance manufacturers, and the resulting ontology, SAREF, was standardized by ETSI in November 2015 (TS 103 264 v1.1.1).

In 2016, ETSI TC SmartM2M requested a Specialist Task Force (STF) to provide input on the evolution of SAREF and create possible extensions in relevant domains of interest. STF 513 was established and developed the first

7 https://www.w3.org/TR/wot-thing-description/

three extensions for SAREF in the energy, environment, and building domains, resulting in SAREF4ENER (TS 103 410-1), SAREF4ENVI (TS 103 410-2) and SAREF4BLDG (TS 103 410-3). STF 513 also developed a second version of SAREF, taking into account feedback received from industry stakeholders since its first release in April 2015. As a result, a new version of SAREF was published in March 2017 (TS 103 264 V2.1.1).

In 2017, the first SAREF-based proof-of-concept solution was demonstrated and implemented on existing commercial products in the energy domain, as part of a second study launched by the EC to ensure interoperability to enable Demand Side Flexibility (SMART 2016/0082) (Daniele and Strabbing, 2018).

A new series of initiatives promoted by ETSI and the EC followed in recent years (e.g. STF 534, STF566, and STF578) in which more extensions have been created for additional domains such as smart cities, agriculture, industry and manufacturing, automotive, eHealth, wearables, and smart lifts, making SAREF a modular framework that comprises SAREF as generic core ontology (TS 103 264 V3.1.1), 11 domain specific extensions (TS 103 410, parts 1-11) and an ontology pattern for systems, connections, and connection points (TS 103 548). The SAREF framework is maintained and evolved under the umbrella of ETSI by an ecosystem of experts from various research organizations, universities, and industry in Europe who collaborate successfully with each other. One of the latest supported initiatives is the development of an open portal for the SAREF community and industry stakeholders, so that they can also directly contribute to the evolution of SAREF[8]. SAREF is currently adopted in a considerable number of European projects that provide applications of semantic interoperability solutions in various domains and are encouraged to contribute with their results and findings to the standardization and evolution process in the ETSI SmartM2M TC.

7.4 Specification and Design of the SAREF Ontology

The general development framework for the SAREF ontology and its extensions (generally referred to as SAREF projects) is specified in the ETSI TS 103 673 technical specification (ETSI, 2020k), and is hosted on the public ETSI Forge portal https://saref.etsi.org/sources/.

7.4.1 A Modular and Versioned Suite of Ontologies

As illustrated in Figure 7.1, the SAREF suite of ontologies is composed of ontologies that define generic patterns such as SAREF4SYST (ETSI, 2020g) (detailed in

8 https://saref.etsi.org/

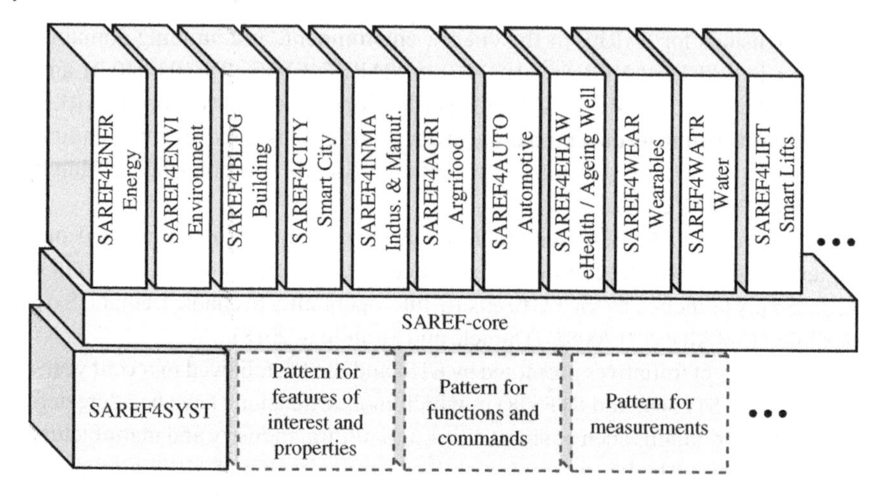

Figure 7.1 The SAREF suite of ontologies with its different modules.

Section 7.6.5), a core ontology SAREF Core (ETSI, 2020j) illustrated in Figure 7.4 (detailed in Section 7.5), and different extensions developed for distinct vertical domains: SAREF4ENER for energy (ETSI, 2020e), SAREF4ENVI for environment (ETSI, 2020f), SAREF4BLDG for smart buildings (ETSI, 2020f), SAREF4CITY for smart cities, SAREF4INMA for manufacturing, SAREF4AGRI for agriculture, SAREF4AUTO for automotive (ETSI, 2020a), SAREF4EHAW for health (ETSI, 2020d), SAREF4WEAR for wearables (ETSI, 2020i), SAREF4WATR for water management (ETSI, 2020h), and SAREF4LIFT for smart lifts (ETSI, 2021).

As SAREF is specified in the ETSI technical documents, it uses *Semantic Versioning*. Each module of the ontology has a distinct version composed of three numbers: a *MAJOR*, a *Minor*, and a *patch*. The increase in MAJOR indicates a break in backward compatibility. The increment in Minor indicates the addition of features. The increment in the patch indicates the correction of a bug.

7.4.2 Methodology

In general, the development of SAREF ontologies follows the Linked Open Terms (LOT) methodology (Poveda-Villalón et al., 2022), which adopts a V-model approach with conditional feedback in some development stages. More specifically, the SAREF development framework defines the different workflows to be followed for new SAREF project versions, SAREF project version development, and SAREF project release. The following roles are defined:

Steering Board member A Steering Board member belongs to the group of persons in charge of steering the development of SAREF, including the SAREF core and SAREF extensions, community participation, and the underlying infrastructure.

Technical Board member A Technical Board member belongs to the group of persons in charge of maintaining the SAREF public forge and the SAREF public portal.

Project leader A project leader is the person in charge of the SAREF project who performs project management tasks.

Ontology developer An ontology developer is a member of the ontology development team who has a great understanding of ontology development and the rights to modify the ontology and interact in the development cycle. Ontology developers create and modify the different development artifacts, provide new requirements to the ontology, and validate whether they are satisfied or not when implemented, and have decision rights about what contributions can be included in the ontology.

Contributor A contributor is a person who is knowledgeable about the ontology domain and who proposes contributions.

Ontology user An ontology user is someone interested in any of the SAREF projects or in proposing a new SAREF project.

Different workflows are established for the creation of an ontology version, the development of an ontology version, and the publication of a project (ETSI, 2020k, Clauses 6.1, 7.1, 8.1). For example, Figure 7.2 illustrates the workflow for the development of project versions that supports the development of SAREF project versions from the SAREF community of users. SAREF project versions may be new versions of SAREF core, new versions of existing SAREF extensions, or initial versions (V1.1.1) of new SAREF extensions. The SAREF project version development workflow is formulated around the use of issues in the corresponding SAREF project issue tracker on the ETSI public forge. This enables us not only to have a single point of interaction for development but also to keep track of the development activity and discussions. Any update in a SAREF project version should be made through a change request, which is posted as an issue in the corresponding repository of the ETSI public forge and assigned an issue number. This includes change requests related to new ontology requirements, defects, or improvements in the ontology specification, in the ontology tests, ontology examples, or ontology documentation. Any contributor can create a new change request or review and discuss existing change requests. Ontology developers should review change requests, propose, and review implementations of accepted change requests. The Steering Board should review change requests. The Project leader is responsible for ensuring that the change requests are approved by SmartM2M and that the implementations of the change requests satisfy the requested change.

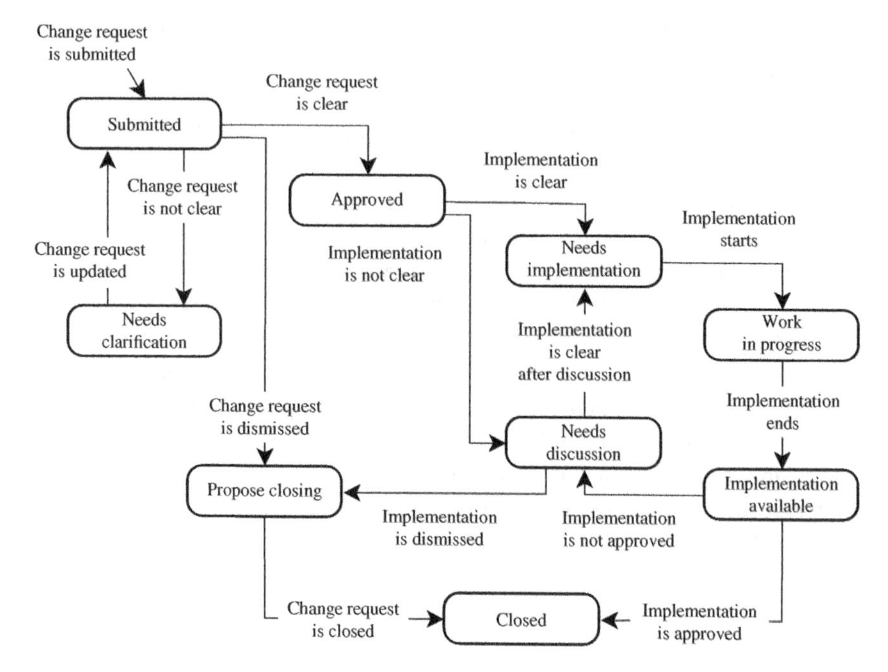

Figure 7.2 The SAREF project version development workflow (adapted from ETSI (2020k), ©ETSI 2020, all rights reserved).

7.4.3 Version Control and Editing Workflow

The sources of the SAREF ontologies are hosted on the public ETSI Forge portal (https://saref.etsi.org/sources/), with four different types of branches: `issue-x` branches to work on an issue, `develop-vx.y.z` branches to work on a version, `prerelease-vx.y.z` branches to work on the final validation of the ontology, and `release-vx.y.z` branches for published versions. Protection rules are defined to prevent ontology developers from directly pushing their changes to `development-vx.y.z` branches or from directly accepting merge requests in `prerelease-vx.y.z` branches.

There are two main practises to identify software versions with git: version tags and release branches. The development of SAREF ontologies uses this second approach, which allows documentation or examples to continue evolving even when an ontology version is published.

7.4.4 Automatization of Requirements and Quality Checks

The requirements for every SAREF ontology project are listed in a specific CSV document with three columns: an identifier, a category, and a requirement expressed as an assertion or a competency question (ETSI, 2020k, Clause 9).

These requirements are then assessed with the Themis tool (Fernández-Izquierdo and García-Castro, 2019).

A set of rules that a SAREF ontology repository must comply with is also defined in the ETSI TS 103 673 technical specification (ETSI, 2020k, Clause 9). The SAREF Pipeline application allows each of these rules to be evaluated with severity level: (a) structure of the repository directory, (b) presence of a defined license file, (c) specification of the ontology requirements, (d) presence of a well-formed file `/saref4[a-z]{4}.ttl/`, (e) declaration of predefined prefixes, (f) presence of an ontology declaration, with a series Internationalized Resource Identifier (IRI) and a version IRI conforming to the naming of the git branch (ex: `develop-v2.1.1`), (g) possible imports of other SAREF ontologies by their version IRI, (h) presence of creators and contributors, (i) naming convention for classes, properties, instances, (j) presence of metadata for each term, (k) the ontology must be in the OWL 2 DL profile, (l) the ontology must be consistent, (m) each class must be satisfiable, (n) no pitfall must be detected by the OOPS! scanner (Poveda-Villalón et al., 2014), (m) presence of tests, (o) presence and quality of examples, and (p) existence of external terms used.

Some of these tests use SHACL shapes (Knublauch and Kontokostas, 2017), others use OWLAPI functionalities after cloning the necessary repositories. The message folder of the application gives a global view of all the errors that can be identified[9]. The SAREF Pipeline can be used with a graphical interface (Figure 7.3a) or a command line (Figure 7.3b). The error report is formatted as markdown, allowing one to quickly open an issue to collaboratively deal with problems (Figure 7.3c). Finally, the application generates different serializations for ontologies and examples, and an HTML documentation inspired by LODE and rewritten with SPARQL-Generate (Lefrançois et al., 2017). See, for example, https://saref.etsi.org/core or https://saref.etsi.org/core/Command.

7.4.5 Continuous Integration and Deployment

We configured Gitlab CI/CD in each SAREF ontology repository to run the SAREF pipeline differently depending on the type of branch where a commit is pushed (issue, develop, pre-release, release), and finally automatically pushes the output files to the SAREF documentation portal https://saref.etsi.org/. Figure 7.3d illustrates the automatic execution of the SAREF pipelines.

7.5 Overview of the SAREF Ontology

Figure 7.4 shows an overview of the main classes of SAREF and their relationships. Then a detailed explanation of each class is presented.

9 https://labs.etsi.org/rep/saref/saref-pipeline/-/tree/master/src/main/resources/messages

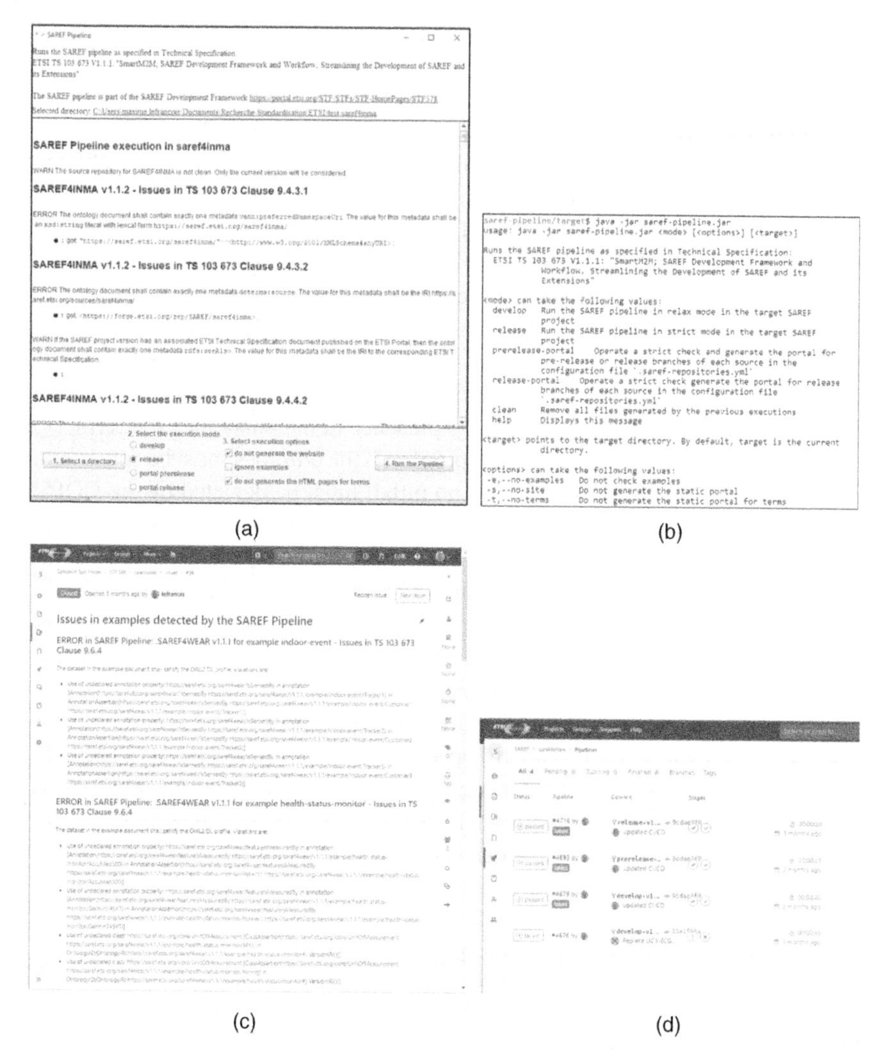

Figure 7.3 The SAREF pipeline checks the compliance of a SAREF project with respect to the ETSI TS 103 673 technical specification (ETSI, 2020k, Clause 9), and generates the public portal. (a) Running the SAREF pipeline with the graphical interface https://saref.etsi.org/sources/saref-pipeline/, (b) Running the SAREF pipeline from the command line https://saref.etsi.org/sources/saref-pipeline/, (c) The output of the SAREF pipeline is formatted in markdown and can be used to create an issue, and (d) Overview of the integration and continuous deployment pipeline: *Snapshot, Staging, Manual release*. Source: https://saref.etsi.org/sources/saref4ehaw/-/pipelines.

7.5.1 Device

SAREF focuses on the concept of device, which is defined as "a tangible object designed to perform a particular task in households, public buildings, or offices. To accomplish this task, the device performs one or more functions." Examples of devices are a light switch, a temperature sensor, an energy meter and a washing machine. A washing machine is designed to wash (task) and to accomplish this task it performs a start and stop function. The saref:Device class and its properties are shown in Figure 7.4.

A saref:Device can have some properties that uniquely characterize it, namely its model and manufacturer (saref:hasModel and saref:hasManufacturer properties, respectively).

SAREF is conceived in a modular way in order to allow the definition of any device from predefined building blocks, based on the function(s) that it performs. Therefore, a saref:Device has at least one function (saref:hasFunction min 1 saref:Function) and can be used for (saref:isUsedFor property) the purpose of offering a commodity, such as saref:Water or saref:Gas. It can also measure properties such as saref:Temperature, saref:Energy, and saref:Smoke. Moreover, a device may consist of other devices (saref:consistsOf property).

The device types that can be represented are actuators (e.g. a saref:Switch that can be further specialized in saref:LightSwitch and saref:DoorSwitch), sensors

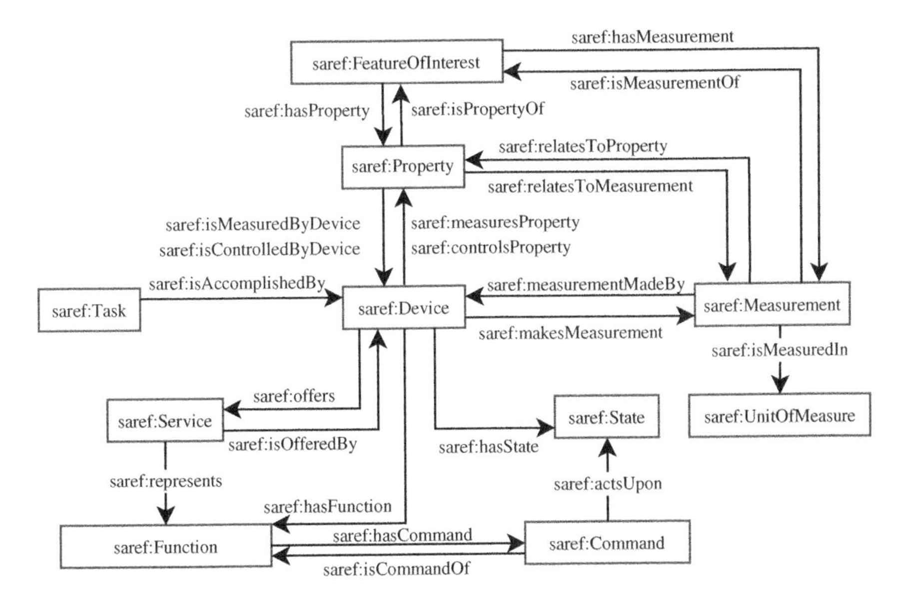

Figure 7.4 Overview of the SAREF ontology (adapted from ETSI (2020j), ©ETSI 2020, all rights reserved).

(e.g. a saref:SmokeSensor and saref:TemperatureSensor), meters, and appliances. Note that there are more types of devices, sensors, and actuators that can be defined to extend SAREF (the device types in Figure 4 represent only some examples that explain the rationale behind SAREF). A description of these types of device is presented in the next clause, in combination with the function they perform. Examples of devices for specific domains are defined in the SAREF extensions.

7.5.2 Feature of Interest and Property

Features of Interest (FOIs) are of high relevance for the IoT domain, including also subdomains such as smart cities and smart agriculture. For example, when generating KPIs as city indicators, such indicators might be related to a given FOI in the city. Additionally, FOIs may have one or more properties to be observed; for example, one can measure the average speed or the CO2 level of a road, the moisture level, or the type of soil of a crop.

SAREF borrows the modeling pattern from the SOSA/SSN ontology, with classes saref:FeatureOfInterest: “any real-world entity from which a property is measured” and saref:Property: “a quality of a feature of interest that can be measured; an aspect of a feature of interest that is intrinsic to and cannot exist without the feature”. A FOI can be linked to its properties using the property saref:hasProperty.

7.5.3 Measurement

The classes saref:Measurement, saref:Property, and saref:UnitOfMeasure allow relating different measurements from a given device for different properties measured in different units, i.e. the saref:Measurement class describes a measurement of a physical quantity (using the saref:hasValue property) for a given saref:Property and according to a given saref:UnitOfMeasure. In this way, it is possible to differentiate between properties and measurements made for such properties and to store measurements for a concrete property in different units of measurement. Furthermore, a timestamp can be added (using the saref:hasTimestamp property) to identify when the measurement applies to the property, which can be used for single measurements or for series of measurements (e.g. measurement streams). Figure 7.4 shows that a saref:Device can measure or control a saref:Property (which may be from a saref:FeatureOfInterest), which in turn relates to a saref:Measurement, which in turn is measured in a given saref:UnitOfMeasure. Note that it is also possible to follow the inverse direction in which a saref:Device makes a measurement in a certain unit of measure (using the saref:makesMeasurement property), and this measurement can be related to a saref:Property (using the saref:relatesToProperty property). A saref:FeatureOfInterest represents any entity of the real world from which a saref:Property is measured.

As an example, the saref:Power and saref:Energy classes can be related to a certain measurement value (using the saref:hasValue property), which is measured in a certain unit of measure, e.g. kilowatt for power (saref:PowerUnit) and kilowatt_Hour for energy (saref:EnergyUnit). Analogously, the saref:Price class can be related to a certain measurement value that is measured using a certain saref:Currency, which is a subclass of the saref:UnitOfMeasure class. Further examples of how to define units of measure can be found in the different SAREF extensions.

The saref:Time class allows one to specify the concept of time in terms of temporal entities (i.e. instants or intervals) according to the existing W3C Time ontology to avoid defining this concept from scratch.

7.5.4 Service, Function, Command, and State

Figure 7.4 shows that a device offers a service (the saref:Service class), which is a representation of a function in a network that makes this function discoverable, registerable, and remotely controllable by other devices in the network. A service shall represent at least one function (saref:represents min 1 saref:Function) and is offered by at least one device that wants (a certain set of) its function(s) to be discoverable, registerable, and remotely controllable by other devices in the network (saref:isOfferedBy min 1 saref:Device). Multiple devices can offer the same service. A service shall specify the device that is offering the service and the function(s) to be represented. For example, a light switch can offer the service of remotely switching lights in a home through mobile phone devices that are connected to the local network (saref:SwitchOnService class). This "remote switching" service represents the saref:OnOffFunction previously described. Note that the concept of service is further elaborated in the oneM2M Base Ontology, to which the reader is referred in order to model the details of a service that are out of the scope of SAREF.

A function is represented in SAREF with the saref:Function class and is defined as "the functionality necessary to accomplish the task for which a device is designed". Examples of functions are saref:ActuatingFunction, which allows transmitting data to actuators; saref:SensingFunction, which allows transmitting data from sensors; saref:MeteringFunction, which allows obtaining data from meters; and saref:EventFunction, which allows contacting other devices.

A saref:Function shall have at least one command associated with it (saref:has-Command min 1 saref:Command). Furthermore, a command can act on a state (saref:actsUpon relation) to represent that the consequence of a command can be a change in the state of a device. Note that a command may act upon a state, but it does not necessarily need to act upon a state. For example, saref:OnCommand acts on saref:OnOffState, but saref:GetCommand does not act on any state, since it only gives a directive to retrieve a certain value.

Depending on the function(s) it performs, a device can be found in a corresponding saref:State. For example, a switch can be in saref:OnOffState, which is also specialized in saref:OnState and saref:OffState. A light switch can be found in saref:OnOffState on which saref:OnCommand and saref:OffCommand will act. Note that SAREF is not restricted to binary states such as saref:OnOffState, but allows us to also define n-ary states (see the saref:MultiLevelState class).

7.6 The SAREF Ontology in the Smart Home Environment

This section describes how the SAREF ontology and its extensions can be applied to smart appliances and the smart home environment. To do so, it will describe the main extensions that are relevant for this environment: those for the energy, water, building, and smart city domains and the extension to represent systems.

As Figure 7.5 presents, the SAREF ontology includes a set of generic classes that can be used in smart homes to represent features of interest, measurements, devices, and their profiles. From these and depending on the aspects of interest, terms from one or multiple extensions can be reused.

The SAREF4ENER extension allows one to represent information on the power profile of any device. It specializes the *Device* and *Profile* classes of SAREF to allow us to include further energy-related information.

The SAREF4WATR extension enables representing information related to the water domain; not only at the device level through water devices (such as water meters) but also at the infrastructure level by describing water assets and infrastructures. The systems underlying these assets and infrastructures can be described in detail using the SAREF4SYST extension.

With the SAREF4BLDG extension, devices (and other building objects) can be included in the context of a particular building. The extension also allows us to represent the topology of a building through its building spaces.

Finally, the SAREF4CITY extension extends the context of features of interest, devices, and their measurements to the smart city. To do so, new spatial features are defined, such as facilities and administrative areas, and city objects can also be represented in them.

This last extension also adds the possibility of representing key performance indicators. In this way, performance measurements and their assessments can be defined for features of interest at any level.

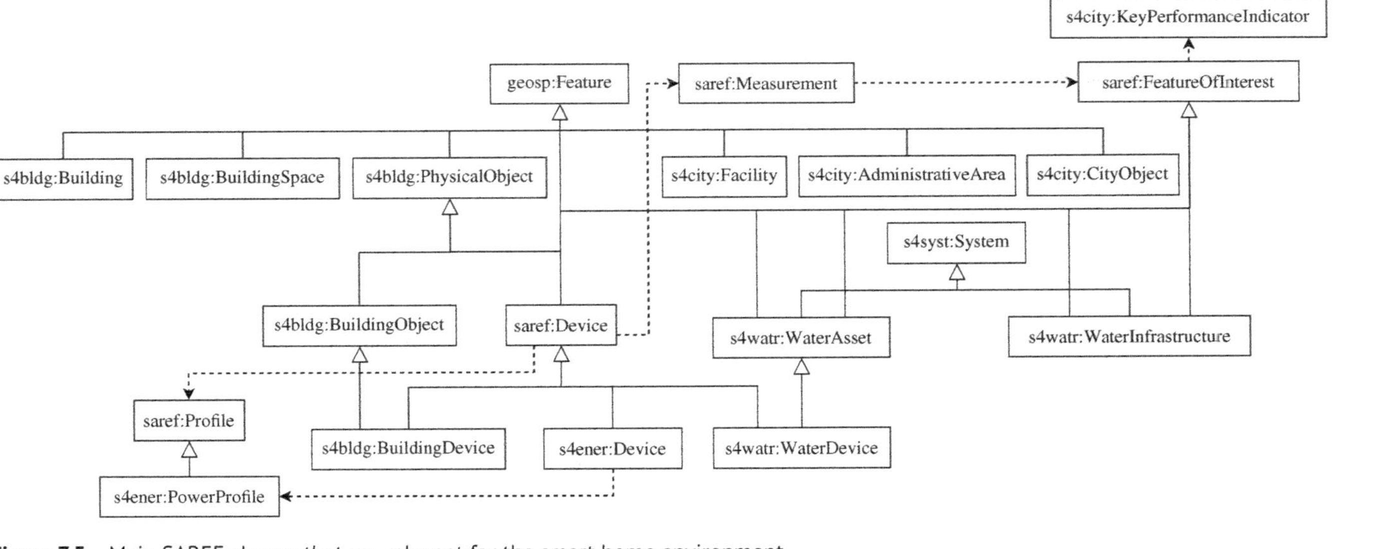

Figure 7.5 Main SAREF classes that are relevant for the smart home environment.

7.6.1 Energy

SAREF4ENER is the first extension of SAREF that was created in 2016 in collaboration with industry associations EEBus[10] and Energy@Home[11] to allow interconnection of their different data models (Daniele et al., 2016). SAREF4ENER is based on EN50631 by CLC TC59 WG7 (EEbus SPINE). SAREF4ENER focusses on demand response scenarios, in which customers can offer flexibility to the smart grid to manage their smart home devices by means of a Home Energy Manager System (HEMS). The HEMS is a logical function for optimizing energy consumption and/or production that can reside either in the cloud or in a home gateway. Moreover, the smart grid can influence the quantity or patterns of use of the energy consumed by customers when energy-supply systems are constrained, e.g. during peak hours. These scenarios involve use cases such as smart energy management of appliances in certain modes and preferred times using power profiles to optimize energy efficiency and accommodate the customers preferences; monitoring and control of the start and status of appliances; reaction to special requests from the smart grid, for example, incentives to consume more or less depending on current energy availability, or emergency situations that require temporary reduction of power consumption.

Figure 7.6 shows the main classes of SAREF4ENER that represent the concepts of "power profile," "power sequence," "alternative," and "slot" that a device uses to communicate its energy flexibility to the HEMS according to the preferences and needs of the consumer.

A s4ener:Device is a subclass of a saref:Device, that is, it inherits the properties of the more general saref:Device and extends it with additional properties that are specific to SAREF4ENER. A s4ener:PowerProfile inherits the properties of the more general saref:Profile, extending it with additional properties that are specific to SAREF4ENER. A power profile is a way to model curves of power and energy over time, which also provides definitions for the modeling of power scheduling, including alternative plans. With a power profile, a device exposes the power sequences that are potentially relevant for the HEMS, for example, a washing machine that wants to communicate its expected energy consumption for a certain day. An alternative group is a collection of power sequences for a certain power profile. For example, the above-mentioned washing machine can offer two alternative plans, a "cheapest" alternative in which the HEMS should try to minimize the user's energy bill and a "greenest" alternative in which the HEMS should try to optimize the configuration to maximize the

10 http://www.eebus.org/en
11 http://www.energy-home.it

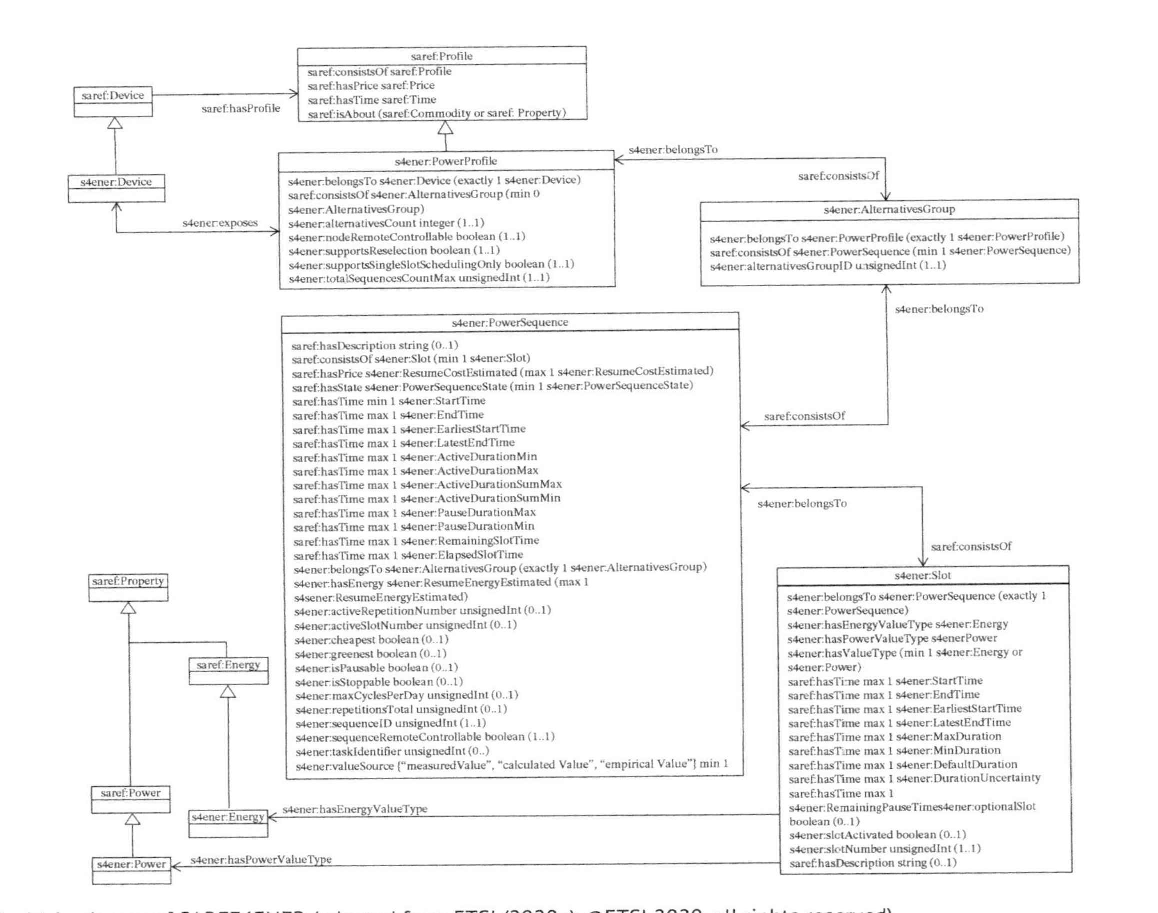

Figure 7.6 Main classes of SAREF4ENER (adapted from ETSI (2020e), ©ETSI 2020, all rights reserved).

availability of renewable energy. An alternative consists of one or more power sequences (s4ener:PowerSequence class). A power sequence is the specification of a task, such as washing or drying, according to user preferences and/or the manufacturer's settings for a certain device. For example, in the "cheapest" alternative mentioned above, the washing machine can ask to allocate two power sequences during the night, while for the "greenest" alternative, it can ask to allocate one power sequence in the morning and one in the afternoon. Of these power sequences, one is allocated for the washing task and cannot be stopped once it started, while the other power sequence is allocated for the tumble drying task and has the flexibility to be paused by the HEMS as long as it finishes within a specified latest end time. A power sequence consists of one or more slots (s4ener:Slot class) that represent different phases of consumption (or production) and their values. In the power sequence allocated for washing, for example, various slots can represent the consumption during the different phases of washing, such as heating the water, washing, and rinsing.

7.6.2 Water

SAREF4WATR (ETSI, 2020h) is the extension of SAREF that provides a common core of general terms for water data orientated to the IoTs. These core terms can be extended to particular water subdomains, for example, to water supply. Figure 7.7 presents the main terms related to water of the SAREF4WATR extension.

The extension specializes devices for the water domain and includes a particular type of water device, a water meter, based on the European M-Bus standard (CEN, 2017a). It also allows for the representation of the tariff that is applied to a water meter, according to the CEN TR 17167:2018 (CEN, 2017b).

The extension also covers a non-exhaustive set of measurable properties that are of interest for this domain: properties of water meters, properties of water flows (based on the European M-Bus standard (CEN, 2017a)), water properties (based on the classification proposed by the World Health Organization (World Health Organization, 2017) and on different EC directives on the quality of drinking water (EC, 1998), bathing water (EC, 2006a), and groundwater (EC, 2006b)) and environmental properties that affect water and the infrastructures that use it.

Water assets and water infrastructures related to different types of water can also be defined. To represent the topology of a water infrastructure or its assets, the GeoSPARQL ontology (Open Geospatial Consortium, 2012) has been reused and, by reusing the SAREF4SYST ontology, the different subsystems of a water infrastructure can be defined. In SAREF4WATR, key performance indicators (KPIs) are intended to be defined for water infrastructures. However, KPIs can also be defined for other features of interest.

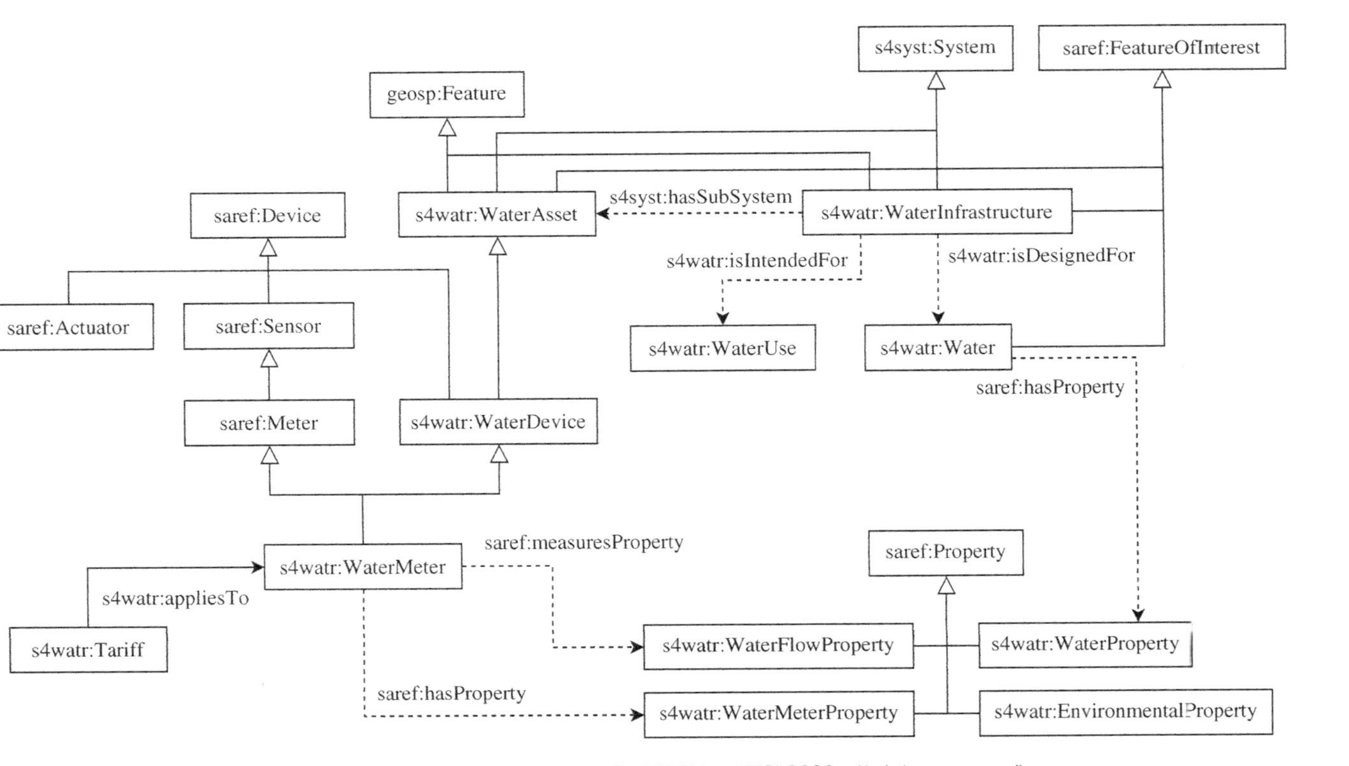

Figure 7.7 Water-related terms of SAREF4WATR (adapted from ETSI (2020h), ©ETSI 2020, all rights reserved).

7.6.3 Building

This section provides an overview of the SAREF4BLDG ontology (ETSI, 2020b), which represents the SAREF extension dedicated to model-building devices based on the International Organization for Standardization (ISO) standard data model Industry Foundation Classes (IFC) (ISO, 2013). The main goal of the SAREF4BLDG ontology is to allow the representation of some IFC features by means of web technologies in combination with SAREF, focussing on the devices, including appliances, described in IFC; and on the location of such devices in buildings.

An overview of the main classes and properties defined in SAREF4BLDG is shown in Figure 7.8.

To reuse the geo ontology modeling for locations, the classes s4bldg:Building, s4bldg:BuildingSpace, and s4bldg:PhysicalObject are represented as subclasses of the class geo:SpatialThing. The s4bldg:Building and s4bldg:BuildingSpace classes are linked to each other through the properties s4bldg:hasSpace and s4bldg:isSpaceOf; which are inverse properties between them. These properties could also be used to define subspaces of a s4bldg:BuildingSpace.

Building spaces can contain physical objects that could represent any type of object or sensors as depicted in Figure 7.8. The property that links the building with the objects is s4bldg:contains.

Finally, the class representing building devices, namely s4bldg:BuildingDevice, is defined as a subclass of both saref:Device and s4bldg:BuildingObject. The device hierarchy extracted from IFC is represented as different subclasses of s4bldg:BuildingDevice and is represented in Figures 7.9a and 7.9b. This hierarchy

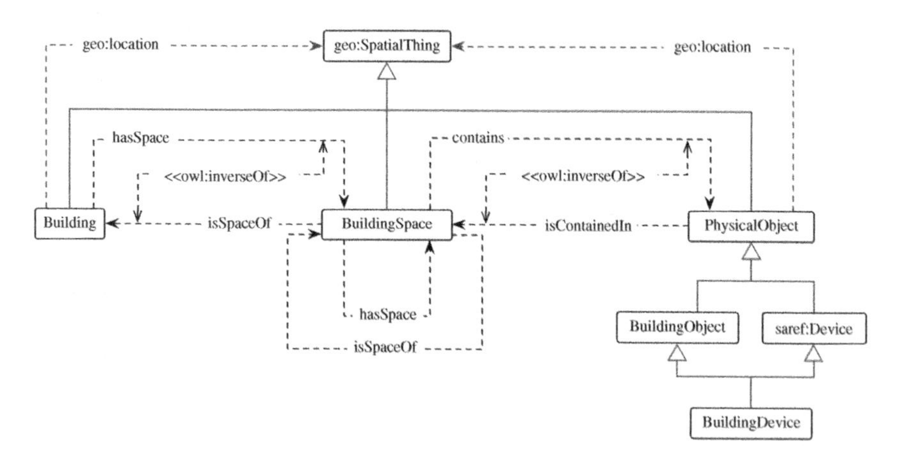

Figure 7.8 Overview of SAREF4BLDG (adapted from ETSI (2020b), ©ETSI 2020, all rights reserved).

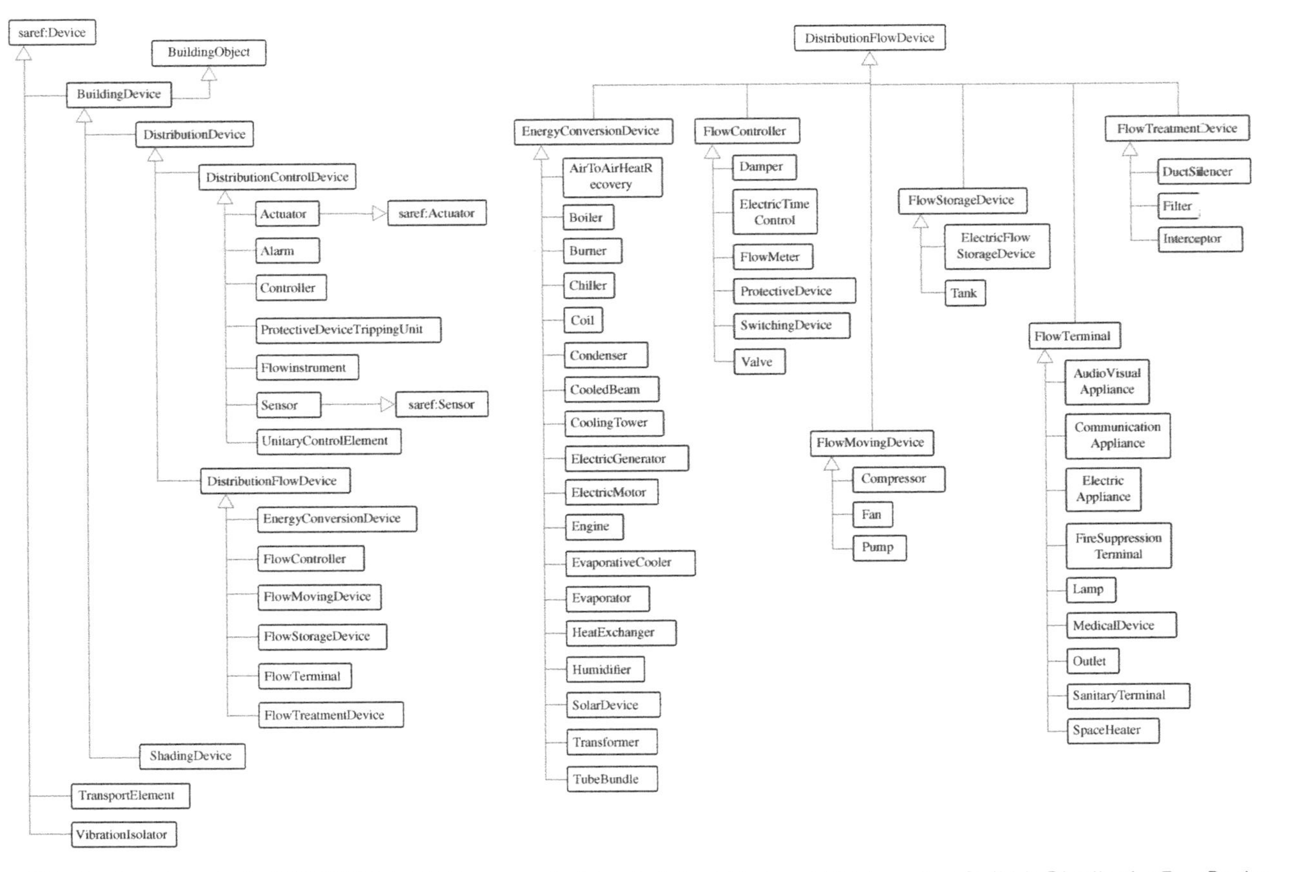

Figure 7.9 Device hierarchy in SAREF4BLDG. (a) Hierarchy of s4bldg:BuildingObject and (b) Hierarchy of s4bldg:DistributionFlowDevice. Source: Adapted from ETSI (2020b), ©ETSI 2020, all rights reserved.

represents the devices defined in IFC reproducing the classification defined in the standard, for example, grouping the devices in s4bldg:DistributionDevice, s4bldg:ShadindDevice, s4bldg:TransportElement, and s4bldg:VibrationIsolator and their inner classifications. More details on the selection of IFC concepts and the ontology development process to build SAREF4BLDG are described in Poveda-Villalón and García-Castro (2018).

7.6.4 City

The SAREF4CITY ontology (ETSI, 2020c) aims to extend SAREF in order to create a general framework for representing smart city data in the IoT domain by identifying the main components. For doing so, different resources have been investigated during the definition of the ontology, for example, ontologies defined by the Open Geospatial Consortium, IoT platforms as FIWARE, European projects and initiatives as the ISA2 program, or the Spanish Federation of Municipalities and Provinces catalogue of vocabularies.

An overview of the SAREF4CITY ontology is shown in Figure 7.10. As can be observed, the main areas represented are as follows: Topology, Administrative Area, City Object, Event, Measurement, Key Performance Indicator, and Public Service.

The topology domain has been represented by reusing the main geographical ontologies, GeoSPARQL and the W3C vocabulary WGS84. The administrative area domain is linked to the topology domain by extending the concept of geosp:Feature with s4city:AdministrativeArea and its subclasses representing cities, countries, districts, and neighborhoods.

The model to represent city objects also relies on the GeoSPARQL topology pattern that allows the connection of city objects with the city or with the parts in which they are located by using the properties geosp:sfContains and geosp:sfWithin inherited from the geosp:SpatialObject class.

Events are modeled by the class s4city:Event that is linked to the agent who organizes them through the property s4city:organizedBy. The facilities in which the events can take place are indicated by the property s4city:takesPlaceAtFacility. The time in which it takes place is represented by the class time:TemporalEntity reused from the W3C Time ontology, and it is indicated by the property s4city:takesPlaceAtTime.

Conceptualization of KPIs involves two main concepts, namely s4city: KeyPerformanceIndicator and s4city:KeyPerformanceIndicatorAssessment.

This distinction is motivated by the need to decouple the definition of a KPI in general terms, for example, the mean air pollution per week, and a particular value of such a KPI, for example, the mean value of air pollution last week in Paris. The relationship between a specific assessment of a KPI (s4city:KeyPerformance

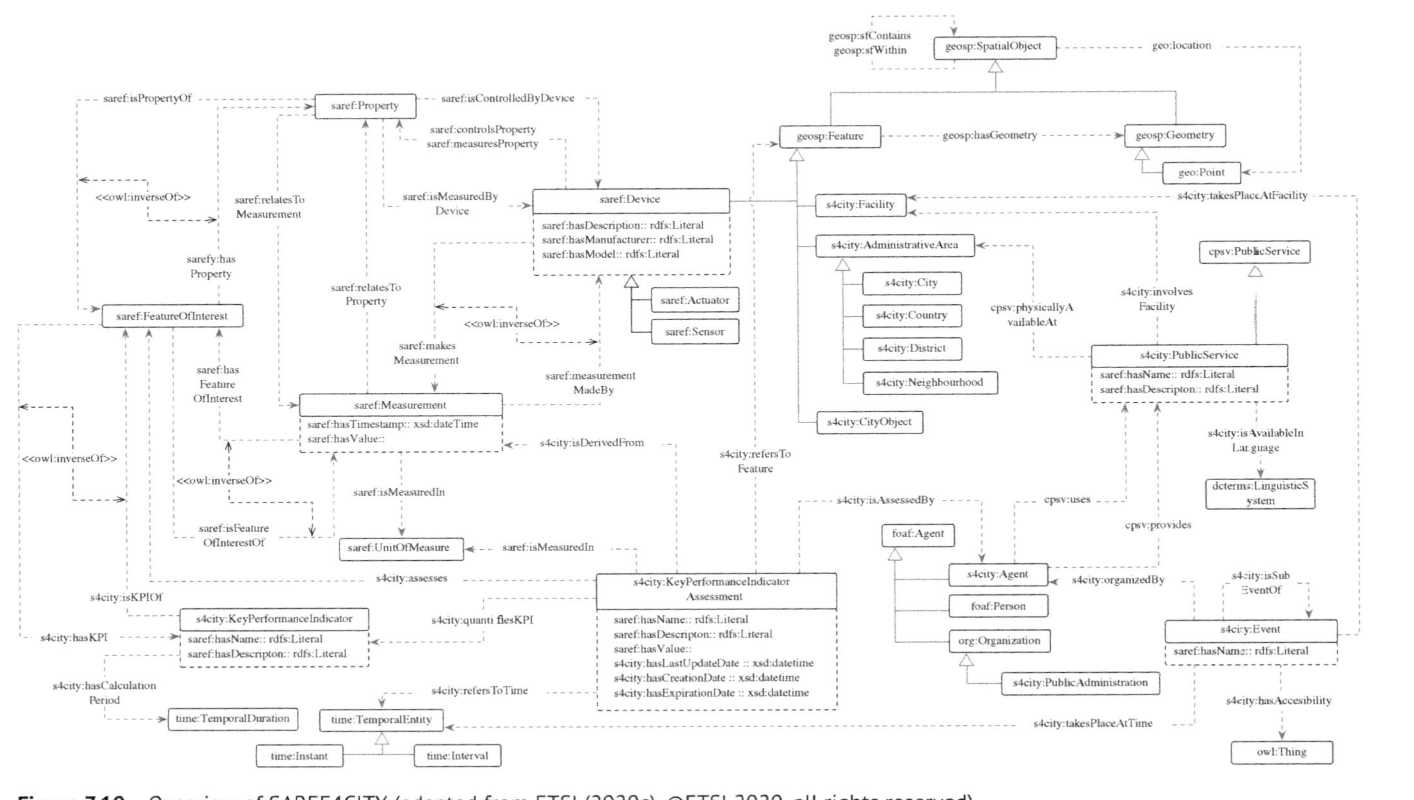

Figure 7.10 Overview of SAREF4CITY (adapted from ETSI (2020c), ©ETSI 2020, all rights reserved).

IndicatorAssessment) and the general KPI definition (s4city:KeyPerformance Indicator) can be established by means of the property s4city:quantifiesKPI.

The property s4city:isKPIOf allows linking from a s4city:KeyPerformance Indicator to the measured saref:FeatureOfInterest. The calculation period of s4city:KeyPerformanceIndicator is provided by the s4city:hasCalculationPeriod property. Some attributes are attached to KPIs, such as their name (using the property s4city:hasName) and a natural language description (s4city:has Description).

Finally, the SAREF4CITY ontology allows representing public services (s4city:PublicService) by extending the reused concept cpsv:PublicService defined in the Public Service vocabulary provided by the ISA vocabularies European initiative. Services can also be linked to the facilities involved by the property s4city:involvesFacility and it is possible to indicate in which administrative area it is provided using the reused property cpsv:physicallyAvailableAt. It is possible to indicate that an agent (s4city:Agent) provides (cpsv:provides) or uses (cpsv:uses) public services and the language in which they are available (s4city:isAvailableInLanguage). The name and description in natural language of public services are represented by the attributes s4city:hasName and s4city:hasDescription, respectively.

7.6.5 Systems

The SAREF4SYST ontology (ETSI, 2020g), shown in Figure 7.11 and inspired by SEAS (Lefrançois, 2017), is the first ontology pattern incorporated into SAREF.

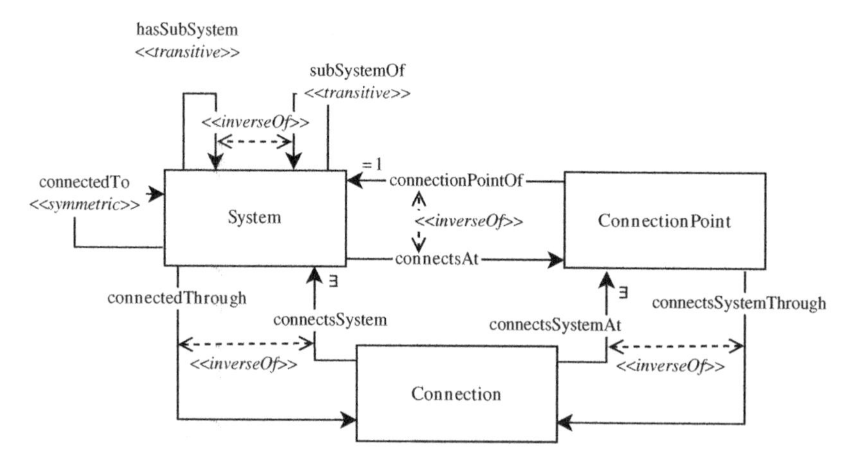

Figure 7.11 Overview of the SAREF4SYST ontology pattern (adapted from ETSI (2020g), ©ETSI 2020, all rights reserved).

It defines an ontology model that can be instantiated for different domains. SAREF4SYST defines systems, connections between systems, and connection points to which systems can be connected. These basic concepts can be used generically to define the topology of entities of interest and can be specialized for multiple domains. For example, to describe areas within a building (systems) that share a boundary (connections). The properties of systems are usually state variables (e.g. agent population, temperature), while properties of connections are usually flows (e.g. heat flow). SAREF4SYST has two main goals: on the one hand, to extend SAREF with the ability to represent the general topology of systems and how they are connected or interact, and, on the other hand, to illustrate how ontology patterns can help ensure a homogeneous structure of the overall SAREF ontology and speed up extension development.

7.7 The SAREF Ontology in Use

As a result of significant standardization efforts such as SAREF, the IoT industry perceives the impact that ontologies can have to enable the missing interoperability. However, most industrial practitioners are not familiar with semantic technology or have an incentive to adopt it, as they believe the learning curve is too steep. Information on ontologies appears to them abstract and scattered over the Web, and thus not easily applicable. The goal of this section is to provide examples of existing SAREF-based implementations in the smart home environment as a support for practitioners to understand in which practical settings the ontologies presented in this chapter can be applied.

The first example is based on a scenario of the Dutch pilot of the H2020 InterConnect project[12] within a household in which smart appliances from different vendors (i.e. Bosch, Siemens, Miele, and Whirlpool) interact with an energy manager in the cloud that takes care of optimizing the energy consumption and production of all home appliances depending on the time of the day, the user preferences, and possible incentives from the grid. In particular, we focus on the specific interaction between smart appliances from Bosch, Siemens or Miele, which are equipped with EEBUS SPINE-IoT communication defined in the EN50631 standard, with an energy manager that makes use of the so-called S2 interface according to the EN50491-12-2 standard. The energy manager is an implementation of the TNO Reflex platform, which organizes the demand and supply of energy, as a tool for aggregation and scheduling of energy flexibility. Smart appliances and energy managers expose their data to a semantic interoperability layer in the format of the SAREF and SAREF4ENER ontologies.

12 https://interconnectproject.eu/wp-content/uploads/2021/03/Interconnect_Netherlands_Residencial_EnglishVersion.pdf

Therefore, SAREF and SAREF4ENER are used as the common language to allow seamless communication in the smart home and enable optimization of energy consumption and production of the home. As an example of a storyline that uses this infrastructure, household owners load their laundry into a washing machine (i.e. an iQ700 Siemens device in the example under consideration) and specify their preference by selecting the latest end time when the laundry needs to be ready. Setting this preference can be done by the user directly using the washing machine display or through the Siemens app available for the mobile phone. The washing machine, which communicates with the energy manager using SAREF and SAREF4ENER, sends the new user demand (in the form of a s4ener:AlternativesGroup that belongs to the washing machine s4ener:PowerProfile), which is received by the Reflex Energy Manager that adapts the plan based on the updated conditions. The Energy Manager is the one who finally decides when the washing machine should start and updates the new start time by sending it to the washing machine.

The second example is based on a SAREF-based implementation in the context of the Greek pilot of the H2020 InterConnect project[13] in which residential buildings are transformed into smart homes with energy meters and sensors installed in the houses. Several companies are involved: GRIDNET, an SME ICT company responsible for transforming 50 homes into smarthomes with smart meters and sensors connected to an IoT gateway and company's cloud services; COSMOTE, a Telco Provider responsible for transforming additional 50 homes to smarthomes with smart meters and sensors connected with an IoT Gateway and companys cloud services; and HERON, a retailer responsible for equipping 200 houses with smart meters and collection of consumption data within the companys cloud services. In addition, AUEB is an academic partner responsible for developing a mobile app that building residents can use to monitor their house consumption and interact with the various services offered by the Greek pilot. This example of SAREF in action focusses on how the four different service providers (i.e. GRIDNET, COSMOTE, HERON, and AUEB) leverage SAREF and its extensions as a common data model to exchange information. GRIDNET has installed Qubino energy meters that communicate wirelessly with Z-Wave technology with an IoT gateway that runs the open-source framework openHAB. COSMOTE uses AEOTEC energy meters with Z-Wave technology that communicate with an IoT gateway that runs the open-source framework Home Assistant. HERON takes advantage of the Wi-fi network of the user to connect directly to Shelly energy meters. An AUEB-developed IoT app presents residents with their energy consumption information regardless of the specific IoT solution

13 https://interconnectproject.eu/wp-content/uploads/2021/03/Interconnect_Greece-Geral_EN-1.pdf

the house has been equipped with (i.e. GRIDNET, COSMOTE, or HERON). The bridging technology for the communication is SAREF, SAREF4BLDG, and SAREF4ENER. As an example of storyline, the user logs into the app and after authentication the app requests historical consumption data, for example, about the last week. A response is generated and forwarded to the mobile app, which is then presented in the user dashboard. The user experience remains the same while using different service providers behind the scenes.

7.8 Lessons Learnt

This section provides some lessons learnt from a long path of development and evolution of the ETSI SAREF ontology for smart applications.

7.8.1 Specification of Ontology Requirements

The ETSI approach for the development of SAREF has been applied in a constrained setting due to the fact that the ETSI STF SAREF projects are usually focused efforts that aim at reaching concrete results in rather short time, compared to European research projects with a typical duration of 3–4 years. As an example, the STF projects carried out in the past years to extend SAREF (e.g. STF 513, STF 534, STF 566, STF 602), in the ontological requirement specification step, have considered an average of three use cases per SAREF extension, as this was sufficient to generate a variety of significant requirements (e.g. an average of 60 competency questions per extension) and, at the same time, keep focus on the scope of the ontology and pace in the ontology development process. On the other hand, it was also necessary to apply the process in larger settings, for example, 112 use cases, 66 services, 166 APIs, and 864 parameters were analyzed in the InterConnect project to enrich the current SAREF suite of ontologies with additional concepts to accommodate new use cases in the smart home and energy domains. In these types of large-scale settings, an initial information overload is unavoidable. For example, this resulted in 350+ requirements in InterConnect, after an initial analysis based on the most relevant use cases and services. However, we could observe that eventually the situation stabilized, i.e. the addition of more use cases did not result anymore in new requirements. Therefore, we learnt that use cases are a useful input to drive the ontology development taking into account concrete needs of the industry and that a limited number of use cases (i.e., 3+ use cases) helps to keep focus and pace, yet, guaranteeing a suitable semantic coverage. However, new concepts may be needed to accommodate new use cases, and the addition of use cases eventually will reach an optimum in which no new concepts will be anymore added to the ontologies, but will provide useful support to validate the already existing concepts.

7.8.2 Stakeholder's Workshops

Since the beginning of SAREF development in 2014, ETSI ontology experts have used to conduct stakeholder workshops in person for the development of SAREF and its extensions with a large number of stakeholders (i.e. 60+ stakeholders). Face-to-face interaction with stakeholders proved to be an extremely successful aspect, especially in increasing the understanding and acceptance of the resulting ontologies (Daniele et al., 2015a). In order to make these workshops manageable and productive, we adopted the best practise to divide the participants into different (small) groups and work in parallel, drawing ontology concepts and relations on a whiteboard. Due to the COVID-19 pandemic in 2019–2021, it has no longer been possible to gather stakeholders in person and ontology workshops had to be conducted entirely online. From this experience we learnt that i) an online workshop helps to involve a larger number of stakeholders compared to the more limited number reachable in a meeting in person, and ii) it is still possible collect ontology requirements (smaller sessions, break-out rooms, etc.), although in a less effective manner compared with face-to-face interaction (results of one day face-to-face are comparable to three days of online workshops). In conclusion, we observe that the desired goals for the SAREF development can also be reached online, but this affects the pace (slower) and the resources (higher) invested to achieve the result compared to the face-to-face setting used in the past.

7.8.3 Tool Support

The development of ontologies with close and active participation of stakeholders is essential and, at the same time, challenging. The learning curve of semantic technology and ontologies for industrial practitioners is steep, as it requires a paradigm shift for traditional software developers, who are not used to thinking in terms of semantics and triples. Moreover, stakeholders often do not have the time nor the incentive to browse through lengthy ontology specifications, and as a consequence, they rather rely on direct consultancy of the (ETSI SAREF) semantic experts, which are a minority compared to the vast majority of traditional software developers, and this is not sustainable. We therefore learnt that it is essential to improve training and communication material that can be reused by the semantic experts across and beyond specific projects, as basis to speed up the adoption of the technology. Moreover, tool support is needed in the near future, starting from visualization tools for nonexperts to quickly browse ontologies and become familiar with them, to intuitive user interfaces that help the same nonexpert users select and focus on only a few parts of interest of the ontologies, especially when they are faced with the challenge of combining different subparts of an extensive framework of ontologies such as SAREF and its extensions (e.g. parts of SAREF, SAREF4ENER, SAREF4BLDG, SAREF4WATR, and SAREF4CITY that are relevant for the smart home environment, as shown in Figure 7.5). Finally, we also learnt that despite the efforts in ETSI for the creation of the user-friendly SAREF portal with the collection of all related technical reports and specifications, it is unclear for the majority of stakeholders how to actively participate to the development and standardize new contributions to SAREF. This chapter serves as an initial tool for stakeholders to find, all in one place, not only the information to become familiar with SAREF and its extensions but also the ETSI workflow for the SAREF projects in order to clarify to stakeholders how they contribute to the further development and standardization of these ontologies.

7.8.4 Ontology Modularization

Two design choices are recurrent in ontology networks: (i) the definition of namespaces for each module, and (ii) the choice of the module in which a term is defined. Choosing a distinct namespace for each module makes it easy to identify from which module a term originates. It also simplifies the publication of ontologies in accordance with the good practise of making a description of each term accessible to its IRI (hash IRIs can be used). However, this approach presents three problems.

First, it is sometimes difficult for a user of these ontologies to remember what the namespace is for each concept. We have, for example, a variety of subclasses of saref:Property spread over the namespaces of the different extensions, depending on where we needed to define them first: saref:Temperature, saref:Humidity, saref:Power, s4ener:PowerMax, s4ener:PowerStandardDeviation, s4inma:Size, saref:Light, s4envi:LightProperty and the following instances of saref:Property: s4envi:Frequency, s4wear:SoundLevel, s4wear:Temperature, s4wear:BatteryRemainingTime, s4watr:Conductivity. An alternative approach would have been to use a single namespace and slash "/" IRIs, and implement redirects from the IRI of each term to the document that describes the ontology where it is defined.

Second, experience shows that it can be relevant to move a term from one module to another. For example, SAREF4CITY V1.1.1 introduced the concept of s4city:FeatureOfInterest, and it was decided during the development of SAREF Core V3.1.1 that this concept should be moved into the core ontology. Therefore, it is now identified by saref:FeatureOfInterest, and the SAREF4CITY implementations had to be modified. This problem would not have arisen if an approach based on a unique namespace and slash IRIs had been adopted.

Finally, what the list of classes and instances of saref:Property shows is that even within the SAREF developer community, the modeling and naming choices are sometimes varied. Therefore, it would seem important to us to rebase the development of SAREF on ontology patterns to harmonize its development.

7.8.5 Ontology Patterns

One of the deliverables of the ETSI STF 556 project is the technical report TR 103 549 entitled "Consolidation of SAREF and its industrial user community, based on the experience of the EUREKA ITEA 12004 SEAS project". This report identifies the implicitly existing patterns in SAREF and that formalization could be appropriate to achieve a consolidated version of the SAREF ontology (ETSI, 2019).

For example, in SAREF Core V2.1.1, measurement, actuation, and metering functions are function types. Usually, a function (e.g. saref:StartStopFunction) has one or more commands to trigger it (e.g. for saref:StartStopFunction, it should be either a saref:StartCommand or a saref:StopCommand). Some commands act on certain states (saref:StartStopCommand acts on a certain saref:StartStopState). It would be advisable, for example, to make sure that all subclasses of the saref:Command class are described in the same way. For example, some subclasses of saref:Command had generic instances, associated with no real action. SAREF also had a command saref:PauseCommand, which was not associated with any function.

Patterns can be instantiated with elements taken in orthogonal dimensions. For example, SAREF4ENER defines s4ener:EnergyStandardDeviation, s4ener:EnergyMax, s4ener:EnergyMin, s4ener:EnergyExpected, s4ener:PowerMax, s4ener:PowerMin, s4ener:PowerExpected, s4ener:PowerStandardDeviation. Manually managing the addition of, for example, a new aggregate type *Average* involves creating many additional properties, such as s4ener:EnergyAverage, s4ener:PowerAverage. A partial solution to this problem is to decouple the dimensions. In the example above, we will specify the property type and the aggregate type.

7.9 Conclusions and Future Work

This chapter provided an overview of how the ETSI SAREF ontology is designed and can be used for smart applications in smart home environments. The design rationale and development framework of SAREF, together with the level of support from the European Commission and ETSI, make it a good candidate for designing interoperable cross-vertical common data spaces with a focus on IoT applications.

SAREF illustrates how the development of an ontology can transition from focused short-term projects mainly involving researchers to large-scale research and development projects with large industrial stakeholders. The SAREF development framework and workflow as specified in ETSI TS 103 673 (ETSI, 2020k), the SAREF pipeline and the public SAREF portal enable SAREF developers to accelerate the development of SAREF and its extensions, as well as the SAREF user community to actively contribute to development. Other keys to SAREF success include participation of stakeholders in regular workshops, good tool support for its development, and the adoption of design best practises including modularization, versioning, and ontology patterns for improved homogeneity.

Acknowledgments

Part of the development of the SAREF suite of ontologies has been funded by European Commission SMART 2013/01077 and 2016/0082 studies; the European Telecommunications Standards Institute Specialist Task Forces 513, 534, 556, 566, 578, and 602; the Horizon 2020 European project InterConnect grant agreement 857237; and the French project ANR-19-CE23-0012-04 CoSWoT. The authors would also like to thank Josef Baumeister (BSH), Jorrit Nutma (TNO), and Donatos Stavropoulos (GRIDNET) for the examples of SAREF in use from the Dutch and Greek pilots of InterConnect.

References

CEN (2017a). EN 13757-2:2017: Communication systems for meters - Part 2: Wired M-Bus communication. Technical report.

CEN (2017b). TR 17167:2018: Communication system for meters - Accompanying TR to EN 13757-2,-3 and -7, Examples and supplementary information. Technical report.

Daniele, L. and Strabbing, W. (2018). Study on ensuring interoperability for Demand Side Flexibility. European Commission.

Daniele, L., den Hartog, F., and Roes, J. (2015a). Created in close interaction with the industry: the smart appliances reference (SAREF) ontology. In: *Proceedings of the Formal Ontologies Meet Industry workshop (FOMI 2015), Lecture Notes in Business Information Processing*, vol. 225 (ed. R. Cuel and R. Young), 100–112. Cham: Springer.

Daniele, L., den Hartog, F., and Roes, J. (2015b). Study on semantic assets for smart appliances interoperability, D-S4 final report. European Commission.

Daniele, L., Solanki, M., den Hartog, F., and Roes, J. (2016). Interoperability for smart appliances in the IoT world. In *Proceedings of the International Semantic Web Conference (ISWC 2016), Lecture Notes in Computer Science*, vol. 9982, 21–29. Cham: Springer.

EC (1998). Council Directive 98/83/EC of 3 November 1998 on the quality of water intended for human consumption. Technical report.

EC (2006a). Directive 2006/7/EC of the European Parliament and of the Council of 15 February 2006 concerning the management of bathing water quality and repealing Directive 76/160/EEC. Technical report.

EC (2006b). Directive 2006/118/EC of the European Parliament and of the Council of 12 December 2006 on the protection of groundwater against pollution and deterioration. Technical report.

ETSI (2016). oneM2M; Base Ontology (oneM2M TS-0012 version 2.0.0 Release 2). ETSI Technical Specification 118 112 v2.0.0.

ETSI (2019). SmartM2M; Guidelines for consolidating SAREF with new reference ontology patterns, based on the experience from the ITEA SEAS project. ETSI Technical Report 103 549 V1.1.1., 07.

ETSI (2020a). SmartM2M; Extension to SAREF; Part 7: Automotive Domain. ETSI Technical Specification 103 410-7 V1.1.1., 07.

ETSI (2020b). SmartM2M; Extension to SAREF; Part 3: Building Domain. ETSI Technical Specification 103 410-3 V1.1.2., 05.

ETSI (2020c). SmartM2M; Extension to SAREF; Part 4: Smart Cities Domain. ETSI Technical Specification 103 410-4 V1.1.2., 05.

ETSI (2020d). SmartM2M; Extension to SAREF; Part 8: eHealth/Ageing-well Domain. ETSI Technical Specification 103 410-8 V1.1.1., 07.

ETSI (2020e). SmartM2M; Extension to SAREF; Part 1: Energy Domain. ETSI Technical Specification 103 410-1 V1.1.2., 05.

ETSI (2020f). SmartM2M; Extension to SAREF; Part 2: Environment Domain. ETSI Technical Specification 103 410-2 V1.1.2., 05.

ETSI (2020g). SmartM2M; SAREF consolidation with new reference ontology patterns, based on the experience from the SEAS project. ETSI Technical Specification 103 548 V1.1.2., 06.

ETSI (2020h). SmartM2M; Extension to SAREF; Part 10: Water Domain. ETSI Technical Specification 103 410-10 V1.1.1., 07.

ETSI (2020i). SmartM2M; Extension to SAREF; Part 9: Wearables Domain. ETSI Technical Specification 103 410-9 V1.1.1., 07.

ETSI (2020j). SmartM2M; Smart Applications; Reference Ontology and oneM2M Mapping. ETSI Technical Specification 103 264 V3.1.1., 02.

ETSI (2020k). SmartM2M; SAREF Development Framework and Workflow, Streamlining the Development of SAREF and its Extensions. ETSI Technical Specification 103 673 V1.1.1.

ETSI (2021). SmartM2M; Extension to SAREF; Part 11: Lift Domain. ETSI Technical Specification 103 410-11 V1.1.1., 07.

Fernández-Izquierdo, A. and García-Castro, R. (2019). Themis: a tool for validating ontologies through requirements. *Software Engineering and Knowledge Engineering*, pp. 573–753.

Gyrard, A., Atemezing, G., Bonnet, C. et al. (2016). Reusing and unifying background knowledge for internet of things with LOV4IoT. *2016 IEEE 4th International Conference on Future Internet of Things and Cloud (FiCloud)*, pp. 262–269. https://doi.org/10.1109/FiCloud.2016.45.

ISO 16739:2013 (2013). *Industry Foundation Classes (IFC) for data sharing in the construction and facility management industries*. International Standardization Organization.

Knublauch, H. and Kontokostas, D. (2017). Shapes Constraint Language (SHACL). W3C Recommendation, W3C, July 20 2017.

Lefrançois, M. (2017). Planned ETSI SAREF extensions based on the W3C&OGC SOSA/SSN-compatible SEAS ontology patterns. *Workshop on Semantic Interoperability and Standardization in the IoT, SIS-IoT*, p. 11p.

Lefrançois, M., Zimmermann, A., and Bakerally, N. (2017). A SPARQL extension for generating RDF from heterogeneous formats. *European Semantic Web Conference*, pp. 35–50. Springer.

oneM2M (2018). oneM2M; Base Ontology (oneM2M TS-0012 version 3.7.1). oneM2M Technical Specification 0012 v3.7.1.

Open Geospatial Consortium (2012). OGC 11-052r4: OGC GeoSPARQL - A Geographic Query Language for RDF Data. Version 1.0. Technical report.

Poveda-Villalón, M. and García-Castro, R. (2018). Extending the SAREF ontology for building devices and topology. *Proceedings of the 6th Linked Data in Architecture and Construction Workshop (LDAC 2018), Vol. CEUR-WS*, Volume 2159, pp. 16–23.

Poveda-Villalón, M., Gómez-Pérez, A., and Suárez-Figueroa, M.C. (2014). OOPS! (Ontology Pitfall Scanner!): An on-line tool for ontology evaluation. *International Journal on Semantic Web and Information Systems (IJSWIS)* 10 (2): 7–34.

Poveda-Villalón, M., Fernández-Izquierdo, A., Fernández-López, M., and García-Castro, R. (2022). LOT: An industrial oriented ontology engineering framework. *Engineering Applications of Artificial Intelligence* 111: 104755.

Vandenbussche, P.-Y., Atemezing, G.A., Poveda-Villalón, M., and Vatant, B. (2017). Linked open vocabularies (LOV): a gateway to reusable semantic vocabularies on the Web. *Semantic Web* 8 (3): 437–452.

World Health Organization (2017). Guidelines for drinking water quality. Fourth edition incorporating the first addendum. Technical report.

8

Scheduling of Residential Shiftable Smart Appliances by Metaheuristic Approaches

Recep Çakmak

Samsun University, Department of Electrical-Electronics Engineering, Faculty of Engineering, Samsun, Turkey

8.1 Introduction

The demand for electrical energy is increasing day by day with the increasing population, urbanization, and the use of electrically powered devices. The energy grid utilities are struggling to provide a balance between demand and supply. On the other hand, the electrical grid which has been installed almost 100 years ago, so it needs a rehabilitation and digitally controlled infrastructure. Although the installation of new power (P) plants and the renewal of the network infrastructure require large investment, the fact that keeping consumers without electricity for one hour after a collapse that may occur in the electricity network due to any malfunction in the existing electricity system causes serious economic losses. In other words, there is a need for financial resources to generate electricity to prevent financial losses that will arise if the existing system is interrupted for any reason. Therefore, it is necessary to ensure the reliability and sustainability of the existing energy grid, as well as the necessity to renew the existing electrical infrastructure and to install new power plants and distributed local energy generation. Meeting the increasing demand and ensuring the stable operation of the network are two important challenges of today's energy grids. Environmentally friendly power sources such as wind, solar, and tidal are preferred for distributed generation. However, these sources provide variable and intermittent power output. For this reason, the electrical grid management paradigm has changed from "supply follows demand" to "demand follows supply" by evolving smart grids.

In smart grid studies, demand-side management (DSM) has a significant importance since DSM contributes to compensating the peak demand which influences the reliability and robustness of energy grids. Furthermore, DSM is effective tool

Energy Smart Appliances: Applications, Methodologies, and Challenges,
First Edition. Edited by Antonio Moreno-Munoz and Neomar Giacomini.

since managing the demand provides a flexible source to complement variable and intermittent renewable power generation. DSM provides grid efficiency and robustness of the grid by compensating peak demand and managing the demand with proper use of renewable power generation. Accordingly, utility companies execute DSM which include demand response (DR), energy efficiency (EE), and conservation programs. The US Department of Energy (DoE) has defined demand response as a tariff or program which provides changes in electricity consumption profiles of the customers via price of electricity or incentive payments (QDR 2006). DSM has been defined and classified into six major types as to daily and seasonal electricity demand by Gellings (1985). Figure 8.1 illustrates these six major DSM objectives (Gellings 1985). In this sense, it can be said that DSM

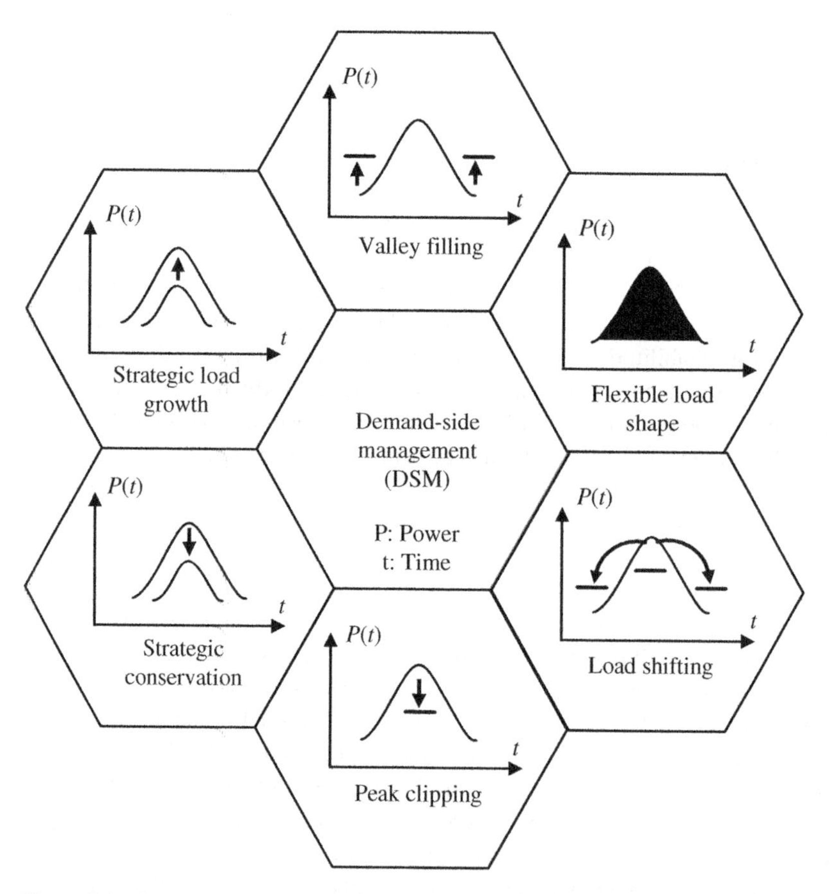

Figure 8.1 Six major DSM objectives proposed by Gellings (1985). Source: Adapted from Gellings (1985).

strives to ensure a smooth and efficient operation of electrical power systems through demand response programs (Bayod-Rújula 2009; Pina et al. 2012).

Demand response (DR), EE, and spinning reserve (SR) are main schemes DSM can utilize. A demand response strategy is a tariff or program which aims to shape power consumption of the customers through incentive payments and/or electrical tariff plans. DR programs can be categorized as price-based and incentive-based programs which include different approaches (Shafie-khah et al. 2016).

DSM and DR studies have been reviewed in some respects such as in terms of utilized optimization strategies (Mariano-Hernández et al. 2021; Javaid et al. 2017), impacts to electrical power systems (Jabir et al. 2018) and also compared according to their performances (Alwan and Abdelwahed 2019). DR programs can be subjected to various classifications according to the arguments and purposes used in the management of consumer loads. In the detailed literature review by Vardakas et al. (2015), DR is classified as follows:

- Price-based DR or incentive-based DR, according to the way it affects the customer behavior.
- Task-based DR or energy management-based DR, according to the decision variable based on load management.
- Centralized DR or distributed DR, in terms of the control of the loads.

The residential time-shiftable loads have great demand response potential and the utilization of residential demand response programs provides efficiency on the electrical grid. For example, in Europe, according to the survey results, if households agree to have their washing machine or dishwasher operated flexibly by an energy manager 5 MW for washing machines and 10 MW for dishwashers, are available as a flexible load on the grid (Stamminger and Schmitz 2017). Another example from the literature presents that flexibility potential for 100 households is 11.45, 9.28, and 16.7 kW for cloth washing machines, dryers, and dishwashers, respectively (Degefa et al. 2018). If this flexible load can be managed within the scope of demand response, the efficiency of the grid can be improved. As a remarkable example, load Losses in the United Kingdom in the low voltage network (Electricity North West [ENW]) constitute 44% of the total losses in the electricity network, and 63% of these low voltage losses are losses related to the consumption profile. The copper losses which are proportional to the square of the current are decreased when peak consumption is reduced. Hence spread across a longer period of the demand through demand curve flatting means that lower losses. Such that, in the event that 10% of the peak consumption in the distribution network in the UK is postponed for four hours, the losses can be reduced by 4 GWh per year (Shaw et al. 2009). So, demand response programs in the residential area have a great importance for the grid management and grid efficiency. The potential of residential demand response can be realized by emerging smart homes. However, if smart time-shiftable appliances are not controlled and scheduled via energy manager, rebound effect may occur which means new peaks happen. For this reason, one of the aims of demand response program which is the peak load reduction through load shifting cannot be achieved (Degefa et al. 2018).

Smart homes which include smart appliances are foreseen to drive major changes in human behavior. The smart homes and smart appliances not only bring great changes to people's behaviors but also create an opportunity to actualize the aforementioned demand response by managing the power consumption profile over a given period of time. The time-shiftable and controllable smart appliances have a great potential to mitigate peak-to-valley power demand gaps in the distribution grids. Also, the smart appliances can improve flexibility of the distribution grids and can decrease the household electricity costs. A smart appliance can be defined as follows. A device which is programmable, connectable to other devices, controllable, and operable autonomously which means programmable itself via artificial intelligence, based on the user preferences and input related to energy price, temperature, lighting conditions, and so on can be defined as the smart appliance. Utility of the smart appliances is emerging as a result of the Internet of Things (IoT) which enables the creation of a system of connected devices that communicate with each other and form an intelligent network to share information through various communication protocols. Essentially, it is foreseen that every smart appliance of the smart home will also be an IoT device in future. By the way, it is noted that it doesn't mean by default that they interoperate. However, in terms of smart grids, the smart appliances are the enablers for demand response programs through their connection and control interface. It will enable not only the residential users which have a smart home to monitor and control any home appliances remotely but also provide a flexible load control opportunity to the grid operators, especially at the distribution grid level.

Smart appliances and IoT devices in households require standardization for them. The European Telecommunications Standards Institute (ETSI) standardized the Smart Applications Reference (SAREF) which is the core model to connect the smart appliances from all domains (SAREF 2022). The intention of SAREF is to enable the link of information coming from different smart appliances, based on different standards instead of replacing the existing standards. So, combined with appropriate software applications, SAREF ontology enables smart appliances in a home to become interoperable even if manufactured by different vendors, as long as all comply to the SAREF ontology. SAREF ontology aims to create a model which provides a consensus to enable matching of existing smart applications domains. Thus, SAREF ontology defines a standard for a domain providing separation and recombination of different parts of the ontology as to specific necessities. This standardization is constituted based on following four essences:

- Reuse and alignment
- Modularity
- Extensibility
- Maintainability

Reuse and alignment refer to adjust and rework of concepts and relationship in the existing devices and platforms. The separation and recombination of different parts of the ontology are provided by its modularity principle. The extensibility of the ontology enables the further improvements and advancements. The maintainability principle of SAREF ontology expedites the defect detection and identification, new requirement accommodation, and management through the changes in parts of SAREF.

Emerging smart appliances with the application of the SAREF ontology and deployment of IoT technologies enable to use time-shiftable smart appliances in demand response programs to realize DSM. The demand response provides the consumers' appropriate response without compromising the comfort to achieve a desired energy demand profile. So, the consumers may save on energy costs and the energy grid achieves technical and economic benefits. However, the management of the demand must be applied by avoiding a rebound effect. The reason for the rebound effect is that when the peak demand shifts to off-peak demand periods thanks to demand response, may create another peak. Thus, requests of the consumers and objectives of the grid utilities must be aggregated and prioritized in an optimal manner, so the scheduling of smart appliances is an optimization problem. Basically, an optimization problem is related to finding the best solution or solution set from all feasible solutions through the consideration of all available constraints. In the literature, there are many studies which use metaheuristic algorithms in DSM. However, most of these studies are focused on the scheduling of the loads for single house. This chapter introduces how metaheuristic algorithms can be utilized in the scheduling of smart appliances applied to DSM. To provide a good understanding of such approach, prominent and modern metaheuristic algorithms are simulated for a case. The effects of the scheduling and a comparison of the algorithms are discussed in terms of convergence performance.

This chapter is organized as follows. First, demand response programs in DSM have been introduced. Then power consumption profiles and usage patterns of the time-shiftable and smart appliances are presented. After that prominent smart metaheuristic algorithms are introduced. Finally, scheduling of time-shiftable smart appliances by smart heuristic algorithms is elaborated and results are presented.

8.2 Demand Response Programs in Demand-Side Management

DSM as an important aspect of the smart grids promises to increase reliability and stability of the grid by shifting demand from the peak period to the low consumption period and therefore reducing the peak demand. DSM programs contain two main operations which are demand shifting by demand response programs and EE by energy conservation programs. Although DSM has two concepts which are EE and demand response, it should be noted that demand response has undeniable benefits in providing EE in distribution grids. The peak load reduction and demand shifting are important for the distribution grid because energy losses are proportional to the square of the current. So, peak load reduction by means of the DSM will improve the efficiency of the grid. There are many potential benefits, which are technical, economic, environmental, and social, through the DSM in the electrical power systems (Jabir et al. 2018).

While DSM aims to reduce consumption and emissions at the demand level, demand response is a subset of DSM which includes non-permanent and short-term load manipulation actions to recast the consumption behavior of the end user. Demand response is defined as "the changes in electric usage by end-use consumers from their normal consumption patterns in response to changes in the price of electricity over time, or to incentivize payments designed to induce lower electricity use at times of high wholesale prices or when system reliability is jeopardized" (QDR 2006).

Smart appliances and smart building concepts provide an opportunity to apply a DSM scheme which is one of the distribution side features of the smart grids. If the comfort of the consumers is not compromised, the consumers give an appropriate response to the demand response programs. So, not only consumers' electricity consumption bills are decreased but also the flexibility and controllability of the consumption provide many benefits to the grid (Prusty et al. 2022).

Demand response programs can be categorized into two main categories which are price-based and incentives-based programs (Albadi and El-Saadany 2007; Shafie-khah et al. 2016). Figure 8.2 shows these demand response programs. These programs survived to get one or more DSM objectives which are described in Figure 8.1. Direct load control gives the distribution system operator (DSO) direct control of the appliances such as central air conditioner is remotely cycled using a switch on the compressor (Hledik and Lee 2021). Interruptible rates create a contract that is limited sheds such that one party to renege on its obligation to provide electricity to the other party a certain number of times over a certain period of time. So, some loads served under interruptible provides ancillary services to the market. Interruptible tariffs provide a discounted price to consumers who agree to interrupt their electric consumption when required as requested by the utility to solve supply shortage or instantaneously in response

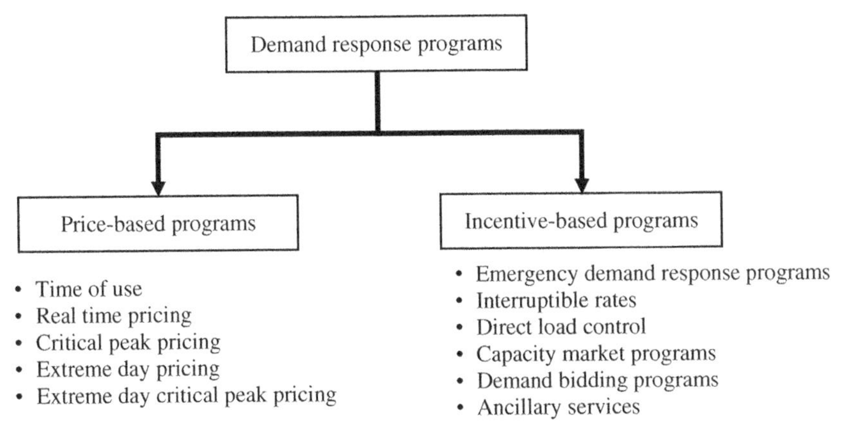

Figure 8.2 Price-based and incentive-based demand response programs. Source: Albadi and El-Saadany (2007) and Shafie-khah et al. (2016).

to a system emergency (Zarnikau 2008; Palensky and Dietrich 2011). On the other hand, emergency demand response programs send the emergency signal to the consumers to get a response such that consumers are paid incentives for their measured load reductions during emergency conditions. Demand bidding programs are also named as buyback programs in which consumers can execute a bidding in electricity wholesale market to reduce consumption. The consumers can get the capacity payment according to their agreement which provides direct load curtailment. The consumers can execute load reduction in the intraday energy market in ancillary services (Contreras et al. 2016).

In price-based demand response programs, consumers can respond to the price structure with changes in energy use, reducing their electricity bills if they adjust the timing of their electricity usage to take advantage of lower-priced periods and avoid consuming when prices are higher. Time-of-use pricing (TOU) offers prices that vary by time period such that they are higher in peak consumption periods and lower in low consumption periods. Real-time pricing (RTP) applies the featured prices day ahead or hour-ahead. These featured prices vary hourly or in 30 minutes intervals. Critical peak pricing (CPP) is combined with TOU pricing and it constitutes price layers that are very high for a few critical hours such as 15 days of the year. Extreme day pricing (EDP) is very similar to critical peak pricing except that in EDP higher prices execute for all hours of the critical day. While critical peak pricing is announced a day before or the day of the critical peak price, the timing of EDP is unknown until the day ahead. Extreme day critical peak pricing is a combination of EDP and critical peak pricing such that it creates the critical peak hours on the extreme days. However, there is no TOU on the other days.

Both incentives-based and price-based demand response programs can be utilized for specific loads or all loads. Also, these demand response programs can be applied as distributed or centralized. The application scheme of the demand response programs determines the smart grid application scheme. The DSO or utility company sends price signals to the smart homes, then the smart energy controller (SEC) of the smart homes schedules and operates the time-shiftable smart appliances in a traditional price-based demand response scheme. On the other hand, this centralized scheme can lead to rebound effects. So, another approach is the decentralized demand response scheme (Ramchurn et al. 2011). Both centralized and decentralized demand response schemes have an optimization problem to schedule the time-shiftable smart appliances as per price or grid objectives and consumer preferences. Consumption of residential sector has reached 25% of the overall electrical consumption in Europe in 2020, and it is foreseen that the consumption rate of the residential sector will reach 29% of the overall energy consumption by 2050 (Bintoudi et al. 2021). It is estimated that the potential of controllable load through smart appliances economically viable in the EU is 40 GW. It means that the shift of this load from peak times to other periods may reduce peak generation in the EU by 10% (Serrenho and Bertoldi 2019). On the energy aspect of smart appliances, these have the capability to receive, interpret, and act on a signal received from an energy provider and adjust its operation according to the settings chosen by the energy consumer. In this context, energy smart appliances support demand-side flexibility such that they can modulate or shift their electricity consumption in response to external signals such as price information, local measurements, or direct control commands.

8.3 Time-Shiftable and Smart Appliances in Residences

The residential appliances which are utilized in demand response programs can be classified into three categories power-shiftable appliances, time-shiftable appliances, and deferrable appliances. Air conditioners and electric water heaters are examples of power-shiftable appliances which can adjust the working power continuously within the predefined range. The operation time of the time-shiftable smart appliances can be shifted in demand response programs through the SEC. Washing machines, dishwashers, and tumble dryers can be considered time-shiftable appliances. Electrical vehicles are in the deferrable appliances categories. The deferrable aspect means that the power consumption can be stopped when the grid load is high and consumer's deferrable load is available to stop and then continue when the grid overall demand is back to low or normal levels (Li et al. 2022).

Figure 8.3 shows representative operating cycles in minute resolution for modern efficient appliances which are cloth washing machine, dishwashing machine,

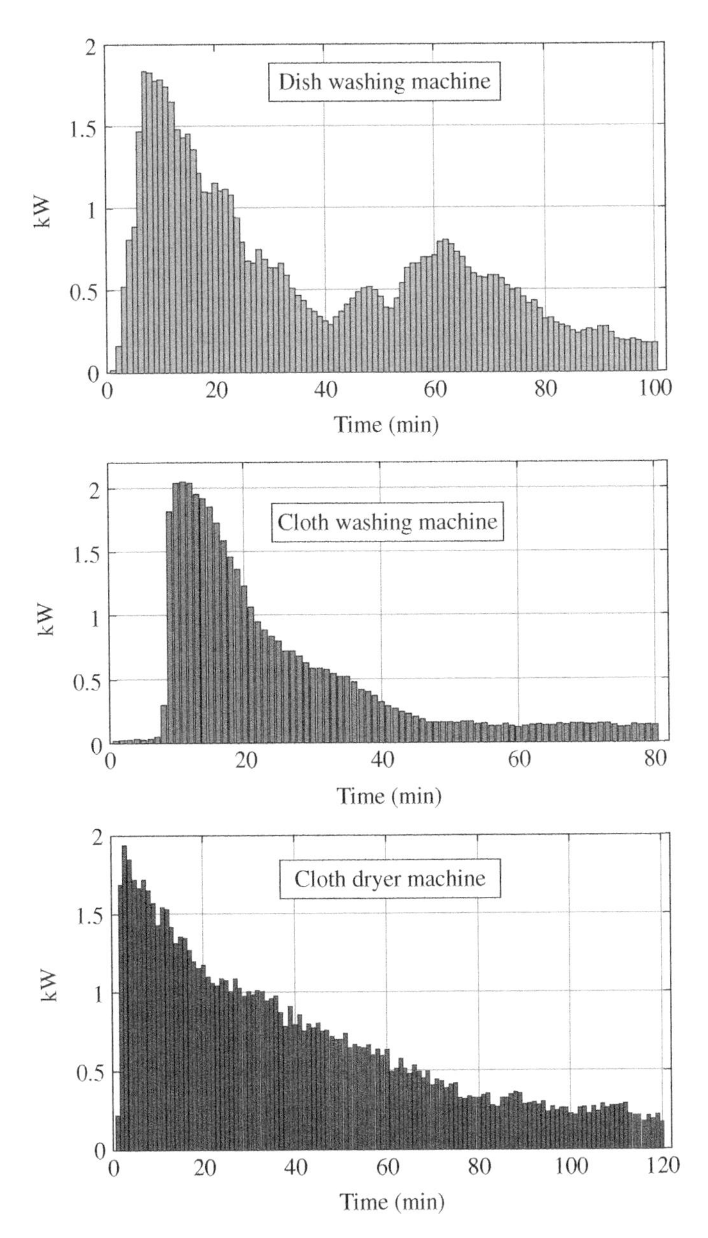

Figure 8.3 Representative operating cycles for dishwashing machine, cloth washing machine, and cloth dryer machine. Source: Degefa et al. (2018)/Reproduced from IEEE.

and cloth dryer machine (Degefa et al. 2018). These appliances are also known as wet appliances, and these have great potential for demand response in terms of demand flexibility (Schofield et al. 2014; Stamminger and Schmitz 2017). Recent public survey study (Çakmak and Altaş 2020) which was done in Turkey shows that these appliances mostly operated in peak consumption times. For example, 47% of the people who attended the public survey expressed that they use the washing machine between 18:01 and 23:59. This time period is the peak consumption period in the grid. Another example from the same public survey is 66% of the respondents express that they use the dishwasher between 18:01 and 23:59. So, these appliances can be utilized in demand response programs through the scheduling. Another example from Europe reports that washing machines and dishwashers of the residential users have 5 and 10 MW flexible load potential, respectively, if 10% of one million household agree to give control of these appliances (Stamminger and Schmitz 2017).

In order to utilize the potential of time-shiftable loads such as dishwashers and washing machines these appliances must be designed with a standard. One of the attempts to standardize this is the SAREF (Smart Applications Reference) to provide controllability, manageability, and monitorability. SAREF mainly focuses on the concept of the device which is designed to do a particular task. For instance, a cloth washing machine has two tasks which are the start and stop functions. SAREF has many classes which are function, command, state, service, profile, and measurement, so that they have relationships with each other (ETSI 2022).

A semantic interoperability in the energy smart appliances which are from different manufacturers can be provided by the SAREF ontology through energy-related information and SEC (Verbeke et al. 2020). This can be applicable by smart machine-to-machine communication thanks to the energy domain (ETSI 2020) and building domain (ETSI 2017) SAREF extensions.

8.4 Smart Metaheuristic Algorithms

Smart metaheuristic algorithms are inspired by nature and are utilized in many areas such as engineering optimization and mathematical problem-solving. This chapter will be elaborately introducing the foremost smart metaheuristic algorithms and their DSM study examples.

Figure 8.4 shows nature-inspired computing techniques in artificial intelligence.

8.4.1 BAT Algorithm

BAT algorithm which inspires echolocation actions of microbats, with varying pulse rates of emission and loudness was invented in 2010 by Xin-She Yang

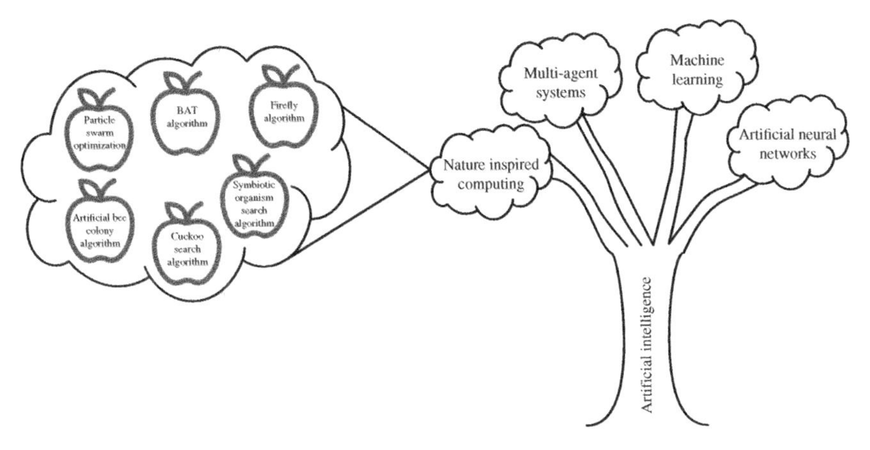

Figure 8.4 Nature-inspired computing techniques in artificial intelligence. Source: Recep Çakmak (co-author).

(2010c). The BAT algorithm is a metaheuristic algorithm, and it is able to find global optimum points by idealizing the echolocation behavior of the bats. BAT algorithm has been used in various studies in the literature such as transport network design problems (Srivastava and Sahana 2019) and controlling of coal gasifiers for optimal response (Kotteeswaran and Sivakumar 2013). Furthermore, BAT algorithm has been utilized in DSM studies recently (Sheng and Gu 2019; Viknesh and Manikandan 2018; Farooqi et al. 2018; Latif et al. 2018).

While x_i indicates a solution candidate for the problem which will be solved by the BAT algorithm and v_i indicates velocity, new solutions, and new velocities are calculated by the following equations for the tth iteration (Yang 2010b).

$$f_i = f_{min} + (f_{max} - f_{min})\beta \tag{8.1}$$

$$v_i^t = v_i^{t-1} + \left(x_i^t - x^*\right) f_i \tag{8.2}$$

$$x_i^t = x_i^{t-1} + v_i^t \tag{8.3}$$

where β is a random number generator in [0–1] range. x^* is the current best solution and it is stored for comparison among all the n bats. A new solution for each bat is created by random walk after selection of current best solution among the local bests as follows:

$$x_{new} = x_{old} + \alpha A^t \tag{8.4}$$

where α is a random number $\alpha \in [-1\ 1]$, while A^t is the average loudness of all the bats for the tth iteration. Additional information and enhanced versions of the original BAT algorithm and their utilization examples could be procured from

Yang and He (2013). The BAT algorithm could be represented by the following pseudocode.

```
Define objective function f(x), x=(x_1,...x_d)^T
Initialize the bat population x_i (i=1,2,...,n) and v_i
Initialize frequencies f_i, pulse rates r_i and the loudness A_i
while (t<maximum iteration number)
 Generate new solutions by adjusting frequency
 Update velocities and locations
 if(rand>r_i)
   Select a solution among the best solutions
   Generate a local solution around the selected best solution
 end if
 Generate a new solution by flying randomly
 if((rand<A_i) & f(x_i) <f(x*))
   Accept the new solutions
   Increase r_i and reduce A_i
 end if
 Rank the bats and find the current best x*
end while
```

In application of BAT algorithm to schedule the residential shiftable smart appliances, the searched best solutions are the optimal starting time of these appliances. While n is the population of the BAT, which is searching for the optimal solutions, d is the dimension of the problem. In the scheduling of residential shiftable smart appliance problem, d represents the number of consumers' scheduled appliances.

8.4.2 Firefly Algorithm (FFA)

Most of fireflies create short flashes in rhythmic form and their flashes are unique to each species. These light flashes are generated via a chemical bioluminescence process, and used for communication, hunting, and warning their enemies. The Firefly algorithm (FFA) was inspired by behaviors of firefly insects and the algorithm was developed and formulated as follows by Xin-She Yang (2009, 2010a, 2010b).

A light intensity $I(r)$ which is a function of r distance from the light source I_s can be calculated by following equation:

$$I(r) = \frac{I_s}{r^2} \tag{8.5}$$

The light is absorbed in the environment according to absorption coefficient $\gamma(\gamma \in [o, \infty))$, and this can be formulated as follows in Gaussian form:

$$B(r) = B_0 e^{-\gamma r^2} \tag{8.6}$$

where $B(r) = B_0$ is the attractiveness of a firefly at r distance and B_0 will be zero at zero distance ($r = 0$).

The distance between ith and jth fireflies can be notated as r_{ij} and it calculated based on Euclidean form as follows:

$$r_{ij} = \|X_i - X_j\| = \sqrt{(x_i - x_j)^2 - (y_i - y_j)^2} \tag{8.7}$$

The action of ith firefly to another one (jth firefly) which is the brighter firefly can be formulated as follows (Yang 2014a):

$$x_i^{t+1} = x_i^t + B_0 e^{-\gamma r_{ij}^2}\left(x_i^t - x_j^t\right) + \alpha \in {}_i^t \tag{8.8}$$

where t is the iteration number, the second term simulates the attraction, and as mentioned above, B_0 will be zero at zero distance ($r = 0$). The third term of Eq. (8.8) is due to randomization and this randomization can be provided by a Gaussian distribution function or Levy flight randomization functions (Yang 2009, 2010a).

The FFA algorithm could be represented by following pseudocode (Yang 2010a):

```
Objective function f(x),  x=(x_1,x_2,...,x_d)^T
Generate an initial population of n fireflies x_i (i=1,2,...,n)
Light intensity I_i at x_i is determined by f(x_i)
Define light absorption coefficient γ
while (t<maximum iteration number)
 for i=1:n
   for j=1:n
   if(I_i<I_j)
     Move firefly i towards j
   end if
    Vary attractiveness with distance r via Equation (8.6)
    Calculate new solutions and update light intensity
   end for j
 end for i
Rank the fireflies and find the current global best x*
end while
```

FFA algorithm is utilized for DSM in the literature by Ishaq et al. (2017) and Debbarma et al. (2020). In application of FFA algorithm to schedule the residential shiftable smart appliances, the searched best solutions are the optimal starting time of these appliances. While n is the population of the fireflies, which are searching for the optimal solutions, d is the dimension of the problem. In the scheduling of residential shiftable smart appliance problem, d represents the number of consumers' scheduled appliances.

8.4.3 Cuckoo Search Algorithm

Some species of Cuckoo birds have aggressive reproduction strategy which is fascinating. These cuckoo birds lay their eggs in communal nests and these eggs hatch before the eggs of the host nests. After hatching the baby cuckoos remove the eggs from the host nest. So, these Cuckoo species reproduce as brood parasitism. However, if a host bird notices the cuckoo's eggs, they either throw off these alien eggs or abandon the nest to make a new one. Brood parasitism behavior of cuckoo birds was imitated in computation form and Cuckoo Search (CS) algorithm was introduced by Xin She Yang and Suash Deb (2009). After discovering the CS algorithm it was upgraded by adopting Lévy Flight (Pavlyukevich 2007) which is an efficient random walk procedure. Key mathematical equations which are utilized in the CS algorithm are presented as follows (Yang 2014b).

The algorithm utilizes a balanced combination between local and global searching by means of random walk-through controlling of the p_a parameter which is a fraction of worse nests that are abandoned and then new ones are built. The local searching is implemented by following equation:

$$x_i^{t+1} = x_i^t + \alpha s \otimes H(p_a - \in) \otimes \left(x_j^t - x_k^t\right) \tag{8.9}$$

where x_j^t and x_k^t are two different solutions which are selected randomly. H indicates the Heaviside function which generates a random number $\in$ from a uniform distribution and s is the step size. After implementation of the local search, Levy flights are utilized in the global search portion of the algorithm. In global random walk by Lévy flights, the new position of a solution member is calculated as follows:

$$x_i^{t+1} = x_i^t + \alpha L(s, \lambda) \tag{8.10}$$

where λ is the Lévy exponent and it is found that $\lambda = 1.5$ and $\alpha = 0.01$ provide sufficient results for most of the problems (Yang 2014b). The basic steps of the Cuckoo Search algorithm can be summarized by following pseudocode (Long et al. 2014).

```
Objective function f(x),  x=(x_1,x_2,...,x_d)^T
Generate an initial population of n host nests x_i (i=1,2,...,n)
while (t<maximum iteration number) or (stop criteria)
  choose a cuckoo randomly
  Generate a solution by means of Levy flight function as follows:
  x_i^{t+1} = x_i^t + αL(s, λ)
  Calculate objective function to evaluate the solution quality
  Choose a nest among n (say, j) randomly
  if(f_i<f_j)
     Replace j by the new solution i
  end
  A fraction (p_a) of worse nests are abandoned
  Build new nests it means new solutions by follows equation
  x_i^{t+1} = x_i^t + αs ⊗ H (p_a − ∈) ⊗ (x_j^t − x_k^t)
  Keep best solution
  Rank the solutions and find the current best
  Update iteration (t= t+1)
end while
```

Cuckoo Search algorithm has been utilized in the literature by Khalid et al. (2017), Aslam et al. (2017), and Cakmak and Altas (2016) for load scheduling/shifting in DSM and home energy management application in demand response. In application of Cuckoo Search algorithm to schedule the residential shiftable smart appliances, the searched best solutions are the optimal starting time of these appliances. While n is the population of the fireflies, which are searching the optimal solutions, d is the dimension of the problem. In the scheduling of residential shiftable smart appliance problem, d represents the number of consumers' scheduled appliances.

8.4.4 SOS Algorithm

Symbiotic Organisms Search (SOS) algorithm is a nature-inspired philosophy which mimics the interactive behavior among organisms in nature. This algorithm represents three symbiotic relationships which are mutualism, commensalism, and parasitism. These relationships are formularized as mutualism phase, commensalism phase, and parasitism phase. The algorithm strives to get optimum solution through performing these phases under a population number, iteration number, and other constraints. The computation phases are described below (Cheng and Prayogo 2014).

In the mutualism phase, it is mimicked that the organisms engage in a mutualistic relationship with the goal of increasing mutual survival advantage in the ecosystem. New candidate solutions for X_i and X_j are calculated based on following equations:

$$X_{i\,new} = X_i + rand\,(0,1) * (X_{best} - Mutual\,Vector^{*}BF_1) \tag{8.11}$$

$$X_{j\,new} = X_j + rand\,(0,1) * (X_{best} - Mutual\,Vector^{*}BF_2) \tag{8.12}$$

$$Mutual\,Vector = \frac{X_i + X_j}{2} \tag{8.13}$$

The commensalism phase computation is described such that organism X_i is updated only if its new fitness is better than its pre-interaction fitness:

$$X_{i\,new} = X_i + rand\,(-1,1) * (X_{best} - X_j) \tag{8.14}$$

In the parasitism phase, a parasite vector is created in the search space by duplicating organism X_i, then it is modified by randomly selected dimensions. In order to provide a host to parasite vector X_j is selected randomly from the ecosystem. The aim of the parasite vector is to replace X_j in the ecosystem. The fitness function is measured for the host-vector and parasite vector, then it is evaluated. If the parasite vector has a better fitness value, it will kill organism X_j and gets its position in the ecosystem or vice versa.

SOS algorithm can be presented by following pseudocode (Verma et al. 2017):

```
Objective function f(x),   x=(x1,x2,...,xd)T
Generate an initial population of n host nests xi (i=1,2,...,n)
while (t<maximum iteration number) or (stop criteria)
 for i = 1: n
    Find the best organism Xbest in the ecosystem
    Perform the Mutualism Phase
       Randomly select one organism Xj such that Xj≠ Xi
       Implement Equation (8.11) and (8.12)
       If objective function value of the modified organisms
better than previous then update the ecosystem.
    Perform the Commensalism Phase
       Randomly select one organism Xj such that Xj≠ Xi
       Implement Equation (8.14)
       If objective function value of the modified organisms
better than previous then update the ecosystem
 Perform the Parasitism Phase
      Randomly select one organism Xj such that Xj≠ Xi
      Generate a parasite vector from Xi
      If objective function value of the modified organisms
better than previous then update the ecosystem
 end for
 The golobal best solution is the optimal solution
end while
```

SOS algorithm has been applied in the literature for DSM and demand response programs-based load scheduling applications by Mukherjee (2018), Çakmak and Altaş (2017), Nasir et al. (2021), and Chatterjee et al. (2020). In application of the SOS algorithm to schedule the residential shiftable smart appliances, the searched best solutions are the optimal starting time of these appliances. While n is the population of the fireflies, which are searching for the optimal solutions, d is the dimension of the problem. In the scheduling of residential shiftable smart appliance problem, d represents the number of consumers' scheduled appliances.

8.5 Scheduling of Time-Shiftable Appliances by Smart Metaheuristic Algorithms

The time-shiftable appliances can be scheduled in two ways. One of them is the scheduling and operation of them through the SEC. Another way is scheduling them optimally thanks to the distributed DSM units. Whatever the approach is, the algorithm needs two things for optimal scheduling. In the SEC-based scheduling approach and distributed DSM the constraints are determined and submitted to the SEC by consumer. These constraints are maximum and minimum starting times of the shiftable appliances. The SECs optimally schedule the time-shiftable appliances based on energy price signal and the constraints of the consumer.

Similarly, in the distributed DSM approach, maximum and minimum starting times of the shiftable appliances are sent to the DSO. The DSO aggregates all consumer's requests which include maximum and minimum starting times of the shiftable appliances, and schedules all requests as to grid objectives which can be peak-to-average ratio (PAR), minimum loss, etc.

In recent days, peak load reduction or reducing the PAR is important for the power systems by DSM reduce to need highly polluting peaker power plants and to decrease losses. However, more important point is to provide match between renewable generation and demand at peak generation times that occurred by distributed renewable-based power plants. Thus, scheduling of time-shiftable appliances which provide demand flexibility is critical to the emerging smart grids. Especially in high PV penetration neighborhood areas, the distributed DSM can be more useful to provide matching generation and consumption in the area. It should be noted that in the distributed DSM approach the requests of the consumers are aggregated in specific time periods such as every five minutes for instance. Figures 8.5 and 8.6 illustrate both scheduling mechanisms.

The scheduling of smart shiftable appliances can be performed in day-ahead request aggregation or daily dynamic request aggregation. In the day-ahead request aggregation, the consumers send their flexible requests of time-shiftable appliances to the DSO. Then the DSO schedules optimally as to grid objective and the consumers' requests which are the constraints of the optimization. In day-ahead scheduling, there is a wide scheduling horizon compared to the daily scheduling mechanism. Figure 8.7 shows the mechanism of both daily and day-ahead scheduling.

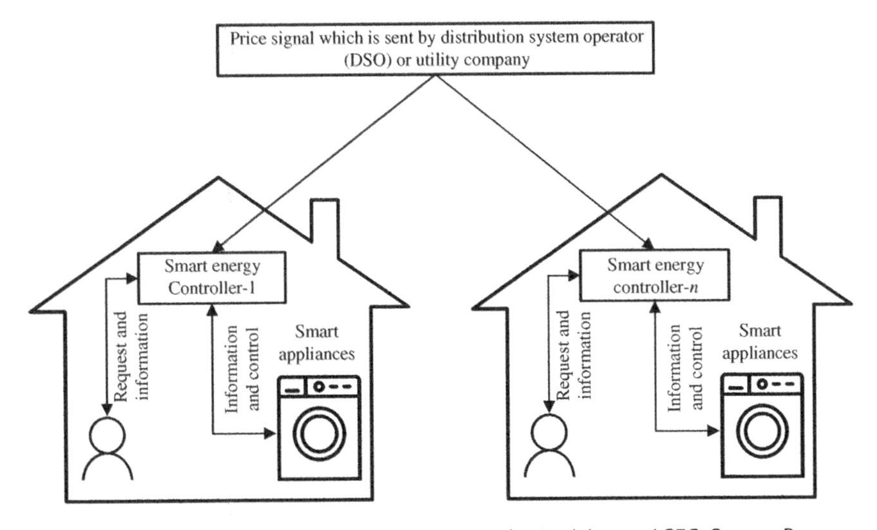

Figure 8.5 Smart appliance scheduling via centralized pricing and SEC. Source: Recep Çakmak (co-author).

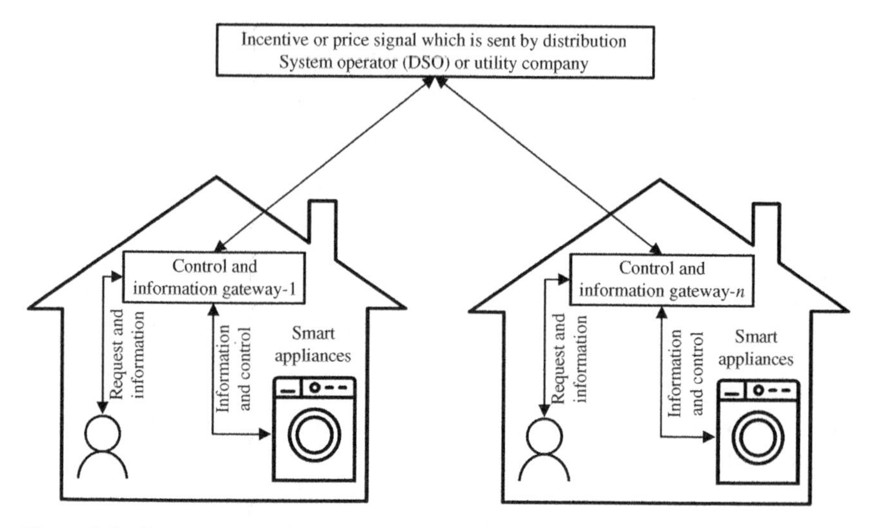

Figure 8.6 Smart appliance scheduling via distributed demand-side management. Source: Recep Çakmak (co-author).

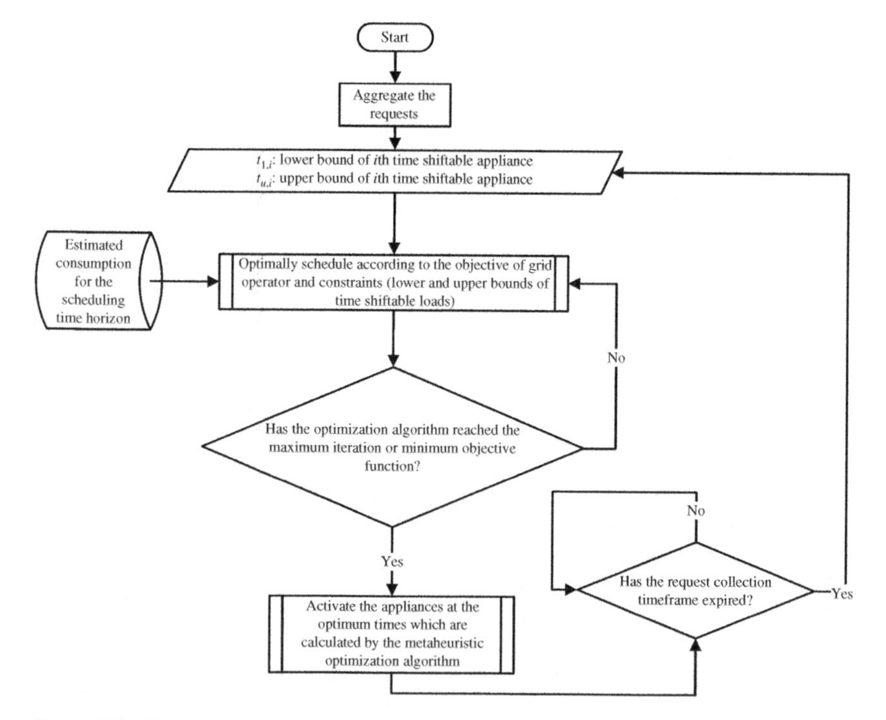

Figure 8.7 Request aggregating and scheduling mechanism. Source: Recep Çakmak (co-author).

In this chapter, it is assumed that there are 150 consumers in a neighborhood area and there is a DSO to schedule time-shiftable appliances. The consumers' time-shiftable appliances are washing machines, dishwashers, and tumble dryers of which the consumption profile is presented in Figure 8.3. It is also assumed that these three appliances are rated all at the same power in each home. It is postulated that the DSO applies a price signal to this neighborhood, and the consumers submit the flexible time range for their time-shiftable appliances. The flexible time range is adopted from Çakmak and Altaş (2017). It is assumed that in each simulation the consumers manipulated day ahead price/incentive signal, then they determine flexible time ranges for their time-shiftable appliances, and each day every consumer uses each time-shiftable appliance once. Both before and after demand response conditions are simulated in this study and obtained results are presented.

The objective function for optimization is utilized in this study as multiplying the average power of the time-shiftable appliances and day ahead time of use price signal. The aim of the smart metaheuristic optimization algorithms is to minimize the objective function. So, both consumers' electricity bills can be reduced, and consumption curve can be manipulated as to DSO objectives which are as far as possible to decrease peak demand and to flatten demand curve. The rates of price signal for day times are created as follows: 0.4US$/kWh for 06:00–09:00, 0.3US$/kWh for 09:00–12:00, 0.2US$/kWh for 12:00–15:00, 0.4US$/kWh for 15:00–18:00, 0.6US$/kWh for 18:00–21:00, 0.4US$/kWh for 21:00–24:00 and 0.2US$/kWh for 00:00–06:00.

The parameters of utilized algorithms and results are given in Table 8.1. In each algorithm, it is aimed to find optimum starting time for the time-shiftable appliances. So, the solution space is searched by the algorithms. According to parameters of the algorithms which are given in Table 8.1, the obtained minimum fitness values are ranked by BAT, Firefly, Cuckoo Search, and SOS. The minimum fitness function is obtained by SOS algorithm. So, it can be said that SOS algorithm is the best algorithm in terms of computation efficiency for this use case. Also, it should be noted that SOS algorithm required minimum parameter which changes the algorithm's performance. One of the disadvantages of metaheuristic algorithms is that they need at least one parameter which affects their performance and should be determined attentively (Yalçin et al. 2020). Thus, SOS algorithm is the best algorithm for time-shiftable load scheduling problem tested in this example in terms of minimum parameter requirements and getting minimum fitness function value.

The scheduled and unscheduled cases comparison is presented in Figure 8.8 showing the SOS algorithm results, and it's obvious that the scheduling of time-shiftable appliances decreased peak load and increased the PAR.

Table 8.1 The parameters of utilized algorithms and obtained results.

Algorithm	Parameters	Obtained minimum fitness function value
BAT	Population size: 20 $A = 0.2$ $r = 0.2$ Iteration: 2,000	243,790
Firefly	Population of fireflies: α: 0.5 γ: 1 Iterations: 2,000	243,249
Cuckoo search	Number of nests: 20 $p_a = 0.25$ Iterations: 2,000	242,260
SOS	Number of organisms: 20 Iterations: 2,000	214,660

Source: Recep Çakmak (co-author).

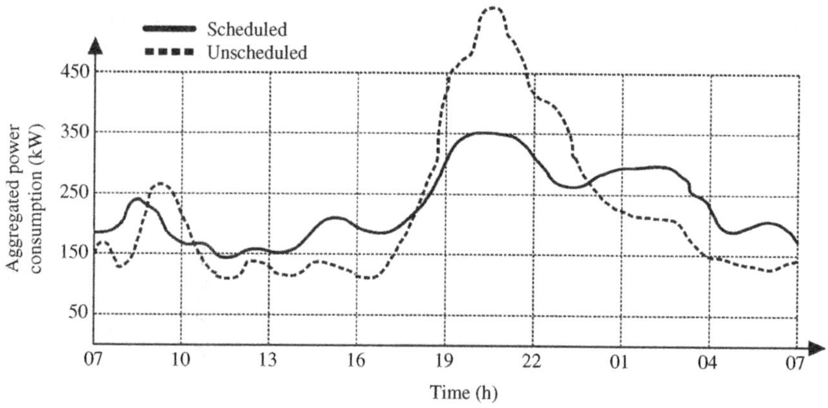

Figure 8.8 Aggregated consumption curves when smart appliances are scheduled and unscheduled. Source: Recep Çakmak (co-author).

The scheduled and unscheduled cases comparison is presented in Figure 8.8. These results were obtained by the SOS algorithm. It is obvious that the scheduling of time-shiftable appliances decreases peak load and the PAR. Peak load reduction has been obtained by 36.3%. Moreover, due to the peak load reduction and demand curve flattening thanks to the scheduling the power losses in the distribution lines have been decreased by 45.2%. So, energy losses along the day would procure that the energy provider lost less energy on the lines after the peak was reduced. These results accentuate the importance of scheduling of time-shiftable energy smart appliances through demand response programs.

Scheduling of smart appliances by metaheuristic approaches would provide not only peak load reduction but also decrease energy losses of the distribution company. Thus, energy-smart appliances and their scheduling have great importance for emerging smart grids in future. Also, optimal scheduling of these appliances can be performed by smart metaheuristic algorithms. Different metaheuristic algorithms which are not use in this study can be applied and compared to each other for future studies.

Symbols and Abbreviations

A	the average loudness of all the Bats in BAT algorithm
d	dimension of the problem which represents the number of consumers' scheduled appliances
n	population size (solution search set) in the nature-inspired algorithms
I	a function of r distance from the light source in Firefly algorithm
r	represents a distance from the light source in Firefly algorithm
p_a	a fraction of worse nests that are abandoned in Cuckoo Search algorithm
s	step size in Cuckoo Search algorithm
H	heaviside function which generates a random number in Cuckoo Search algorithm
α	step scale coefficient in a random number generator
λ	Lévy exponent in Cuckoo Search algorithm
γ	light absorption coefficient in Firefly algorithm
x_i	solution candidate which represents optimal starting time of the scheduled appliance
CS	Cuckoo search
CPP	critical peak pricing
DR	demand response
DoE	The United States Department of Energy
DSM	demand-side management
DSO	distribution system operator
EE	energy efficiency
EDP	extreme day pricing
ETSI	The European Telecommunications Standards Institute
FFA	Firefly algorithm
IoT	internet of things

P	power
PAR	peak-to-average ratio
RTP	real-time pricing
SR	spinning reserve
SAREF	smart applications reference
SEC	smart energy controller
SOS	symbiotic organisms search
TOU	time-of-use pricing

Glossary

Metaheuristic Advanced heuristic algorithms which can search for optimal points without knowing the problem.

Optimization A solution procedure which aims to find the optimal solutions to the objective function under constraints.

Ontology An explicit specification of a conceptualization.

Smart Energy Controller A controller which manages energy smart appliances according to consumer requests, price/incentive signals from the electricity utility, past, current, and future energy usage thanks to artificial intelligence techniques.

Distribution System Operator An entity responsible for distributing and managing energy from the generation to the end users.

Control and Information Gateway A device which controls household appliances, measures the consumption, and provides bi-directional communication between consumer and electric utility.

Time-shiftable appliance A household appliance whose operating time can be shiftable.

Demand Response Changing of electricity consumption profile of consumers through price or incentive.

Demand-Side Management Modification of consumer demand level and profile through various strategies such as price, financial incentives, and education.

References

Albadi, M.H. and El-Saadany, E.F. (2007). Demand response in electricity markets: an overview. In: *2007 IEEE Power Engineering Society General Meeting*, 1–5. IEEE.

Alwan, H.O. and Abdelwahed, S. (2019). Demand side management – literature review and performance comparison. In: *2019 11th International Conference on*

Computational Intelligence and Communication Networks (CICN) (3–4 January 2019), 93–102. Los Alamitos, CA, USA: IEEE Computer Society Conference Publishing Services (CPS).

Aslam, S., Bukhsh, R., Khalid, A. et al. (2017). An efficient home energy management scheme using cuckoo search. In: *International Conference on P2P, Parallel, Grid, Cloud and Internet Computing*, 167–178. Springer.

Bayod-Rújula, A.A. (2009). Future development of the electricity systems with distributed generation. *Energy* 34: 377–383.

Bintoudi, A.D., Bezas, N., Zyglakis, L. et al. (2021). Incentive-based demand response framework for residential applications: design and real-life demonstration. *Energies* 14: 4315.

Cakmak, R. and Altas, I.H. (2016). Scheduling of domestic shiftable loads via cuckoo search optimization algorithm. In: *2016 4th International Istanbul Smart Grid Congress And Fair (ICSG)*, 1–4. IEEE.

Çakmak, R. and Altaş, İ. (2017). Optimal scheduling of time shiftable loads in a task scheduling based demand response program by symbiotic organisms search algorithm. In: *Saudi Arabia Smart Grid (SASG)*, 1–7. IEEE.

Çakmak, R. and Altaş, İ. (2020). Türkiye'deki Müşterilerin Akıllı Şebekelerde Yük Kaydirmaya Dayali Talep Tarafi Yönetimine Tepkileri. *Karadeniz Fen Bilimleri Dergisi* 10: 395–416.

Chatterjee, S., Dey, S. and Dasgupta, M. (2020). A solution to demand response with soft-computing techniques. In: *2020 Fourth International Conference on Inventive Systems And Control (ICISC)*, 122–126. IEEE.

Cheng, M.-Y. and Prayogo, D. (2014). Symbiotic organisms search: a new metaheuristic optimization algorithm. *Computers and Structures* 139: 98–112.

Contreras, J., Asensio, M., De Quevedo, P.M. et al. (2016). Chapter 4 – Demand response modeling. In: *Joint RES and Distribution Network Expansion Planning Under a Demand Response Framework* (ed. J. Contreras, M. Asensio, P.M. De Quevedo, et al.), 33–40. Academic Press.

Debbarma, S., Kumar, K.K.P., Soren, N. et al. (2020). Day ahead demand side management using firefly algorithm. In: *AIP Conference Proceedings*, 050068 (ed. Xiao-Zhi Gao, Ranjan Kumar Ghadai, Kana Kalita, et al.). AIP Publishing LLC.

Degefa, M.Z., Sæle, H., Petersen, I. and Ahcin, P. (2018). Data-driven household load flexibility modelling: shiftable atomic load. In: *IEEE PES Innovative Smart Grid Technologies Conference Europe (ISGT-Europe)*, 1–6. IEEE.

ETSI (2017). SmartM2M; Smart Appliances Extension To SAREF; Part 3: Building Domain.

ETSI (2020). SmartM2M; Smart Appliances Extension To SAREF; Part 1: Energy Domain.

ETSI (2022). Smart Applications Reference Ontology, and Extensions. https://saref.etsi.org/index.html (accessed 6 January 2023).

Farooqi, M., Awais, M., Abdeen, Z.U. et al. (2018). Demand side management using harmony search algorithm and BAT algorithm. In: *Advances in Intelligent Networking and Collaborative Systems* (ed. L. Barolli, I. Woungang and O.K. Hussain), 191–202. Cham: Springer International Publishing.

Gellings, C.W. (1985). The concept of demand-side management for electric utilities. *Proceedings of the IEEE* 73: 1468–1470.

Hledik, R. and Lee, T. (2021). Chapter 9 – Load flexibility: market potential and opportunities in the united states. In: *Variable Generation, Flexible Demand* (ed. F. Sioshansi), 195–210. Academic Press.

Ishaq, A., Ayub, N., Saba, A. et al. (2017). An efficient scheduling using meta heuristic algorithms for home demand-side management in smart grid. In: *International Conference on Intelligent Networking and Collaborative Systems* (ed. Leonard Barolli, Isaac Woungang, and Omar Khadeer Hussain), 214–227. Springer.

Jabir, H.J., Teh, J., Ishak, D., and Abunima, H. (2018). Impacts of demand-side management on electrical power systems: a review. *Energies* 11: 1050.

Javaid, N., Naseem, M., Rasheed, M.B. et al. (2017). A new heuristically optimized home energy management controller for smart grid. *Sustainable Cities and Society* 34: 211–227.

Khalid, A., Zafar, A., Abid, S. et al. (2017). Cuckoo search optimization technique for multi-objective home energy management. In: *International Conference on Innovative Mobile and Internet Services in Ubiquitous Computing* (ed. Leonard Barolli and Tomoya Enokido), 520–529. Springer.

Kotteeswaran, R. and Sivakumar, L. (2013). A novel BAT algorithm based re-tuning of PI controller of coal gasifier for optimum response. In: *Mining Intelligence and Knowledge Exploration* (ed. R. Prasath and T. Kathirvalavakumar), 506–517. Cham: Springer International Publishing.

Latif, U., Javaid, N., Zarin, S.S. et al. (2018). Cost optimization in home energy management system using genetic algorithm, bat algorithm and hybrid bat genetic algorithm. In: *2018 IEEE 32nd International Conference on Advanced Information Networking And Applications (AINA)* (16–18 May 2018), 667–677. Los Alamitos, CA, USA: IEEE Computer Society Conference Publishing Services (CPS)

Li, S., Cao, D., Huang, Q. et al. (2022). A deep reinforcement learning-based approach for the residential appliances scheduling. *Energy Reports* 8: 1034–1042.

Long, W., Liang, X., Huang, Y., and Chen, Y. (2014). An effective hybrid Cuckoo search algorithm for constrained global optimization. *Neural Computing and Applications* 25: 911–926.

Mariano-Hernández, D., Hernández-Callejo, L., Zorita-Lamadrid, A. et al. (2021). A review of strategies for building energy management system: model predictive control, demand side management, optimization, and fault detect & diagnosis. *Journal of Building Engineering* 33: 101692.

Mukherjee, V. (2018). Day-ahead demand side management using symbiotic organisms search algorithm. *IET Generation Transmission and Distribution* 12: 3487–3494.

Nasir, T., Bukhari, S.S.H., Raza, S. et al. (2021). Recent challenges and methodologies in smart grid demand side management: state-of-the-art literature review. *Mathematical Problems in Engineering* 2021: 1–16.

Palensky, P. and Dietrich, D. (2011). Demand side management: demand response, intelligent energy systems, and smart loads. *IEEE Transactions on Industrial Informatics* 7: 381–388.

Pavlyukevich, I. (2007). Lévy flights, non-local search and simulated annealing. *Journal of Computational Physics* 226: 1830–1844.

Pina, A., Silva, C., and Ferrão, P. (2012). The impact of demand side management strategies in the penetration of renewable electricity. *Energy* 41: 128–137.

Prusty, B.R., Arun, S.L., and De Falco, P. (2022). Demand response in smart buildings. In: *Control of Smart Buildings: An Integration to Grid and Local Energy Communities* (ed. A. Tomar, P.H. Nguyen and S. Mishra), 121–131. Singapore: Springer Nature Singapore.

QDR, Q. (2006). *Benefits of Demand Response in Electricity Markets and Recommendations for Achieving Them*. US Department of Energy.

Ramchurn, S., Vytelingum, P., Rogers, A. et al. (2011). Agent-based control for decentralised demand side management in the smart grid. In: *Proceedings of 10th International Conference on Autonomous Agents and Multiagent Systems – Innovative Applications Track (AA- MAS 2011)*, Tumer, Yolum, Sonenberg and Stone (eds.). Taipei, Taiwan: International Foundation for Autonomous Agents & Multiagent Systems (IFAAMAS), 5–12.

SAREF, E. (2022). Smart Applications Reference Ontology, And Extensions. https://saref.etsi.org (accessed 6 January 2023).

Schofield, J., Carmichael, R., Tindemans, S. et al. (2014). Residential consumer responsiveness to time-varying pricing. Report A3 for the "Low Carbon London" LCNF project: Imperial College London.

Serrenho, T. and Bertoldi, P. (2019). *Smart Home and Appliances: State of the Art. Energy, Communications, Protocols, Standards*, 2–36. Brussels: JRC Technical Reports.

Shafie-Khah, M., Heydarian-Forushani, E., Osório, G.J. et al. (2016). Optimal behavior of electric vehicle parking lots as demand response aggregation agents. *IEEE Transactions on Smart Grid* 7: 2654–2665.

Shaw, R., Attree, M., Jackson, T., and Kay, M. (2009). The value of reducing distribution losses by domestic load-shifting: a network perspective. *Energy Policy* 37: 3159–3167.

Sheng, S. and Gu, Q. (2019). A day-ahead and day-in decision model considering the uncertainty of multiple kinds of demand response. *Energies* 12: 1711.

Srivastava, S. and Sahana, S.K. (2019). Application of BAT algorithm for transport network design problem. *Applied Computational Intelligence and Soft Computing* 2019: 9864090.

Stamminger, R. and Schmitz, A. (2017). Load profiles and flexibility in operation of washing machines and dishwashers in Europe. *International Journal of Consumer Studies* 41: 178–187.

Vardakas, J.S., Zorba, N., and Verikoukis, C.V. (2015). A survey on demand response programs in smart grids: pricing methods and optimization algorithms. *IEEE Communication Surveys and Tutorials* 17: 152–178.

Verbeke, S., Aerts, D., Reynders, G. et al. (2020). *Final Report on the Technical Support to the Development of a Smart Readiness Indicator for Buildings*. Brussels/Belgium: European Commission.

Verma, S., Saha, S., and Mukherjee, V. (2017). A novel symbiotic organisms search algorithm for congestion management in deregulated environment. *Journal of Experimental & Theoretical Artificial Intelligence* 29: 59–79.

Viknesh, V. and Manikandan, V. (2018). Design and development of adaptive fuzzy control system for power management in residential smart grid using BAT algorithm. *Technology and Economics of Smart Grids and Sustainable Energy* 3: 19.

Yalçin, E., Çam, E., and Taplamacioğlu, M.C. (2020). A new chaos and global competitive ranking-based symbiotic organisms search algorithm for solving reactive power dispatch problem with discrete and continuous control variable. *Electrical Engineering* 102: 573–590.

Yang, X.-S. (2009). Firefly algorithms for multimodal optimization. In: *Stochastic Algorithms: Foundations And Applications* (ed. O. Watanabe and T. Zeugmann), 169–178. Berlin/Heidelberg: Springer.

Yang, X.-S. (2010a). Firefly algorithm, stochastic test functions and design optimisation. *International Journal of Bio-Inspired Computation* 2: 78–84.

Yang, X.-S. (2010b). *Nature-Inspired Metaheuristic Algorithms*. Luniver Press.

Yang, X.-S. (2010c). A new metaheuristic BAT-inspired algorithm. In: *Nature Inspired Cooperative Strategies For Optimization (NICSO 2010)* (ed. J.R. González, D.A. Pelta, C. Cruz, et al.), 65–74. Berlin/Heidelberg: Springer.

Yang, X.-S. (2014a). Chapter 2 – Analysis of algorithms. In: *Nature-Inspired Optimization Algorithms* (ed. X.-S. Yang), 23–44. Oxford: Elsevier.

Yang, X.-S. (2014b). Chapter 9 – Cuckoo search. In: *Nature-Inspired Optimization Algorithms* (ed. X.-S. Yang), 129–139. Oxford: Elsevier.

Yang, X.-S. and He, X. (2013). BAT algorithm: literature review and applications. *International Journal of Bio-Inspired Computation* 5: 141–149.

Yang, X. and Suash, D. (2009). Cuckoo search via Lévy flights. In: *2009 World Congress On Nature & Biologically Inspired Computing (NaBIC)* (9–11 December 2009), Coimbatore, India: IEEE, 210–214.

Zarnikau, J.A.Y. (2008). Chapter 8 – Demand participation in restructured markets. In: *Competitive Electricity Markets* (ed. F.P. Sioshansi), 297–324. Oxford: Elsevier.

9

Distributed Operation of an Electric Vehicle Fleet in a Residential Area

*Alicia Triviño, Inmaculada Casaucao, and José A. Aguado**

Escuela de Ingenierías Industriales, University of Málaga, 29071, Málaga, Spain

9.1 Introduction

The interest for renewable energy sources is gaining popularity. Three main reasons explain this fact. First, they offer nondependency with fuel-based energy sources, which is a relevant limitation for geopolitical reasons. Second, society is more conscious about the risks that pollution causes on the health and the consequences it has on the climate change. Third, in some cases, these sources can be accessed locally with affordable facilities so their installations offer an attractive solution to reduce the electricity bills.

Power grids are facing new challenges due to the inclusion of more renewable energy sources. Moreover, they also have to deal with an increment on the consumption, which is expected to concentrate in urban sites as 65% of the population is foreseen to live in cities by 2050. One of the consumers of this energy will be electric vehicles (EVs). Government and public institutions are fostering the use of EV as a sustainable mode of transportation. The estimated fleet of EVs is forecast as 145 million for 2030 (International Energy Agency, 2017).

Thus, serving all the EVs will increase the electricity demand. According to Gryparis et al. (2020), and following a linear regression model, electricity demand will increase by 15.98% in the European Union countries by 2050, assuming an increase in electricity production of 90 million MWh during the period 2030–2050. The problem is even more serious if we consider that most people follow a similar behavior as they charge their EVs at home once they are back from work, and this happens at a narrow interval for most people. If not correctly managed, providing energy to these elements will require a large investment on

* Corresponding Author: Alicia Triviño; atc@uma.es

Energy Smart Appliances: Applications, Methodologies, and Challenges,
First Edition. Edited by Antonio Moreno-Munoz and Neomar Giacomini.

electrical infrastructure to cope with the energy demand. In addition, phenomena such as line overloading in both primary and secondary distribution systems, transformers overloading, line losses, voltage drops, voltage unbalances, and power quality worsening will take place more frequently (Faddel et al., 2018). However, EVs are a flexible load so their charging can be scheduled when it is more convenient for the power grid, given the consumers' need is met. In this way, they can also constitute as an important asset for the electrical grids to cope with the uncertainty and intermittency of renewable energy sources.

It will be essential to abandon the Plug and Charge strategy (i.e., the EV is charged as soon as the vehicle is plugged into the wall) and to equip EVs with smart and autonomous charging systems to decide when to perform this task. The use of EVs as controllable loads could report significant advantages to the grid such as balancing balance the load by shaving peak, filling valley according to the grid demand, perform frequency or voltage regulation and even reduce the total harmonic distortion.

The design for these scheduling algorithms should be particularly tailored for home scenarios, as most vehicles will be charged at households (Dudek, 2021). On-site renewable energy sources and limited electric installations are some key features that must be taken into account. Other parameters such as the users preferences, their behavior, or the electricity prices are also of interest.

In the near future, the use of EVs will not be restricted to mere controllable loads but they will also participate in the grid as energy providers. Since EVs are averagely parked for almost 22 hours a day, their batteries can be a valuable asset for delivering energy when required. V2X (vehicle-to-everything) stands out for this new paradigm. Four of the main cases are represented in Figure 9.1. The most

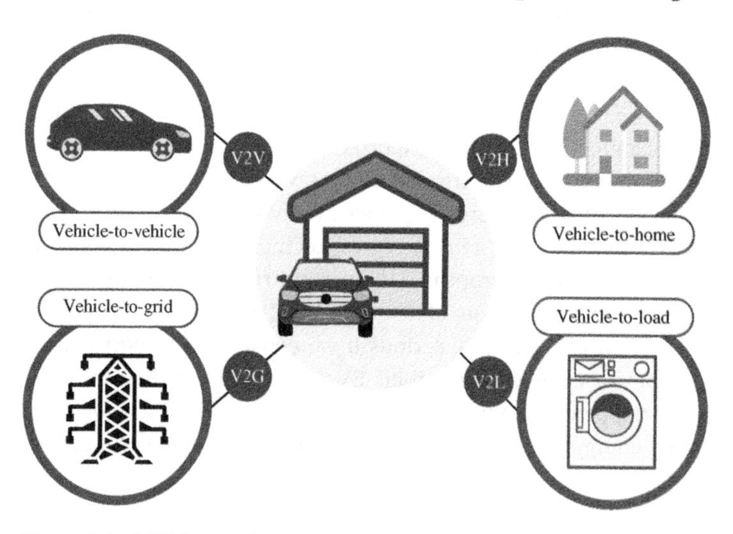

Figure 9.1 V2X interactions.

widespread case is Vehicle to Grid (V2G), where the electrical energy stored in the car's batteries is fed into the grid. In vehicle-to-vehicle (V2V) structure (Sousa et al., 2018), there is an exchange of power between EVs, so that if one of the vehicles does not have access to a charging station, it can extract energy from another nearby vehicle. On the other hand, vehicle-to-home (V2H) is a structure that enables the transmission of power from the EV battery to the home for consumption. Finally, the vehicle-to-load (V2L) mode of operation should be mentioned. In this case, energy stored in EV batteries can be used to power domestic devices and appliances directly, without using the electrical distribution of the house. It should be noted that all four operating modes work in bi-directional mode, where the power flow can be transmitted in both directions.

EVs are said to be prosumers (producers or consumers). With this dual role, we can differentiate four types of charging, as illustrated in Figure 9.2. When Plug-in and Charging mode is taking place, the EV starts charging at the moment it is plugged-in. In Delayed Charging mode, in contrast to the first one, the EV starts charging a specific time after being plugged-in. Both modes perform charging at constant power and without interruptions. If the charging power is variable over time, being adjusted by a control algorithm according to external requirements and conditions, such as the price of electricity, the charging mode used is Smart Charging. Similarly, if in addition to the aforementioned, the EV can transfer its energy stored in the battery in such a way that it serves to alleviate the demand for power from the grid by energy-demanding devices, in this case, it refers to V2X charging.

In this chapter, we review the main techniques to control the charge and discharging of EVs. The algorithms to decide about when to do these tasks are called the dispatching strategies. We proceed with two approaches. First, as individual entities, we will study how the EVs can be managed to get electrical benefits at home. Second, we will explain how aggregators can control a set of EVs in different homes to increase the potential ways to participate in services of the electrical market.

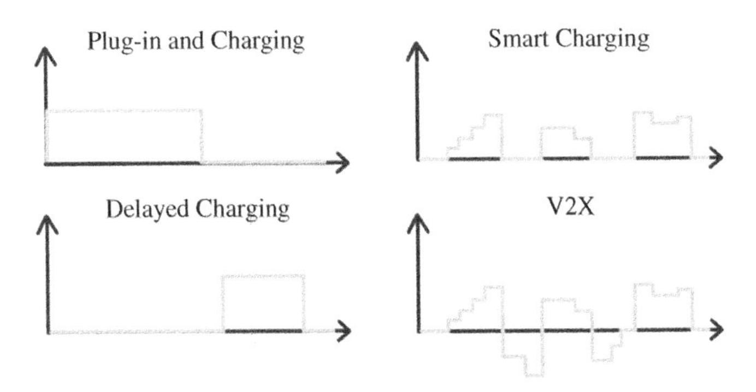

Figure 9.2 Charging modes.

The rest of the chapter is structured as follows. In Section 9.2, we review the types of charging stations. The focus in this section is on to those that can be installed at homes. In Section 9.3, services that the EVs can provide to the grid and to the home energy management are described. This study is performed for charging and discharging operations. Section 9.4 includes a review about the main strategies proposed for the control of a fleet of EVs, especially those applied for a domestic scenario. Section 9.5 explains two distributed dispatching algorithms for the charging and discharging operations of a set of EVs in a residential area. Finally, Section 9.6 presents the main conclusion of this chapter.

9.2 EV Charging Stations

According to Society of Automotive Engineers (SAE) Standard J1772v (SAE International, 2012), three different EV charging levels are defined:

- Level 1. It allows charging via a 120 V AC (alternating current) plug, which is equivalent to a typical household plug. Therefore, installing any special electrical equipment beforehand is not necessary for charging the vehicle.
- Level 2. A 240 V and 40 A AC plug is used to charge devices.
- Level 3. In this level, a DC (direct current) technology called CHAdeMO is used to implement the charging process. 480 V is the output voltage.

International Electrotechnical Commission (IEC) provides a different classification of EV battery charging techniques in its standard 61851 (IEC 61851-1:2017, 2017). The Commission names four modes in particular:

- Mode 1. The charge in this mode is performed at a maximum voltage of 250 V and a maximum current of 16 A, limiting the power to 3.7 kW. This connection can be supported by a typical household socket. It should not be used for prolonged periods of time due to overheating.
- Mode 2. In a similar way to mode 1, mode 2 is supported by a conventional plug, but with the inclusion of a protection system. The purpose of this protection is to verify the charging and discharging power, as well as to provide a communication scheme. The maximum current value allowed in this mode is 32 A. In mode 2, the connection to the grid is made via a Schuko type connector, while in the vehicle is usually a Mennekes connector.
- Mode 3. Compared to the previous modes, mode 3 requires a specific installation connected to the grid. This installation is called a Wallbox, which incorporates protection and control functionalities, as well as communication between the EV and the charging station. This technique is suitable for installation in smart grids.

- Mode 4. Similarly to mode 3, mode 4 systems incorporate control and protection mechanisms, however, they include a Wallbox that converts AC into DC, which will be the method in which the vehicle is charged. These types of chargers are recommended for fast charging, with a maximum current of 400 A and a power limit of 240 kW. Due to the main characteristics of these chargers, they become a very expensive solution.

If the EV to be charged at home is conventional in terms of battery characteristics and charging power, mode 2 can be used. However, although it does not require a special installation, the charging speed would be slow and the charging power is limited to approximately 3.7 kW. However, in a domestic context, mode 3 is the most commonly used method for charging EVs. With the implementation of this mode, the EV is connected to the AC grid by means of equipment exclusively dedicated to charging the EV. Moreover, in this type of installation, the homeowner will know in advance the vehicle model that will be charged in the installation, so the charging point will supply the appropriate cable, adapted to the vehicle.

Another aspect to consider when describing a charger is identifying where the energy comes from. According to this criterion, we can define three types of chargers:

- Grid-connected, which only receives energy from the main power system.
- Off-grid, which is connected to a self-powered home so that the energy is extracted from renewable energy resources captured by the domestic installation. Energy storage systems are required in the installation.
- Hybrid. The home is connected to the grid but it also gets energy from renewable energy sources. A control system will decide from where the energy should be extracted according to criteria such as energy cost, environmental impact, or user's preferences.

In the commercial sector, there are numerous solutions available. For grid-connected type, the Wallbox brand can be highlighted, as they offer chargers with a charging capacity from 7 kW up to 22 kW, such as the Wallbox Pulsar Plus model. In the same way, and with the same power range, Circutor or Orbis manufacturers offer a different alternative with their model Circutor eNEXT and orbis Viaris Uni, respectively. Circutor develops one of its sectors on domestic charging solution, as well as EVBOX with its model EVBOX Elvi and Simon with Simon plug and drive.

Wireless charging is gaining attention recently as an EV domestic charger (Triviño-Cabrera et al., 2020). It is a mature technology now that avoid risks associated with the physical connections plugged versions have higher voltages and current present in the cables. For the execution of the dispatching strategies, wireless charging allows the participation of EVs without any manual intervention. Using this method, it is expected that the availability of EVs to perform some services increase due to the possibility of offering automated scheduling without the need of ensuring that the EV is plugged in. Wireless charging will provide more efficient energy management at homes.

The most mature implementation of wireless chargers for EVs relies on magnetic-resonant technology including commercial devices for domestic stations as WiTricity, Qualcomm Halo, or Plugless Power. It has to be noted that wireless chargers must adapt their electronics and control to provide V2X services (Triviño et al., 2021).

9.3 EV Services

As previously explained, EVs will be an active asset, which may provide services to the grid or to improve the energy management at home. Next, we will review how the EVs can contribute to these two sectors.

9.3.1 Ancillary Services

EVs are identified as valuable assets for the provision of several grid services. Among them, the following stand out:

- Frequency regulation (Peng et al., 2017), by the injection or consumption of active power to/from the grid.
- Voltage support. The conventional way to alleviate local voltage issues in distribution networks is by means of forcing the injection or consumption of reactive power. There are research available demonstrating that charging stations can be modified to provide reactive power.
- Reducing current harmonics. The battery chargers are operated as variable impedance, which works as an appropriate filtering for the reduction of harmonics.

9.3.2 Domestic Services

Providing power grids with advanced algorithms and control is an established trend which also will be present at home. Individual homes or buildings are expected to count on smart home energy management systems, which will execute effective load management (not only for traditional domestic appliances but also for EVs and Energy storage systems) while accommodating renewable energy sources (Hou et al., 2019).

One of the main applications for which V2H is becoming widespread is the use of the EV as battery storage inside the home (Tuttle et al., 2013). Typically, this application is found in smart homes, where renewable power generation is available, such as solar panels, wind turbines, etc. All the energy produced by the renewable source can be used directly for consumption or, in case of power overproduction, it can be stored in batteries. In a home with an integrated V2H system, this energy excess can be stored in the battery of the EV. In this system, in the event of low energy availability from the renewable source due to external agents, such as unfavorable weather conditions, or a demand that exceeds the power produced by the smart home's generation system, the power can be extracted from the EV's own battery and be fed back into the house for consumption. Moreover, energy can be stored during the hours when the price of electricity is cheaper, so that the energy stored in the vehicle's battery can be used during the hours when the price of electricity is high.

By using the battery of the EV as a power source, the energy stored can be used to power different devices or specific applications within the home. For example, control algorithms can be designed where all the energy from the EV is used to power the heating and cooling system of the home. Similarly, the energy stored in the EV can be used for household electrical applications when electricity prices are high. These applications can be for lighting, as well as for supplying appliances such as dishwashers, microwaves, or televisions. It must be taken into account that in order for the EV battery to supply the power demanded by the devices or applications, a control algorithm must be designed and implemented to regulate the operation of the system.

On the other hand, and as can be deduced from the above information, the fact that the battery of the EV can be used as a power source, allows one of its applications within the V2H to be a backup system, as long as a minimum charge according to the user's minimum transportation needs. Therefore, in the event of a general or local power outage or power failure, the EV can serve as a source of electricity (Kosinka et al., 2020).

9.4 Dispatching Strategies for EVs

Smart home energy management system (HEMS) is an essential component of the future homes to make a more efficient use of energy. Basically, it analyzes the domestic assets and the energy demands to decide how and when the power is going to be consumed or delivered. Figure 9.3 illustrates the interaction of the HEMS with other components at home.

The decisions of the HEMS could attend to several goals, but the most popular ones are reducing the electricity bills, accommodating renewable energy sources or having a self-sufficient house.

However, the use of renewable energy, introduces variability into the system due to their fluctuating nature, as they are dependent on external factors, such

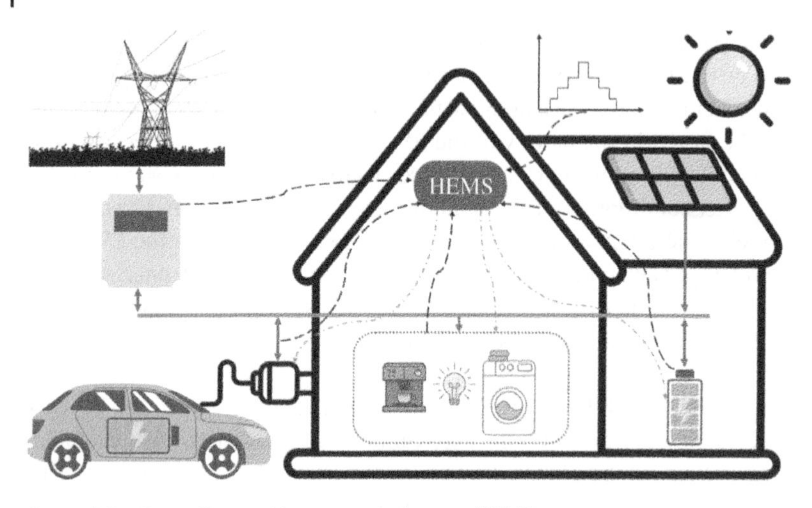

Figure 9.3 Home Energy Management System (HEMS).

as weather conditions, making them unpredictable sources. Thus, there is one parameter in energy systems that play an important role: energy flexibility. According to the International Energy Association, Power system flexibility is the ability of a power system to reliably and cost-effectively manage the variability and uncertainty of demand and supply across all relevant timescales (IEA, 2018). In the household, there are devices that can promote energy flexibility, such as washing machines or dishwashers, as their activation can be shifted to different times, depending on the system's requirements.

In general, HEMS systems are governed by algorithms that perform the actions automatically. These algorithms can be of different types, as well as focused on different priority objectives, such as cost minimization, renewable energy utilization, adjustments in the degradation model of storage systems, etc.

There are proposals where different algorithms are designed for a HEMS. Forootani et al. (2022) proposed an advanced satisfaction-based HEMS using deep reinforcement learning (DRL). The main objective of that approach is to reduce the electricity cost and user dissatisfaction, by determining ahead of time the 24-hour energy consumption of each home appliance. A different approach is described in Song et al. (2022), where it is proposed a multi-objective optimization model focused on satisfying users' need, by developing an algorithm that considers running costs, satisfaction index and peak-valley balance index. The work in Tantawy et al. (2022) develops a HEMS, including photovoltaic sources and battery storage units (BSU) and controlling the electrical loads with four different algorithms: genetic algorithm (GA), particle swarm optimization (PSO), whale optimization algorithm (GA), and sine cosine algorithm (SCA).

9.4.1 Classification of EV Dispatching Strategies

The way in which the EV exchanges power with the grid is normally performed by an element that transmits the setpoint to the EVs connected to the system. The setpoint consists of the power that the EV has to consume or produce. This element can be directly a grid operator (Figure 9.4a), as suggested in Datta et al. (2018) or an aggregator (Figure 9.4b), which acts as an intermediary between the power grid and EVs, as proposed in Mohiti et al. (2019). In general, the tendency is to opt

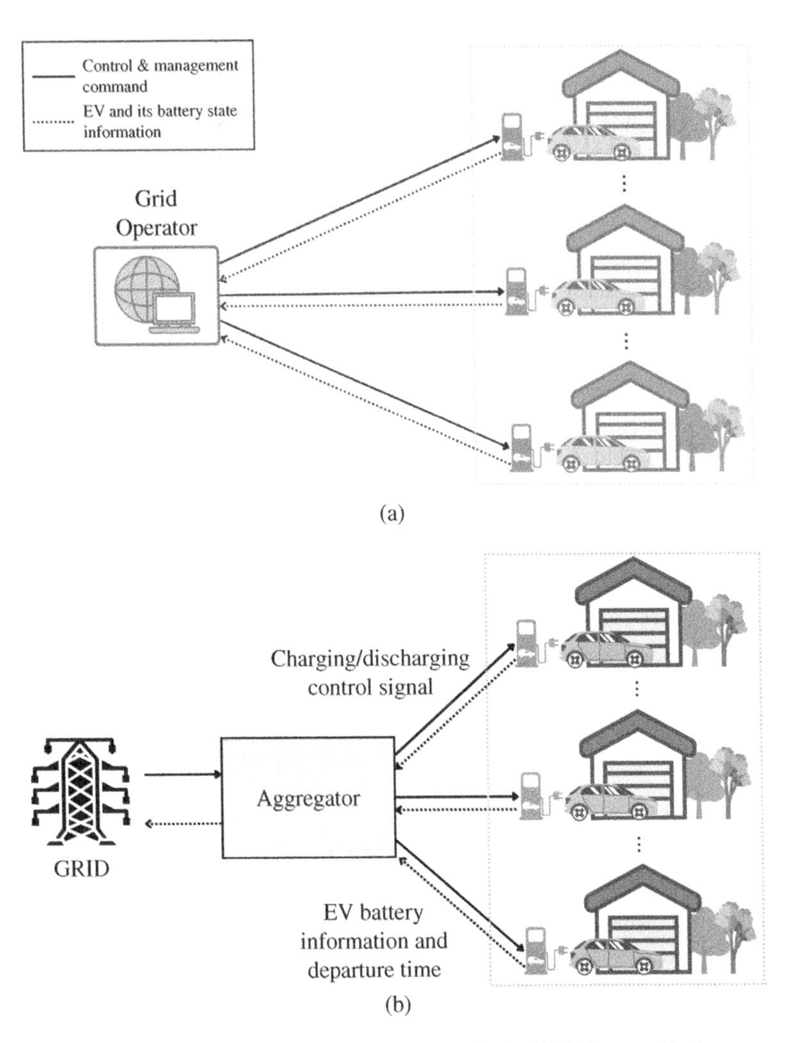

Figure 9.4 Grid-controlled and aggregator methods. (a) Grid-controlled system. (b) System with aggregator.

for the second option. Thus, an aggregator can manage charging and discharging in three different ways: centralized, decentralized, and hierarchical control structures (Aghajan-Eshkevari et al., 2022).

Under the centralized setting, the aggregator, after aggregating the information and charging requirements of each one of the EVs in the feeder under management, determines the charging and discharging parameters of each of them at each instant of time, taking into account the grid operation point and following an objective function. That is, the aggregator aims for the best global operation from a centralized position. In this case, the EV owner does not have control of the vehicle's charge. One of the disadvantages of this system is the possible failure of the system in the event of an optimization problem without a global solution, so it would be advisable to install a backup system. Likewise, it is worth noting that centralized approaches do not have a high scalability, since the difficulty of solving the problem increases with the number of vehicles in the system.

In a decentralized setting, it is vehicle owners who decide whether the vehicle will be charged or not, mainly on the basis of electricity price criteria. In other words, each charging station resolves its own charging and discharging optimization problem, and individual information is shared among others to achieve overall system balance. The aggregator or system operator can contribute with price strategies, as well as proposing shifting loads from peak to off-peak hours in order to reduce costs.

Finally, the hierarchical control structure combines features of centralized method and decentralized method. In this case, control is divided into two levels. On the first level, there is a central controller that has control over the aggregators, while on the second level there are aggregators that control a group of EVs. The problem with this method is that a failure in the central controller would affect the entire system. For this reason, there are other variants of this structure in which the central controller is eliminated, and the aggregators communicate with each other without the need for all the information to end up at a centralized controller. In this way, if one of the aggregators fails, only the group of EVs controlled by that aggregator would be affected, and not the rest.

9.5 Proposed Distributed EV Dispatching Strategy

In this section, we propose two generic decentralized EV dispatching algorithms for a residential area. In this area, it is expected that a disperse number of EVs will be present, each one in different homes. In every residence, the HEMS will be responsible for controlling the energy flow inside the home, that is, it will control when to make use of the grid connection or when benefit from the energy extracted from its own renewable sources.

To increase the benefits the controlled charging of EVs will bring, they can be grouped into an aggregator. With the proposed algorithms, it is aimed to show how the aggregators work for two main scenarios: (i) a set of EVs to be charged but there are electrical restrictions and (ii) a set of EVs coordinated to provide energy to the grid.

In the first scenario, the aggregator will consider the power restriction of the area in which the EVs under control are. Specifically, it is assumed that all the electrical installations in the houses are connected to the distribution network by means of a single central transformer. The maximum power that the transformer can provide is P^{trans}. Running several estimations of the domestic loads, we can conclude that only a percentage of this maximum power can be assigned to the EVs. This percentage is time-dependent. Thus, we define P^{tot} as the maximum power that the transformer can support for the charging of the EVs. Figure 9.5 illustrates the scenario we consider.

It is necessary to define at which power each EV will be charged. This type of operation requires the modeling of the EV's owners satisfaction i when charging at a power level p_i with the following term u:

$$u_i = \frac{P_i^{max}}{SoC_i} ln(p_i + 1) \tag{9.1}$$

where P_i^{max} represents the power of the home charger and SoC_i is the State of Charge of the EV i. With this Equation, we want to model that chargers with higher power are expected to be used with greater power flows too. In addition, if the SoC

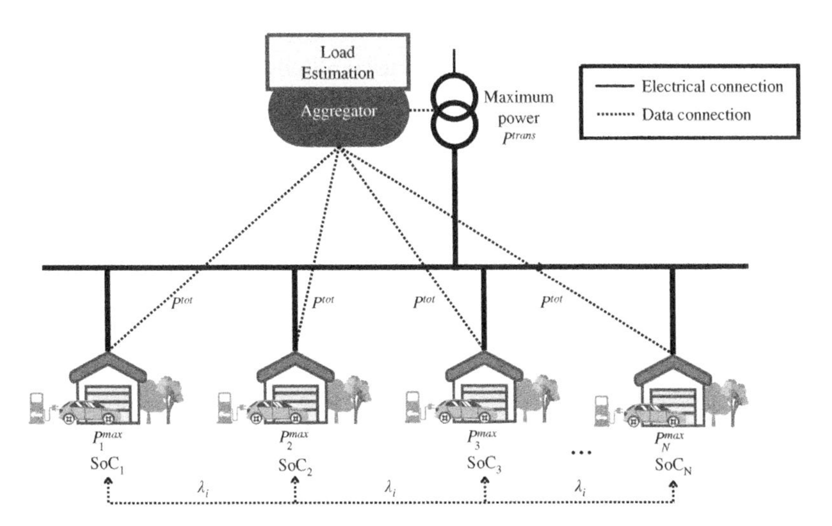

Figure 9.5 Basic scenario tested for the distributed management of EV charging.

is low, the EV has a higher demand for charging sooner. It is also assumed there are N vehicles.

The objective function of the general problem is:

$$\max \sum_{i=1}^{N} u_i \tag{9.2}$$

subject to

$$\sum_{i=1}^{N} p_i < P^{tot} \tag{9.3}$$

With this objective function, the assignation of the charging power is based on maximizing the satisfaction of all EV's users. We can define the Lagrangian function of the previous problem as:

$$\mathcal{L} = u_i + \lambda_i \left(\sum_{i=1}^{N} p_i - P^{tot} \right) \tag{9.4}$$

The optimality condition for the Lagrangian is

$$\mathcal{L}p_i = \frac{P_i^{max}}{SoC_i \left(p_i + 1\right)} + \lambda_i = 0 \tag{9.5}$$

Thus, the values for λ are:

$$\lambda_i = -\frac{SoC_i \left(p_i + 1\right)}{P_i^{max}} \tag{9.6}$$

The proposal fits in the group of distributed dispatching. With an iterative algorithm, it is possible to converge to a valid solution. In this case, the solution is considered as valid when an equilibrium is reached. For such formulation, the equilibrium conditions require that all Lagrange multipliers are the same, that is,

$$\lambda_1 = \lambda_2 = \cdots = \lambda_N \tag{9.7}$$

The proposed algorithm has the following phases:

- Phase 1: Initialization.
  It starts by assigning the maximum power supported by each domestic charger to its corresponding EV, that is:

  $$p_i = P_i^{max} \quad \forall i \in \{1, N\} \tag{9.8}$$

  Then, every λ_i must be updated according to Equation 9.6.
- Phase 2: Checking the power limits
  With the power assignation, it is verified if the maximum power of the transformer has been exceeded, by calculating δ_p:

  $$\delta_p = \sum_{i=1}^{N} p_i - P^{tot} \tag{9.9}$$

If δ_p is positive, it means that the power has been exceeded and a new power assignation must be performed. If δ_p is negative, it means that the power has not been exceeded. The ideal situation would be:

$$\epsilon_p < \delta_p < 0 \tag{9.10}$$

- Phase 3: Lambda update
  As the equilibrium point is reached when all λ_i are the same, their values are updated according to the difference between these values. In particular,

$$Diff_i = \sum_{j=1}^{N} (\lambda_j - \lambda_i) \tag{9.11}$$

  For this operation, the EVs need to exchange their λ_i. It is important to note that this data contains private information such as the SoC or the features of the domestic charger. In this way, the algorithm can work in a distributed way while preserving private information.
  The update of λ considers two main conditions. First, it will use the deviations of the previous values for this variable for all the EVs in the same group, as the target is to reach equilibrium. Second, it will also use the potential margin to modify the powers assigned considering the power limits imposed by the transformer.

$$\lambda_i = \lambda_i + \frac{Diff_i}{N} - 0.09 * \delta_p \tag{9.12}$$

  Once each EV has computed its λ_i, they can determine the consequent p_i with Equation 9.6, so that:

$$p_i = -\frac{p_i^{max}}{SoC_i * \lambda_i} - 1 \tag{9.13}$$

- Phase 4: Verifying the equilibrium conditions
  If $Diff_i < \epsilon_\lambda$, then the power assignation has been performed successfully and the algorithm ends. If not, the EVs will return to Phase 2.

The algorithm was tested with an aggregator controlling 5 EVs in a residential area. Each domestic charger has a different maximum charger power and their EVs are with a different SoC. The transformer has a maximum 10 kW assigned to the charge of the EVs. The data of the results are summarized in Table 9.1 and also include the final power at which the EVs will be charged and their λ_i.

As can be observed, all the EVs get the same value for λ_i. This is the equilibrium condition to decide when to finish the iterative algorithm. The number of iterations to reach this point is 10, as represented in Figure 9.6.

The algorithm can also be adapted for V2G operations. In this new formulation, the aggregator decides to provide P^{obj} to the grid, and it will be extracted from the

Table 9.1 Data for the distributed algorithm for the charging management of domestic EVs.

	EV_1	EV_2	EV_3	EV_4	EV_5
P_i^{max} [kW]	3.7	3.7	11	7	3.7
SoC_i	0.5	0.8	0.9	0.4	0.6
λ_i	3.19	3.19	3.19	3.19	3.19
p_i [kW]	1.3	0.44	2.82	4.47	0.93

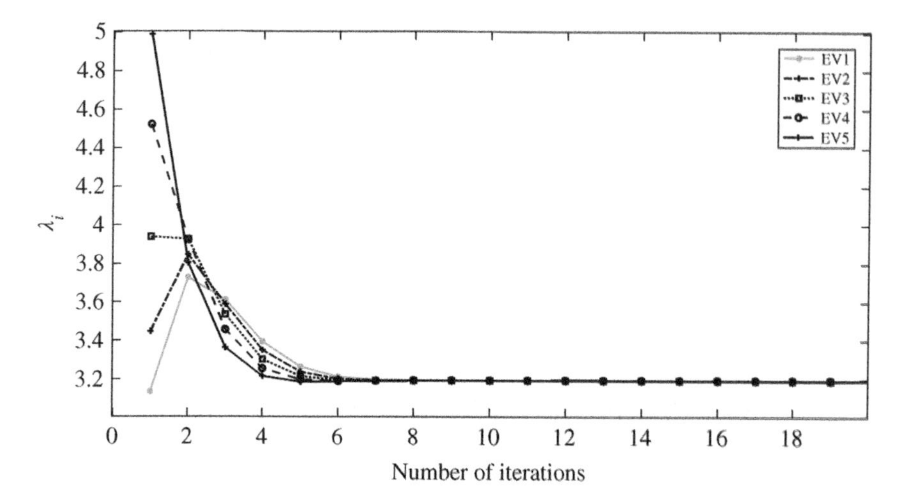

Figure 9.6 λ_i values for every iteration for the distributed charging algorithm.

set of EVs it is managing. The aggregator informs the EVs about this goal and the EVs will decide the power that each one delivers in a distributed way. Figure 9.7 illustrates the problem to be solved.

Each user must determine their interest to participate in the V2G transaction. This interest is modeled by variable u_i:

$$u_i = \frac{P_i^{max}}{penalty_i} ln(p_i + 1) \tag{9.14}$$

where $penalty_i$ represents the domestic costs for an EV to participate in the V2G service. If the EV is part of the V2G service, it cannot provide domestic services, which may be of interest as well. This parameter is a number between 0 and 1.

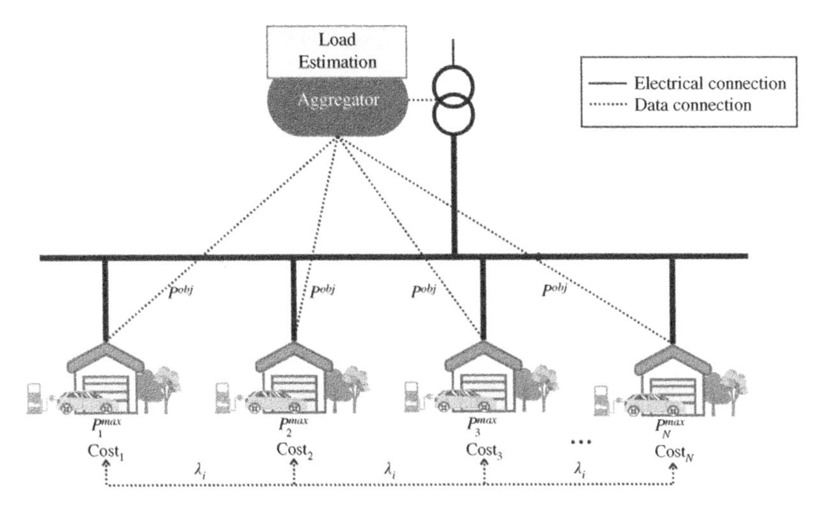

Figure 9.7 Scenario with home PV and storage tested for the distributed management of V2G operation.

The goal of the algorithm is to maximize the benefits for all the EV users. It is possible to formulate a similar problem to the one expressed in Equation 9.2, where the restriction is:

$$\sum_{i=1}^{N} P_i < P^{obj} \tag{9.15}$$

In a similar way to the previous formulation, we could proceed with the definition of the Lagrangian function and an analogous resolution. Reaching the equilibrium point means that the iterative algorithm has achieved a valid solution. The equilibrium is also associated with equivalent values for the Lagrangian multipliers. The algorithm for the distributed V2G is defined with the following phases:

- Phase 1: Informing about the V2G service
  The aggregator decides that the fleet is going to contribute to the V2G services with a discharge of P^{obj} W. It sends this command to all the EVs it is managing.
- Phase 2: Initialization phase
  Each EV computes their own penalty function ($penalty_i$) to participate in this operation for the next interval. This computation may be based on the prediction of the home consumption, the current charge of the home electrical storage, the estimation of the future use of the EV as a domestic service or the irradiation that the PV (Photovoltaic) panels are receiving, among other factors. It is important to note that the penalty function is a dynamic value.

The Lagrangian multiplier associated with this penalty function is:

$$\lambda_i = -\frac{penalty_i\,(p_i + 1)}{P_i^{max}} \tag{9.16}$$

- Phase 3: Lambda update
  The EV transmits its Lagrangian multiplier to the other EVs in the fleet. Based on the received data, it can assume. Then, $Diff_i$, as in Equation 9.11 is calculated. If the difference is low, the algorithm ends and the power assignation is assumed to be the final one. If not, the λ values are updated as in Equation 9.12 where δ_p reflects how far the current power assignation is from the power demanded by the objective function, that is:

$$\delta_p = \sum_{i=1}^{N} p_i - P^{obj} \tag{9.17}$$

After the update of the λ values, each EV changes its assigned power as:

$$p_i = -\frac{P_i^{max}}{penalty_i * \lambda_i} - 1 \tag{9.18}$$

This second algorithm has been tested for the configuration summarized in Table 9.2 to offer a total power of 10 kW.

As can be observed, the algorithm is correctly executed. The total power provided by the EVs is 9.99 kW, which is close to the objective power set by the aggregator. In Figure 9.8, we can see how the λ_i values have been updated for every iteration, reaching to a common constant of 4.69. Only 6 simulations are necessary to reach the equilibrium point, which makes this algorithm appropriate for the scenario considered.

Table 9.2 Data for the distributed algorithm for the discharging management of domestic EVs.

	EV_1	EV_2	EV_3	EV_4	EV_5
P_i^{max} [kW]	3.7	3.7	11	7	3.7
$penalty_i$	0.4	0.3	0.5	0.4	0.4
λ_i	4.69	4.69	4.69	4.69	4.69
p_i [kW]	0.97	1.63	3.69	2.73	0.97

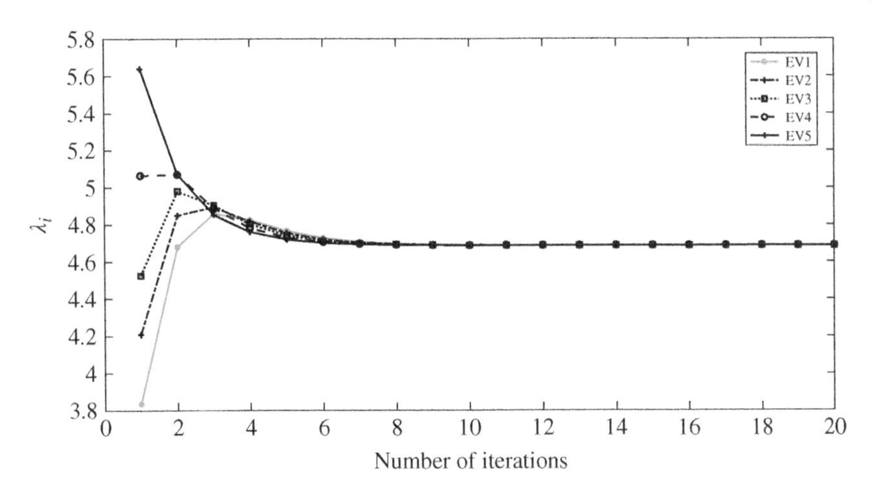

Figure 9.8 λ_i values for every iteration for the distributed discharging algorithm.

9.6 Conclusions

This chapter has presented a review of existing charging stations types that can be installed at homes. Services to the distribution grid and home energy services provided by EV were also presented.

It was proposed two (single and V2G) decentralized optimal charging strategies for an EV aggregator in the context of residential applications. The goal of both strategies is to obtain the maximum optimal energy management of the EVs controlled by an aggregator. Each EV defines its interest to participate in the charging and discharging processes. For the charging, the SoC and the electrical features of the domestic chargers are considered to optimize schedule for each EV while satisfying maximum power limits. Alternatively, the V2G operation establishes the convenience for contributing to the objective power based on the expected home demands, the capacity for extracting renewable energy or by the SoC. On this aspect, it was formulated the control problem as an optimization problem consisting of a convex objective function subject to linear local constraints.

A novel algorithm was proposed to address a decentralized solution which solves the problem iteratively by sending control signals (Lagrange multipliers) to each subproblem and receive their decision variables to update the control signals. The decentralized algorithm allows all EV chargers to update their charging/discharging actions in parallel with no extra communications among involved chargers.

The simulation results confirm that the both proposed strategies are computationally efficient and convergence to the global optimum.

Acknowledgments

This work was supported in part by the Spanish Ministerio de Ciencia e Innovacion (MICINN) project PID2019-110531-RA-I00 from the "Proyectos de I+D+i - RTI Tipo A" program. This work was also partially supported by Junta de Andalucia (Spain) Project Ref: P20 01164.

References

Aghajan-Eshkevari, S., Azad, S., Nazari-Heris, M. et al. (2022). Charging and discharging of electric vehicles in power systems: an updated and detailed review of methods, control structures, objectives, and optimization methodologies. *Sustainability* 14 (4). https://doi.org/10.3390/su14042137.

Datta, U., Kalam, A., and Shi, J. (2018). Electric vehicle (EV) in home energy management to reduce daily electricity costs of residential customer. *Journal of Scientific and Industrial Research* 77: 569–565.

Dudek, E. (2021). The flexibility of domestic electric vehicle charging: the electric nation project. *IEEE Power and Energy Magazine* 19 (4): 16–27. https://doi.org/10.1109/MPE.2021.3072714.

Faddel, S., Al-Awami, A.T., and Mohammed, O.A. (2018). Charge control and operation of electric vehicles in power grids: a review. *Energies* 11 (4). https://doi.org/10.3390/en11040701.

Forootani, A., Rastegar, M., and Jooshaki, M. (2022). An advanced satisfaction-based home energy management system using deep reinforcement learning. *IEEE Access* 10: 47896–47905. https://doi.org/10.1109/ACCESS.2022.3172327.

Gryparis, E., Papadopoulos, P., Leligou, H.C., and Psomopoulos, C.S. (2020). Electricity demand and carbon emission in power generation under high penetration of electric vehicles. A European Union perspective. *Energy Reports* 6: 475–486. https://doi.org/10.1016/j.egyr.2020.09.025. Technologies and Materials for Renewable Energy, Environment and Sustainability.

Hou, X., Wang, J., Huang, T. et al. (2019). Smart home energy management optimization method considering energy storage and electric vehicle. *IEEE Access* 7: 144010–144020. https://doi.org/10.1109/ACCESS.2019.2944878.

IEA (2018). *Status of Power System Transformation 2018*. https://doi.org/10.1787/9789264302006-en. https://www.oecd-ilibrary.org/content/publication/9789264302006-en.

IEC 61851-1:2017 (2017). " — IEC Webstore." https://webstore.iec.ch/publication/33644 (accessed 28 December 2022).

International Energy Agency (2017). *Global EV Outlook 2017*. https://doi.org/10.1787/9789264278882-en. https://www.oecd-ilibrary.org/content/publication/9789264278882-en.

Kosinka, M., Slanina, Z., Petruzela, M., and Blazek, V. (2020). Control system for V2H applications. *2020 20th International Conference on Control, Automation and Systems (ICCAS)*, pp. 916–921. https://doi.org/10.23919/ICCAS50221.2020.9268383.

Mohiti, M., Monsef, H., and Lesani, H. (2019). A decentralized robust model for coordinated operation of smart distribution network and electric vehicle aggregators. *International Journal of Electrical Power & Energy Systems* 104: 853–867. https://doi.org/10.1016/j.ijepes.2018.07.054.

Peng, C., Zou, J., and Lian, L. (2017). Dispatching strategies of electric vehicles participating in frequency regulation on power grid: a review. *Renewable and Sustainable Energy Reviews* 68: 147–152. https://doi.org/10.1016/j.rser.2016.09.133.

SAE International (2012). *SAE Electric Vehicle and Plug in Hybrid Electric Vehicle Conductive Charge Coupler*. www.sae.org/standards/content/j1772_201210/ (accessed 28 December 2022).

Song, Z., Guan, X., and Cheng, M. (2022). Multi-objective optimization strategy for home energy management system including PV and battery energy storage. *Energy Reports* 8: 5396–5411. https://doi.org/10.1016/j.egyr.2022.04.023.

Sousa, T.J.C., Monteiro, V., Fernandes, J.C.A. et al. (2018). New perspectives for vehicle-to-vehicle (V2V) power transfer. *IECON 2018 - 44th Annual Conference of the IEEE Industrial Electronics Society*, pp. 5183–5188. https://doi.org/10.1109/IECON.2018.8591209.

Tantawy, E.S.F., Amer, G.M., and Fayez, H.M. (2022). Scheduling home appliances with integration of hybrid energy sources using intelligent algorithms. *Ain Shams Engineering Journal* 13 (4): 101–676. https://doi.org/10.1016/j.asej.2021.101676.

Triviño-Cabrera, A., González-González, J.M., and Aguado, J.A. (2020). *Wireless Power Transfer for Electric Vehicles: Foundations and Design Approach*. Springer. ISBN 3030267059.

Triviño, A., Gonzalez-Gonzalez, J.M., and Castilla, M. (2021). Review on control techniques for EV bidirectional wireless chargers. *Electronics* 10 (16). https://doi.org/10.3390/electronics10161905.

Tuttle, D.P., Fares, R.L., Baldick, R., and Webber, M.E. (2013). Plug-in vehicle to home (V2H) duration and power output capability. *2013 IEEE Transportation Electrification Conference and Expo (ITEC)*, pp. 17. https://doi.org/10.1109/ITEC.2013.6574527.

10

Electric Vehicles as Smart Appliances for Residential Energy Management

Indradip Mitra[1], Zakir Rather[2], Angshu Nath[2], and Sahana Lokesh[1]

[1]*Deutsche Gesellschaft für Internationale Zusammenarbeit (GIZ), New Delhi, India*
[2]*Indian Institute of Technology Bombay, Powai, Mumbai, Maharashtra, India*

10.1 Introduction

In order to address the climatic issues, transition of transportation sector from conventional fuel-based vehicles to electric vehicles (EVs) is being experienced across the world, with ambitious targets of EV penetration set by different countries/regions. One of the critical emergence due to transition to EVs is the shifting of vehicle refueling from commercial spaces to the user residence. It has been observed that EV users prefer to charge their vehicles at home due to convenience and low cost of charging, provided they have adequate parking and charging infrastructure. The US Department of Energy claims that 80% of EV charging is done at home (Blonsky et al. 2021).

As distribution feeders are generally sized based on the residential loads, so inclusion of EV charging which can effectively double the power requirement of each household, would also lead to significant stress on the distribution system (Cazzola et al. 2018). Therefore, for secure and stable EV integration, determining the hosting capacity of the distribution feeders should consider all the critical constraints, such as thermal limit of the distribution lines, transformers, etc., voltage profile and power quality issues, distribution losses, and protection aspects. Currently, from the distribution operator's perspective, EVs are largely an unknown type of load, with potential variations both spatially and temporally. The different types of technical impacts of EV integration on the distribution network are summarized in Figure 10.1.

While EV charging load introduces different challenges in residential energy management and power demand, grid operation, and management, the controllable nature of EV loads also provides opportunity to help mitigate such

Energy Smart Appliances: Applications, Methodologies, and Challenges,
First Edition. Edited by Antonio Moreno-Munoz and Neomar Giacomini.

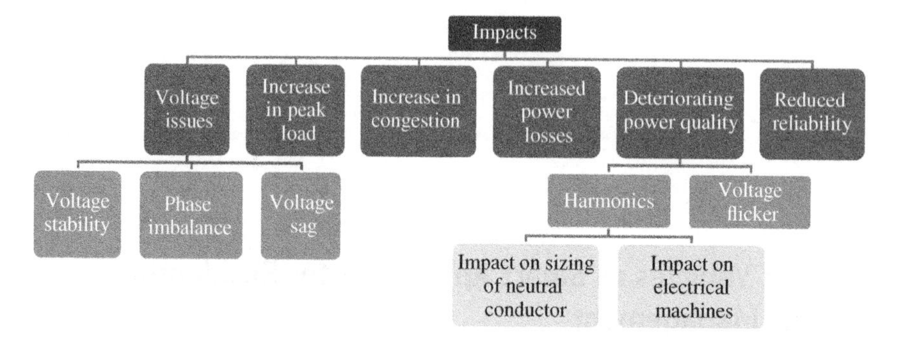

Figure 10.1 Impact of high EV charging load on the distribution network. Source: Rather et al. (2021a)/NITI Aayog/CC BY-SA 2.5 IN.

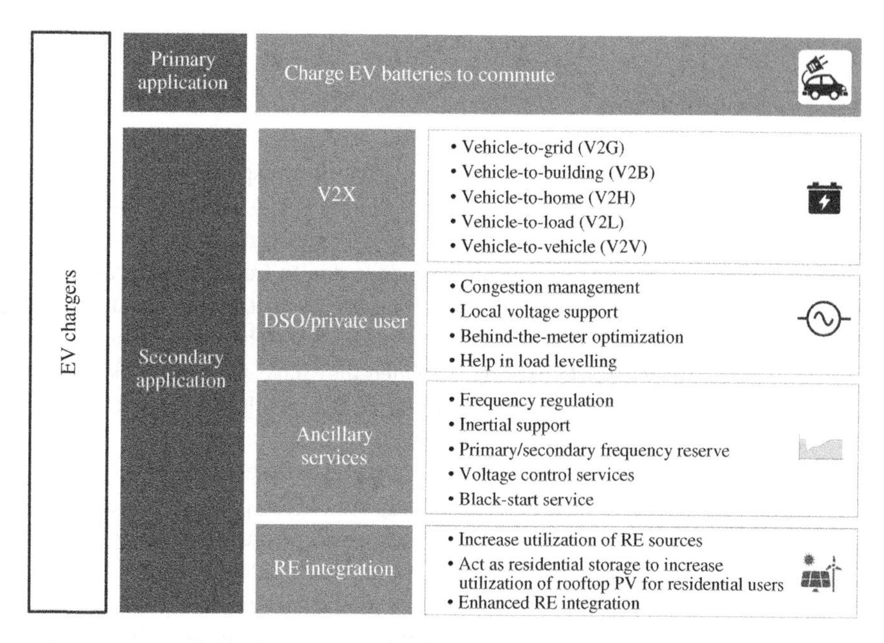

Figure 10.2 Applications of EV charger. Source: Rather et al. (2021a)/NITI Aayog/CC BY-SA 2.5 IN.

challenges. The charging power of EVs can be controlled or their time of charging can be altered, either by molding the user's behavior or using smart charging technologies.

The potential applications that can be provided by EV charging infrastructure are summarized in Figure 10.2. In this chapter, the different applications and usability of EVs as smart residential loads would be explored.

10.2 EV Charging Standards and Charging Protocols

EV charging standards are regulated by different regulatory bodies around the world. Institute of Electrical and Electronics Engineers (IEEEs), International Organisation for Standards (ISO), International Electrotechnical Commission (IEC), and Society of Automotive Engineers (SAEs) are leading regulatory bodies that provide standards for every entity in the EV ecosystem. These standards cover the complete spectrum of design and construction of EV charging stations, safety considerations, communications as well as testing. A brief description of the standards is provided in Section 2.1.

10.2.1 EV Charging Standards

10.2.1.1 IEC 61851

IEC 61851 covers conductive alternating current (AC) and direct current (C) charging for on-board and off-board charging up to 1000 V AC and 1500 V DC (IEC 2017). It covers mechanical, electrical, communication, and performance requirements of electric vehicle supply equipment[1] (EVSE) including different characteristics and operating modes, connection specifications between EV and EVSE and safety requirements (Rather et al. 2021a). This standard classifies EVSE into four different charging modes as described in Figure 10.3. Besides difference in allowable voltage and current levels, the modes of charging are predominantly categorized based on the protection, controllability, and type of connection between the EV and EVSE.

- In *mode* 1, the EV is directly connected to a standard electrical outlet without any supplementary pilot wire or auxiliary contacts, or any form of overcurrent or thermal protection.
- In *mode* 2, the EV is connected to a standard electrical outlet through an In Cable Control Box (ICCB) which enables some level of controllability of EV charging.
- In *mode* 3, charging the EV is charged through an EVSE which is permanently connected to the AC supply network. It has extended controllability and protection functionalities.
- In *mode* 4, charging is similar to mode 3 in that here too the EV is charged through an EVSE permanently connected to the grid, however here the EVSE converts the AC input from the mains to DC and feeds the DC output to the EV.

1 The EVSE is a physical interface that enables the EV to be connected to the electrical mains. Depending on the level of sophistication of the EVSE, there are different protection and control equipments embedded within an EVSE, that enables a controlled charging of the EV.

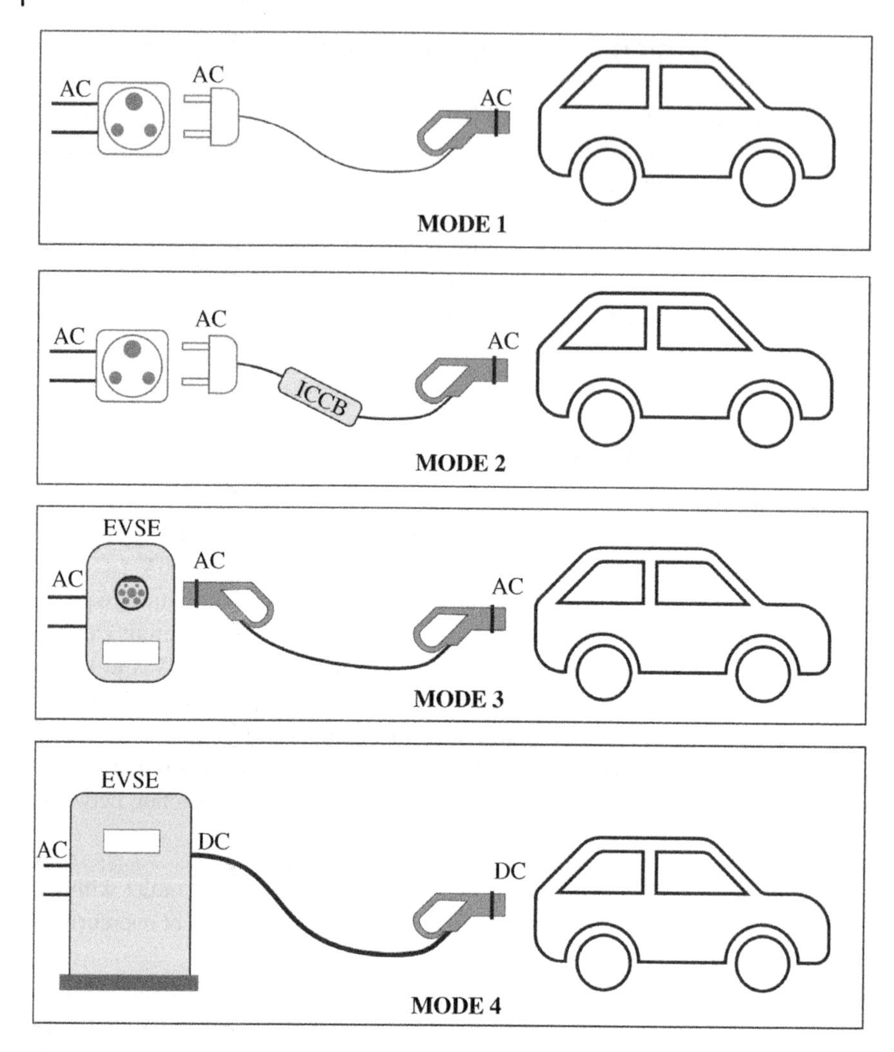

Figure 10.3 Modes of charging as defined by IEC 61851. Source: Angshu Nath based on details from IEC 61851 (IEC 2017).

10.2.1.2 SAE J1772

This standard also provides the specification of AC and DC chargers, but while the IEC 61851 is more prevalent in Europe, the SAE J1772 has been widely adopted across the United States and Japan (Rather et al. 2021a). The standard covers the details on control pilot pin,[2] its working, and the signal flow for controlling the

2 The control pilot is used for communication between the EVSE and the EV.

Table 10.1 Charging levels as defined in SAEJ1772.

	SAE Standard		
Parameter	**Level 1**	**Level 2**	**Level 3**
Voltage (V)	120	240	200–600
Current (A)	16	80	400
Power (kW)	1.92	19.2	240

Source: Rather et al. (2021a)/NITI Aayog/CC BY-SA 2.5 IN.

EV charging. This standard also provides information on charging levels based on current and voltage levels as given in Table 10.1.

10.2.1.3 GB/T 20234

The Guobaio (GB/T) standard, issued by the Standardization Administration of China, contains guidelines for plug, socket, and connector for EV conductive charging (Rather et al. 2021a). Followed mostly in China, this standard also provides testing methods and specifications for conductive charging up to 690 V AC, 250 A, and 1000 V DC, 400 A (ChineseStandard.net 2016).

Some other relevant standards related to EV charging are provided in Table 10.2.

10.2.2 Charging Protocols for EV Charging

10.2.2.1 Type 1 AC Charger

Developed using the SAE J1772 and the IEC 62196 standard, type 1 charger is used for AC charging of EVs, predominantly in USA and Japan.

Table 10.2 Other relevant standards on EV charging.

Standard	Details
SAE J1773	Specifies the requirements for inductive charging
IEC 61980	Defines the practices for wireless transfer of power from the charging plates to the EV for supply rating up to 1000 V for AC and 1500 V for DC
IEC 60364	Applicable for safety and protection of EV and EV charging
IEEE 1547	It provides requirements and information about the operation, safety, and testing of interconnected Distributed Energy Resources (DER) with the electrical grid up to 10 MVAR at the point of common coupling

Source: Rather et al. (2021a)/NITI Aayog/CC BY-SA 2.5 IN.

10.2.2.2 Type 2 AC Charger

Also referred to as Mennekes, type 2 AC charger is also used for AC charging of EVs, mostly within Europe. It is the official charging plug within the European Union since 2013. It has since been adopted as the recommended connector in some countries outside of Europe, including New Zealand.

10.2.2.3 CHArge de MOve (CHAdeMO) Protocol

CHArge de MOve (CHAdeMO) provides DC charging for EVs that ensures seamless communication between charging point and vehicle. It is currently the only charging protocol that allows for bi-directional charging capability.

10.2.2.4 Combined Charging System (CCS)

Combined charging system (CCS) protocol is used for DC charging of EVs. Based on interoperability with either type 1 charger or type 2 charger, the CCS charger is either classified as a CCS1 or a CCS2 charger. CCS1 is a combination of type 1/SAE J1772 and DC ports, whereas CCS2 is a combination of type 2/Mennekes and DC port. These two CCS connectors follow SAE J1772, IEC 62196 protocols for charging purposes. These connectors are country-specific, which means that North America, Central America, Taiwan, and Korea use the CCS1 connector whereas CCS2 is used in Europe, South America, Arabia, Australia, and some other countries.

10.2.2.5 Tesla Charging Protocol

Unlike other vehicle original equipment manufacturers (OEMs), Tesla has developed its own proprietary charging protocol for charging their vehicles. It is not openly available for adaptation by other OEMs. Tesla has also developed a Supercharger, that allows for fast charging of its vehicles.

10.3 Communication Protocols Used in EV Ecosystem

Besides standards on EV charging, utilization of EVs for smart charging needs a well-developed communication infrastructure. Different communication protocols in the EV ecosystem have been described below.

10.3.1 Open Charge Point Protocol

Open Charge Point Protocol (OCPP) is a vendor independent open-source protocol between the EVSE and the Charge Point Operator (CPO)/Charging Management System (CMS). It provides a common language for communication between the EVSE and the CMS/CPO. Any charger that is OCPP compliant can be configured so as to run any OCPP-compliant software. It exchanges "transaction event" containing the start of a transaction, the stop of a transaction, meters values, and status notification. OCPP 2.0 supports ISO 15118 with added features of plug and charge and smart charging (Rather et al. 2021a).

10.3.2 Open Automated Demand Response (OpenADR)

Open automated demand response (OpenADR) enables smart charging features to the EVs. It enables an open and interoperable solution for demand response by communicating different price signals, resource capacity availability, demands from the distribution network operator, load demands from the client, etc. (OpenADR Alliance n.d.). This communication link connects the distribution system operator (DSO) with the energy management system and generally works on Zigbee and Home Area Network (HAN) (Rather et al. 2021a).

10.3.3 Open Smart Charging Protocol (OSCP)

Open smart charging protocol (OSCP) communicates predicted capacity of the network to supply the system whereas clients indicate their charge requirements to the central system. The communication link is established between the CPO, eMobility service provider (eMSP), and the DSO. OSCP is used for smart charging by communicating capacity, managing the grid, and handling the capacity budget (Rather et al. 2021a).

10.3.4 IEEE 2030.5

This is an Internet Protocol (IP) based adoption of Smart Energy Application Profile 2.0/SEP2 which provides a smart grid solution via internet-connected devices (Rather et al. 2021a). This standard is originated at Zigbee smart Energy Protocol V1 and became the IEEE standard in 2013. It provides load control, meter data exchange, exchanging tariff information, energy flow reservation, and management of distributed energy resources (DERs) functionalities.

10.3.5 ISO/IEC 15118

One of the critical communication standards used today in the EV ecosystem, this standard provides high-level communication between EV and EVSE as compared to IEC 61851. It defines the requirements for smart charging, metering, and billing for wireless charging. It is provided by the collaborative work of ISO and IEC for providing general requirements and uses cases for conductive and wireless high-level communication between EV and supply equipment (Neaimeh and

Table 10.3 Smart charging and communication protocols and the supporting roles.

	Smart charging protocols			Charge point – CMS	EVSE – EV
Protocol	OSCP	Open ADR	IEEE 2030.5	OCPP	ISO/IEC 15118
Version	1.0	1.1	2	1.6	—
Smart charging	✓	✓	✓	✓	✓
Roaming	—	—	—	✓	—
Reservation of charge point	—	—	—	✓	✓
Registration handling	—	✓	✓	—	—
Controls charge point	—	—	—	✓	—
Grid management	✓	✓	✓	✓	—
EV charging	—	—	—	—	✓
Billing and transactions	—	—	—	✓	—
Authorization of charging	—	—	—	✓	✓

Source: Rather et al. (2021a)/NITI Aayog/CC BY-SA 2.5 IN.

Andersen 2020). Additional roles of ISO/IEC 15118 series are, to provide network and application protocol requirements, physical and data link layer requirements for high-level communication between EVSE (Rather et al. 2021a).

The open-source protocols for EV smart environment have four major functionalities

- smart charging,
- roaming,
- communication between the central system and charge point, and
- communication between EV and charge point.

For achieving the above-mentioned major functionalities of the EV smart environment, different protocols are implemented to address the same functionality. Various protocols for achieving the same functionality have some typical overlapping roles. The number of protocols for specific functions with their roles is mentioned in Table 10.3.

10.4 Residential EV Charging Infrastructure

Compared to other loads in the residence, an EV charging load may be several magnitudes higher. Individual loads in a modern residence are typically rated less than 5 kW. Even cumulative residential loads are typically in the range of 1–7 kW depending on the size of the residence (Chuan and Ukil 2015). Comparatively, a single four-wheeler EV charger can draw 3–22 kW, varying from model to model.

So, integration of EV charging into the building network is critical and needs proper planning. Most distribution networks around the world have employed a very hands-on attitude for integration of EV chargers into the network mandating strict regulations and permissions.

10.4.1 Prerequisites to Installation of EV Charge Point

Prior to installation of a charge point in the residence, the user in consultation with a competent authority should check for the following,

- **Permission for supply:** Although not an issue for homeowners with single-family dwelling residences, for people residing in multi-unit dwellings, prior to installation of a charge point should get the necessary permissions from the relevant property managers. Besides permission from the property managers, the EV users would also need to get the approvals for installation of EV chargers from their local energy provider.
- **Adequacy of supply:** Depending on the capacity of the existing supply and the connected loads to the property, the competent authority should determine whether there is sufficient margin for addition of an EV charge point, with or without a CMS. If the supply is inadequate, the potential solutions include,
  - Installing a lower-capacity charger
  - Implementing load management strategies, either controlled automatically or through manual control
  - Upgrading the capacity of the power connection to the property
- **Existing earthing arrangements:** Another critical information that needs to be checked is the existing earthing circuit in the residence. This information would feed into determining the earthing arrangement needed for the charging unit.

10.4.2 EV Charger Connection Requirements and Recommendations

As EV loads are typically of much higher capacity compared to traditional residential loads, integration of EV charging to the residential circuit is typically much more complicated than simply plug-and-play. Interconnection of EV charging infrastructure is regulated by the respective distribution network operators, and most of them have specific protocols and procedures.

10.4.2.1 United Kingdom

In the United Kingdom, to install an EV charger at home, the EV user needs to consult one of the charge point installers[3] who makes an assessment of

---

3 A list of approved charge point installers is maintained by the UK Office for Zero Emission Vehicles, https://assets.publishing.service.gov.uk/government/uploads/system/uploads/attachment_data/file/1089450/residential-installer-authorisations-log.csv/preview.

the property to determine whether it has the available power capacity to accommodate an EV charging load (Rather et al. 2021b). The installer will also check the premises and if there is a power source close to vehicle parking space. This information is then sent to the grid operator who cross-checks if the distribution network can accommodate the increase in demand.

In case of vehicle-to-grid (V2G) connection, the following protocols are needed in the United Kingdom (MacLeon and Cox 2018).

- From the point of view of the electricity network, an EV exporting power is a generator connected in parallel to the network. For the generator to be connected it has to meet a set of technical requirements.
- In the UK, all distribution network operators are required to comply with the GB Distribution Code, which is maintained by the Distribution Code Review Panel and approved by Ofgem. Engineering Recommendation (EREC) G98 and EREC G99 are applicable for all new installations.
- EREC G99 underlines the connection procedure for connecting micro-generation to LV network, where the aggregated power is less than 50 or 17 kW per phase.
- EREC G98/1 updates the connection requirements of fully type-tested micro-generators (up to and including 16 A per phase) to the LV network. It also covers the connection procedures of multiple micro-generators in a close geographical location.
- EREC G99/1 covers the connection of generating units having an extended range than micro-generators.

10.4.2.2 The Netherlands

Liander, which is one of the largest Distribution network operators in the Netherlands, has the following procedure for installation of private EV chargers for its residents (Liander 2021),

- The customer has to first contact a specialized company to provide the charging station.
- These companies will first analyze the current contracted load and capacity of the residential distribution panel and determine if the existing electrical connection needs to be upgraded.
- To upgrade the electrical connection, the customer has to pay Liander the necessary upgrade fee.
- After the necessary upgradation to the electrical connection, the specialized company will install the EV charger.

In case of bidirectional chargers, the requirements for connection are however more stringent. ElaadNL which is acknowledged innovation center for smart charging initiated by the Dutch grid operators has laid out a set of draft requirements for connection of EV chargers with bidirectional charging capabilities.

- The CPO which has bidirectional chargers installed should report the location of the vehicle-to-everything (V2X) chargers to the local DSO via the platform Energieleveren (www.energieleveren.nl/). This site maintains a registry of all distributed generations installed in the Dutch grid.
- The standards for connection to the distribution grid that needs to be followed for bidirectional chargers are
  - NEN-EN 50549-1:2019: Requirements for generating plants to be connected in parallel with distribution networks – Part 1: Connection to an LV distribution network – Generating plants up to and including type B
  - VDE-AR-N 4105: Power Generating Plants in the Low Voltage Grid
- The EVSE should have proper physical signs that it is capable of bidirectional charging.
- The charging should be controllable via a central system.
- The V2X system must be equipped with means to automatically disconnect from the grid during power outage (anti-islanding).
- DSO requirements
  - For production units below 800 W
    - The power factor in the transfer point of a connection may be between 0.9 capacitive and 0.9 inductive
    - Protection for undervoltage that responds within two seconds at 80% of nominal voltage
    - Protection of overvoltage that responds within two seconds at 110% of overvoltage
    - A protection for frequency variation that responds within two seconds at frequencies outside of 48–51 Hz.
  - Others (>800 W)
    - The production unit is able to stay connected to the grid and in operation within the following bands
      - In the frequency band of 47.5–48.5 Hz during 30 minutes
      - In the frequency band of 48.5–49 Hz during 30 minutes
      - In the frequency band of 49–51 Hz for unlimited duration
      - In the frequency band of 51–51.5 Hz during 30 minutes
    - The unit should be able to activate frequency response in which
      - The frequency threshold value is adjustable between 50.2 and 50.5 Hz
      - The droop is adjustable between 4% and 12%
      - The default setting of droop is 5%

- The electricity production unit may reduce its active power at a frequency of 49.5 Hz with a gradient of 10% of the maximum capacity at 50 Hz per frequency drop of 1 Hz.
- The production unit should be able to resynchronize to the grid if
  - The voltage is between 0.9 and 1.1 pu
  - The frequency is between 49.9 and 50.1 Hz
  - The minimum time the voltage and frequency are in between the values mentioned is 60 seconds.
- A production unit with maximum capacity higher than 11 kW connected to the LV grid should be at least equipped with
  - A measuring device for the current
  - A signaling function whether the electricity production unit is connected in parallel to the grid.
- The protection should be
  - A protection for undervoltage that responds within two seconds at 0.8 pu voltage and within 0.2 seconds at 0.7 pu voltage
  - A protection for overvoltage that responds within two seconds at 1.1 pu voltage
  - A maximum current/time protection
  - A protection for frequency variation that responds within two seconds at frequencies outside of 47.5–51.5 Hz.

Table 10.4 General requirements.

Requirements for the vehicle side charging interface	ISO 17409: safety requirements for connection to an external electric power supply
Requirements for charging infrastructure and charging interface	IEC 61851-1: general requirements of electric vehicle conductive charging system
	IEC 62196-1: general requirements for connectors for EV charging
	IEC 60364-7-722: installation of low voltage systems requirements for the power supply of electric vehicles
Electromagnetic compatibility	IEC 61851-21-1: electromagnetic compatibility requirements for on-board charging devices for electric vehicles in order to establish a conductive connection to an AC or DC power supply
	IEC 61851-21-2: electromagnetic compatibility requirements for off-board charging systems for EV

Source: German National Platform for Electric Mobility (NPE) (2017).

Table 10.5 Standards for charging interface.

Wired charging	IEC 61851-1, IEC 61851-23, IEC 62196-1, IEC 62196-2, IEC 62196-3, ISO 17409: Combined Charging System (CCS) for AC and DC charging
	IEC 625752: mode 2 charging cable including IC-RCD safety device
Wireless charging	IEC 61980: infrastructure requirements
	IEC 61980-2: charge controller, positioning of vehicles
	ISO 15118-1, ISO 15118-2, ISO 15118-8, IEC-61980-2: communication for wireless charging
	ISO 19363: safety requirements and charge controller requirements for vehicle side charging interface
Communication	ISO 15118: communication interface between vehicle and charging infrastructure and also for wireless charging
	ISO 15118-3: requirements for the physical layers and data link layers for wired communication
	ISO 15118-8: requirements for the physical and data link layers for wireless communication
	ISO 15118-2: requirements for grid and application protocols for V2G

Source: German National Platform for Electric Mobility (NPE) (2017).

10.4.2.3 Germany

In Germany, installation of private charging points has the following requirements,

- There is an obligation for all users to register their charging unit with their network operator.
- Each connection point must have its own residual current device (RCD)
- EV chargers connected to the German low-voltage grid have to comply with VDE-AR-N 4100 standard.

Germany has also released a roadmap of standards and specifications, which includes standards for EV charging interface (both wired and wireless) and information and communication technology (Rather et al. 2021b) (German National Platform for Electric Mobility (NPE) 2017) (Tables 10.4 and 10.5).

10.5 Impacts of EV Charging

10.5.1 Impact on Electricity Distribution Network

The energy requirement of an EV which determines the charging behavior of the user varies from user to user based on the average distance traveled. Moreover, depending on the travel schedule, the time of charging requirement will vary. Though EV users can potentially save on energy expenses if the EV charging is primarily done during off-peak demand hours (time of use [ToU] tariffs), it may not always be feasible for the user due to their travel requirements. So, the challenges related to EV charging on the network stem from primarily two different reasons,

- **The temporal variability of EV charging:** As the EV load is a new type of load for the grid operator, it is difficult to accurately forecast the charging load. The difficulty is further amplified due to the different travel behavioral patterns of EV users. Also, the charging load demand is different for residential chargers and public charging stations. Therefore, depending on whether the distribution network has a public charging station, the daily load curve would change.
- **Charging characteristics of the EV:** With the continuous development of high energy density batteries, the charging power of the batteries has also steadily been increasing. However, the actual charging power of the EV is dependent on other factors too, such as AC/DC charging,[4] the State of Charge (SoC) level of the battery, the rated charging capacity of the EV on-board charger (if AC charging), the rated charging capacity of the EVSE.

Due to the above-mentioned factors, the resulting impact on the distribution network due to the addition of EV loads is presented here.

10.5.1.1 Voltage Issues

High penetration of EVs can potentially lower the voltages at the distribution network, resulting in imbalance in the three phases of the network or even leading to voltage instability. Some of the main voltage-related issues are listed here.

Voltage Sag Distribution networks, generally due to the high resistance to inductance (R/X) ratios of the distribution lines, are more susceptible to voltage sags due to high power drawn which may even breach the nominal voltage operating zones. As EV charging entails a higher power demand compared to other residential loads, high penetration of EV will significantly increase the power demand in low voltage grids, which can potentially lead to undervoltage issues. In addition to this, EV charging may coincide with other loads in the system, as EV users are likely to charge their vehicles in the evening period when they return to their homes. This coincidence between EV charging period with the peak demand period may further aggravate the voltage sag issue. As given in Eq. (10.1), as the current flowing

---

4 Even though the vehicle may have DC charging capability, the charger that the EV is connected to may be an AC charger. So, having an EV with DC charging capability does not guarantee that charging would always be DC and vice-versa. Also, as AC and DC charging would both have different charging power in the same EV, so by charging the EV in either mode would add a different amount of load in the network.

through the feeder increases due to higher load demand, the receiving end voltage V_r is lowered from the sending end voltage V_s due to voltage drop in the impedance of the feeder:

$$V_r = V_s - IX_l \tag{10.1}$$

Voltage Stability Issues The relation between active power and voltage of a bus is represented by the power-voltage curve as shown in Figure 10.4. It signifies the voltage change with increasing active power. Based on the line resistance and reactance, each bus has a critical voltage that corresponds to the maximum active power that can be drawn from the bus, and any further increase in load at the bus will lead to voltage collapse. The ratio of change in voltage due to change in active power drawn from the bus is termed as Voltage Sensitivity Factor (VSF) (Deb et al. 2018). A high VSF means that even for small changes in active power drawn, there is a large drop in voltage at the bus and vice versa. Therefore, systems that are operating at high VSF regions are more susceptible to reach instability. With the addition of higher amount of loads into the system would move the normal operating point of the feeder toward higher VSF, without any additional reactive power support.

Phase Imbalance Unique to India, the 2 and 3 W EV sector has seen a massive growth and is expected to dominate national sales. However, these 2 and 3 W are generally charged using single-phase chargers. If these chargers are not uniformly

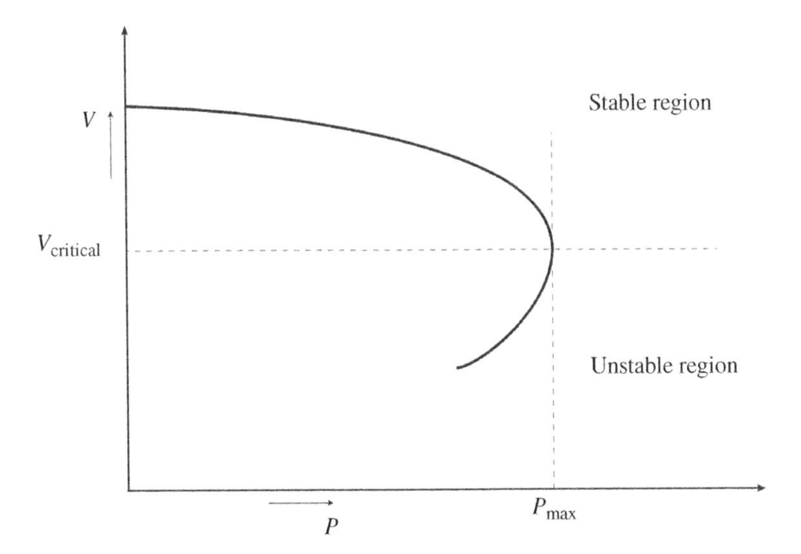

Figure 10.4 Typical power vs. voltage curve. Source: Deb et al. (2018)/MDPI/CC BY 4.0.

distributed among the phases of a distribution feeder, it may lead to unbalanced phase voltages and current loading. Unbalanced operation of the network results in higher losses in the network, voltage issues and is detrimental to overall system health. For example, if power demand in one phase is higher than the other two phases, there will be voltage drop in the phase with high demand, while potentially leading to overvoltage in other two phases, due to shifting of the neutral point in a three-phase four-wire distribution system (Weckx and Driesen 2015).

10.5.1.2 Increase in Peak Load

Uncontrolled and noncoordinated EV charging could increase the peak load on the grid if the time of charging of the vehicles coincides with the existing peak demand of the feeder. If a larger share of EV users start charging their EVs during peak demand periods, it would lead to significant burden in the distribution system, which could contribute to feeder congestion, potential overloading of the transmission system, and distribution network assets like transformers, cables, etc. This extra load would also lead to increase in generation and may also increase the electricity price (Klettke et al. 2018). It could also stress the system with the increased requirement of ramp limits.

10.5.1.3 Congestion

Since coincidence of EV charging with the network peak load, may further stress the system, the increased EV load can result in overloading of different assets of the distribution network, such as distribution transformers, cables, fuses, etc. Such overloading can significantly reduce the lifespan of this equipment, while simultaneously reducing the efficiency of the energy transmission system.

10.5.1.4 Losses

Power losses in the distribution network generally refer to the I^2R losses of the overhead lines/underground cables. Therefore, total power loss in the system is given by

$$P_t = \sum P_i = \sum I_i^2 R_i \tag{10.2}$$

where P_i is the loss of each line, I_i is the current flowing through each line, and R_i is the resistance of each line. So, with added current flowing through the lines due to the extra EV charging load, the loss in the system also increases making the supply of power less efficient. Besides, unbalanced loading can lead to uneven losses among the three phases. The amount of losses in the system increases particularly during peak periods.

Case Study: Impact of EV Charging on an Ideal Distribution Feeder The following case study has been carried out to determine the impact of EV charging on an ideal distribution feeder. As shown in Figure 10.5, the distribution feeder has a voltage

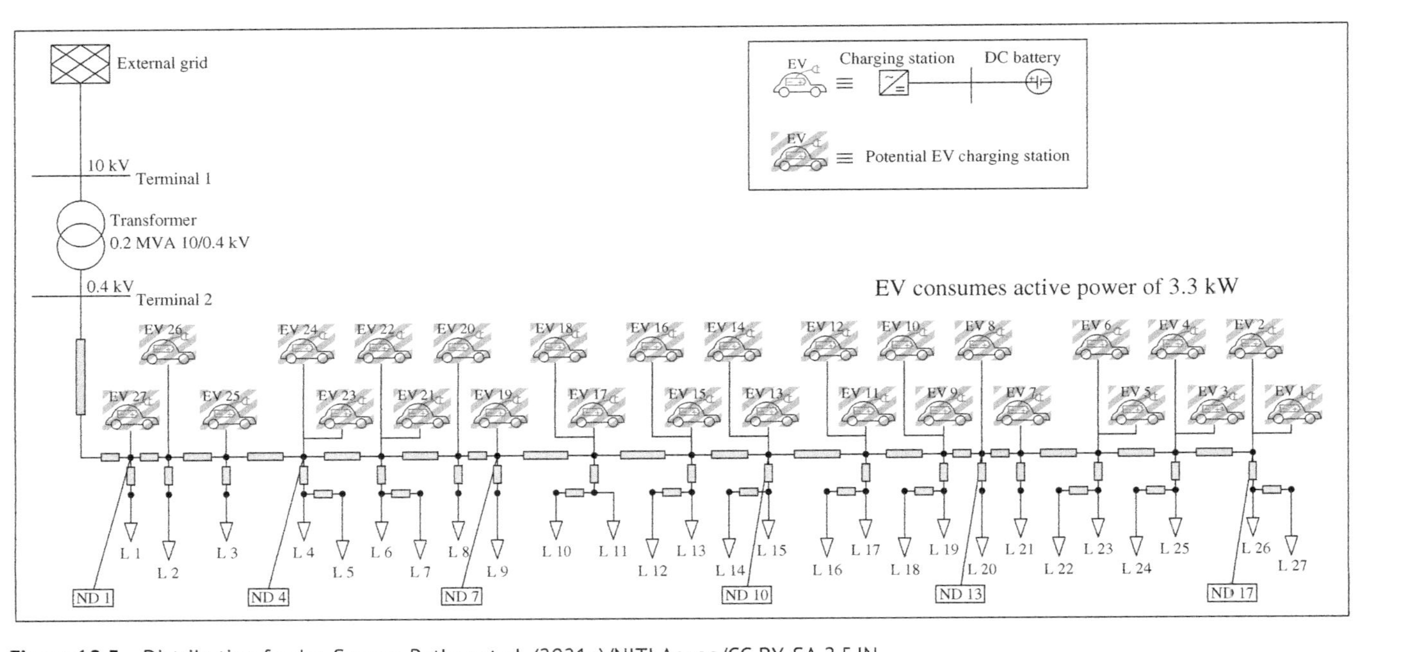

Figure 10.5 Distribution feeder. Source: Rather et al. (2021a)/NITI Aayog/CC BY-SA 2.5 IN.

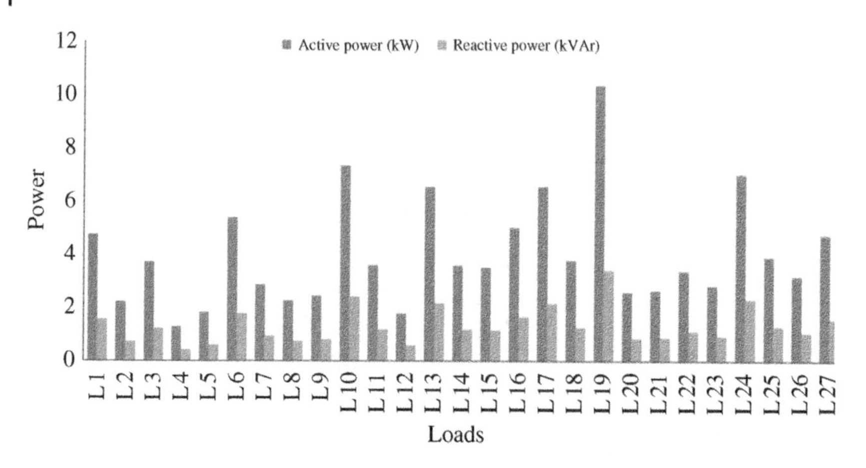

Figure 10.6 Static loads in the feeder. Source: Rather et al. (2021a)/NITI Aayog/CC BY-SA 2.5 IN.

level of 0.4 kV and is connected to the external grid through a 0.2 MVA 10/0.4 kV transformer. There are 27 different residences along the feeder with each residence having an EV. The active and reactive power drawn by the connected loads of the feeder have been given in Figure 10.6. All the EVs are considered to have a battery of 24 kWh. The residential charging option has been assumed to be a three-phase 11 kW AC charger.

Using the above-mentioned assumptions, the impact on the feeder voltage, hosting capacity, and power losses have been analyzed under different EV charging load levels.

To monitor the impact of EV penetration on the distribution feeder, the feeder voltage has been measured at different nodes along the length of the feeder. The location of these nodes has been shown in Figure 10.7, which shows the variation of voltages at the different nodes of the feeder when the number of EVs charging simultaneously have been increased from 0 to 27. When no EVs were charging the voltage at the start of the feeder (ND 1) had the highest voltage level at around 1.008 pu and the voltage at the rear end of the feeder (ND 17) had the lowest voltage level at 0.98 pu. The voltage along the entire feeder length went further down as the number of EVs charging simultaneously was increased, with a higher drop in voltage seen for ND 17 than for ND 1.

Next, the impact of voltage at ND 17 is monitored, for two different EV connection orders under different penetration levels. In the first case, the EVs are connected from the rear end of the feeder, i.e., the first EV is connected to the final node in the feeder (ND 17), the second EV is connected to the penultimate node (ND 16) and so on and the final EV is connected to the starting node of the feeder (ND 1). In the second case, the EVs are connected from the start of the feeder, i.e.

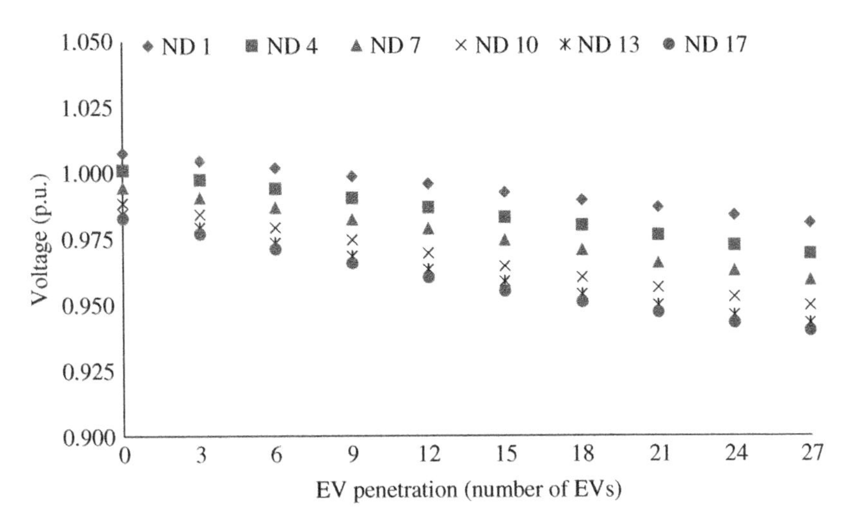

Figure 10.7 Feeder voltage at different nodes under different EV penetration levels. Source: Rather et al. (2021a)/NITI Aayog/CC BY-SA 2.5 IN.

the first EV is connected to ND 1, the second EV is connected to ND 2, and so on until the final EV is connected to the final node. For both cases, the voltage at ND 17 is monitored to check if there is any difference in feeder voltage if the order of EV loading is altered.

As seen in Figure 10.8, if no EV is connected then the voltage at ND 17 is the same for both cases. But as the number of EVs increases, the voltage is worse for

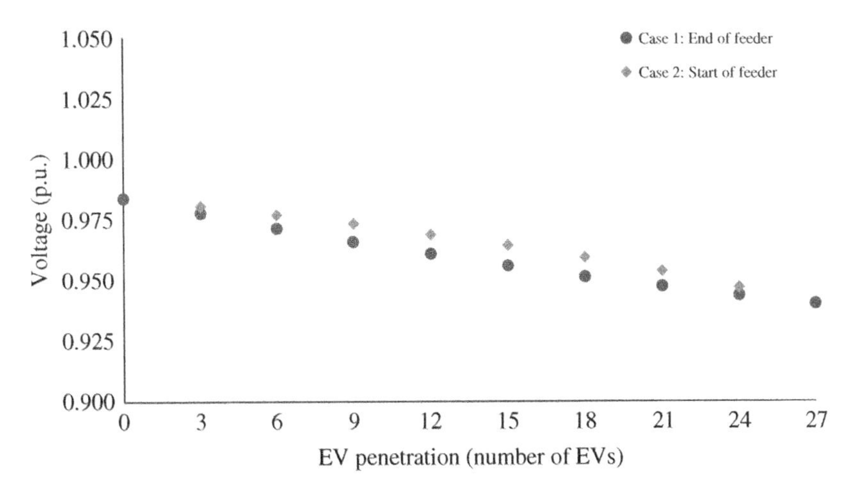

Figure 10.8 Voltages at ND 17 for different levels of EV penetration. Source: Rather et al. (2021a)/NITI Aayog/CC BY-SA 2.5 IN.

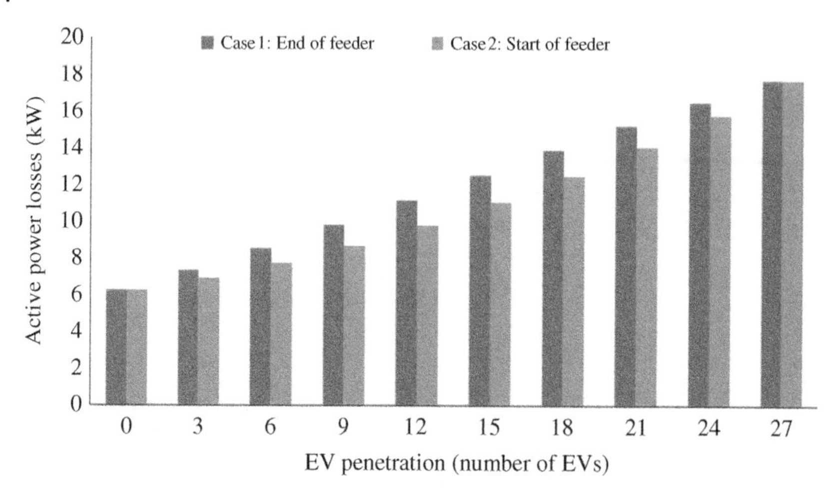

Figure 10.9 Total active power losses for different penetration levels of EVs. Source: Rather et al. (2021a)/NITI Aayog/CC BY-SA 2.5 IN.

case 1, where the EVs are connected from the end of the feeder than for case 2. This is because, when the EVs are connected from the end of the feeder, the current drawn flows through the entire length of the feeder, which increases the voltage drop across the feeder. If the EVs are connected from the start of the feeder, then no EV charging current flows through the feeder, due to which the voltage drop is reduced.

The active power losses due to increasing EV penetration in the feeder have been shown in Figure 10.9. Here, the active power loss is more for cases where the EV loading is done for the tail end of the feeder which can be attributed to the increase in the length of line, i.e. higher resistance, that the current has to flow through, which increases the I^2R losses.

10.6 Smart Charging for Home Charging

Also referred to as dumb charging, in uncontrolled charging, the EVs start charging as soon as the user plugs-in their EVs to the charger. However, in smart charging, EV charging is controlled, either by controlling the charging power or shifting the time of charging. The scheduling of charging is done with consideration of residential load demand, energy cost, grid parameters/constraints as well as the needs of the EV owner. Since smart charging involves controlling the charging based on external parameters, it requires observability and communication between the different entities. The different stakeholders/entities generally involved in smart charging include,

- EV user
- Distribution system operator (DSO)
- Charging management system (CMS)
- Aggregator

Of the different stakeholders, the CMS schedules the charging based on the objectives of the EV user's transportation needs, DSO requirements, and cost savings maximization for the EV user. A standardized communication infrastructure establishes the necessary communication requirements for reliable information flow between the different involved entities and devices, while the smart meter ensures that the financial aspect of employing smart charging is properly accounted for.

The objective of smart charging varies according to stakeholders and their requirements. Minimization of charging cost, maximization of user's travel requirements, and load leveling are the common objectives from EV owner's perspective. Overall system cost minimization, demand response, maximization of renewable energy (RE) utilization, and providing grid support services are the major objectives of the system operator. Smart charging is performed using different strategies which are the combination of information flow and decision-making ability between aggregator and EV owner.

The benefits of smart EV charging for a residence include, but are not limited to,

- Deferring of maximum contract demand for the household, thereby saving in capacity upgradation cost.
- Minimization of EV charging cost
- Maximization of use of local energy generation resources
- Cost savings by participating in demand response and ancillary services for the grid if used as Unidirectional controlled charging (V1G).
- Addition revenue opportunities by participating in demand response and ancillary services for the grid if used as Bidirectional controlled (V2G) charging

10.6.1 Type of Smart Charging

Depending on the level of control over the charging power, smart charging can be categorized into different types as given in Table 10.6 and Figure 10.10.

The utilization of ToU tariffs is a passive smart charging methodology. Here, the EV charging is not directly controlled by the CMS/home energy management system, instead, the EV users are incentivized to adapt their charging behavior, by charging their vehicles during off-peak periods during which the price of electricity is generally lower. In *dynamic pricing*, the price signals have higher temporal

Table 10.6 Different types of smart charging.

Description	Control over charging power	Possible uses	Maturity	Examples
Basic control	On/off	• Grid congestion management • Voltage Support • RE integration • Demand response	Partial market deployment	• My Electric Avenue commercial pilot in the United Kingdom
Unidirectional controlled (V1G)	Increase and decrease in real-time the rate of charging based on external control signals or time-based tariffs	• Grid congestion management, • RE integration, • Voltage Support, • Demand Response, • Ancillary service	Partial market deployment	• Charging of EVs controlled based on availability of green energy by Jedlix, Netherlands • Electric Nation smart charging commercial trial in the United Kingdom • AgileOctopus dynamic half-hourly tariff in the United Kingdom (beta product)
Bidirectional V2G	Reaction to grid conditions or response to control signal from the energy management system or time-based tariffs	• Grid congestion management, • RE integration, • Voltage support, • Demand response, • Ancillary service	Partial market deployment	• Commercial pilot in Denmark for use of V2G services in providing frequency regulation service • Blue Bird Electric Buses used V2G technology to earn revenue by energy arbitrage in Illinois, USA
Bidirectional V2X (V2H/V2B)	Integration of bidirectional charging and home/building energy management systems	• Behind-the-meter optimization • Micro-grid optimization	Partial market deployment	• V2H functionality is being provided by select EV models released (or soon to be released) by Ford, GMC, Porche

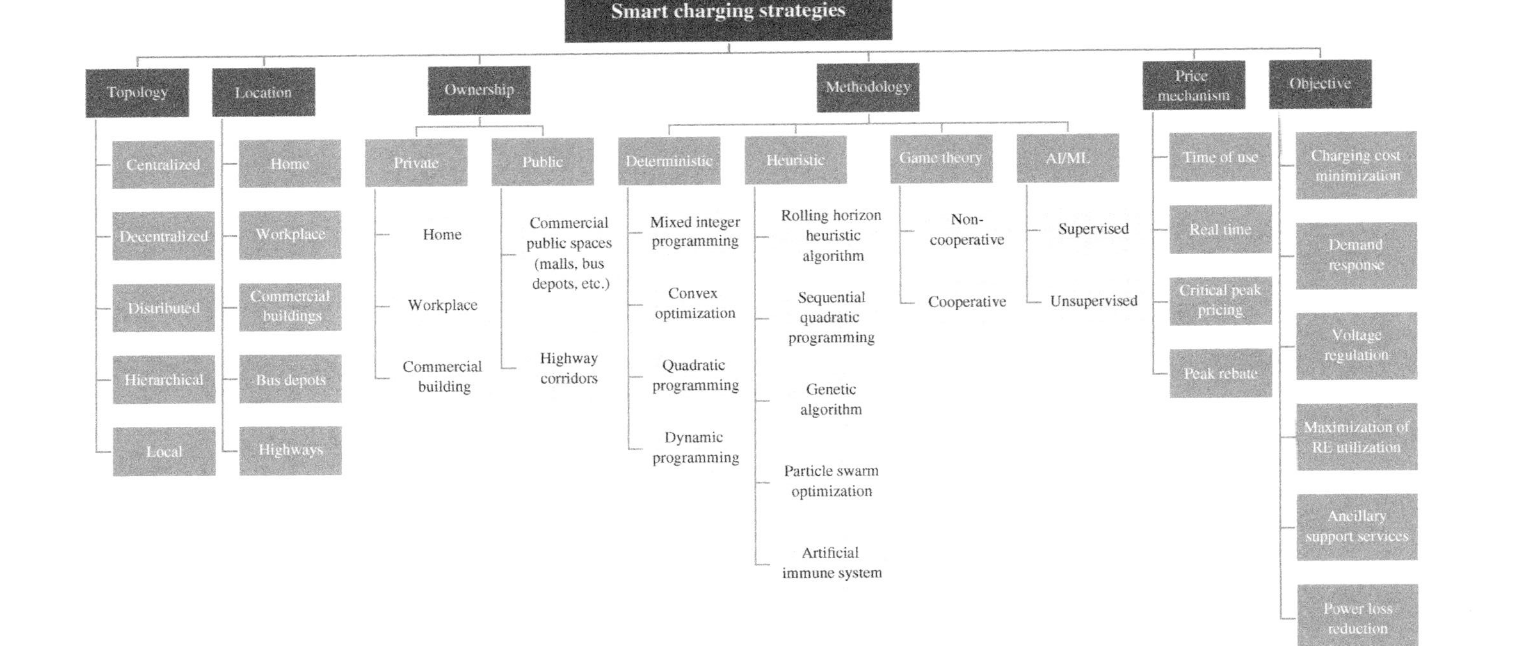

Figure 10.10 Classification of smart charging strategies. Source: Adapted from Rather et al. (2022).

resolutions (15 minutes/30 minutes/1 hour, etc.) which enable finer control over the selection of the optimal charging period.

In *basic control*, there is no fine control over the charging power of the EV, instead depending on the loading of the distribution transformers/unwanted grid operation parameters/total load of the household, the charging of the vehicles can be either switched on/off. This control is automated by a centralized CMS.

In *unidirectional controlled (V1G) charging*, the charging power is increased/decreased depending on different grid conditions based on the communication signal from the CMS. *Bidirectional vehicle-to-grid* (V2G) and grid-to-vehicle (G2V) are similar to V1G, however here, apart from controlling the charging power consumption, power can be even fed back to the grid. However, implementation of V2G is restricted to only EV models and EV chargers that are capable of providing bidirectional charging capability. Similar to V2G, *vehicle-to-load/building/home* (V2X) can be utilized for behind-the-meter optimization as well as optimization of small micro-grid.

10.6.2 Requirements for Smart Charging

To enable smart charging there are technical and regulatory requirements as well as definition of the roles of the different players involved.

On the technical front, smart charging requires chargers with smart charging capability that are able to regulate the charging power based on some signals from the CMS, which needs a well-developed communication infrastructure (hardware infrastructure as well as well-defined interoperable communication protocols). There is also a need for smart meters for accurate billing of energy consumption. From the regulatory perspective, there also needs to be adequate regulations that help enable smart charging.[5] These regulations include both technical as well as commercial market regulations. The technical regulations need to specify the minimum functionalities needed in the charging points prior to their installation. The commercial regulations need to specify the specific electric market products and services that can be provided by the EVs and their payment structure. These products may include demand response services, ancillary services, etc.

Further, there need to be fiscal incentives so that the users participating in smart charging are properly remunerated (by minimization of charging costs in case of unidirectional charging). The different requirements for smart charging are given in Table 10.7.

---

5 The Government of UK published an electric vehicle smart charge points regulation on 3 February 2022, mandating that all private charge points sold in the UK sold after 30 June 2022, would need to comply with smart charging functionality as mentioned in the regulation.

Table 10.7 Requirements to enable residential smart charging.

Technical requirements	Hardware	Private smart charging points
		Smart meters
		Smart grid infrastructure for grid observability
	Software	An energy management software that runs the algorithm to implement smart charging by taking real time inputs from the EVs, residential load, tariff, and the grid condition (for more details please refer to Chapter 6)
	Information and communication technology (ICT)	Interoperable communication protocols for communication across different entities. For example, OpenADR (Chapter 5)
		Interoperable standards for communications including hardware requirements
Regulatory requirements	Electricity market	EVs through aggregators should be allowed to participate in the electricity market
		Create fiscal incentives to incentivize smart charging (lower charging cost for unidirectional smart charging and revenue earning for bidirectional smart charging)
	Distribution system	Time-based EV tariff needs to be promoted.
Stakeholders roles	State institutions	Create sponsored projects to kickstart smart charging
		Help in financing the projects
	E-mobility market	Incentivize customers to participate in smart charging programs through different schemes
		Make smart chargers easily accessible to EV users

10.6.3 Additional Smart Charging Enablers

Apart from the requirements mentioned in Section 6.2, there are some additional enablers for smart charging as given below:

1. **Consumer behavior:** The success of Smart charging will be determined by the interest of EV consumers to participate in smart charging solutions. This may vary among EV users, as some users can be technology enthusiasts willing to try and test new technologies, while a portion of the users may be skeptical of participating in smart charging either due to range anxiety, cybersecurity concerns, or for economic reasons. The active participation of users will be possible if there are incentives for the users to participate in smart charging.
2. **Big data and artificial intelligence:** Smart charging is a data-intensive service, where large amounts of data from the entire fleet of EVs and the

measurements from the distribution network need to be properly analyzed so that the accurate state of the system may be painted in real-time. The service also needs to consider the complexity of electricity markets and participate in these markets based on the current state of the system. Big data analytics will thus play a significant role in the proper management of smart charging solutions. Along the same lines, artificial intelligence tools will become necessary for complex decision-making considering a myriad of objective functions and constraints. Digitalization will play a key role in the optimization between transport services and grid services. Data analytics will enable matching the mobility demand with the power supply and load patterns and further help determine optimal charging point locations.

In the electrical network, the generation of electricity has to be equal to the load demand for each instant. However, integration of stochastic generation resources have introduced variability in the energy generation. Similarly, the introduction of EV which are comparatively high power loads en masse will create stochasticity in the loads. The stochasticity of generation and loads necessitates the use of tools that can identify underlying trends and patterns to help reduce such stochasticity. Artificial intelligence and machine learning can significantly help in this regard, as these can help study the charging behavior of users and help make better predictions on the expected load pattern.

For example in the case study presented in Section 5.1, AI/ML can be used to determine the charging pattern of the EVs by studying the user behavior, the time of day, and other relevant parameters. The CMS can use the expected charging profile of the different EVs in the network to create an optimized charging schedule that minimizes the grid impact without severely affecting the grid. If a CMS is deployed to manage the charging of the EVs in the particular feeder and the CMS is based on an AI-based algorithm, then over time, the CMS would be able to predict the expected load at different nodes in the feeder by learning about the consumers charging behavior, i.e. the load profile at the different nodes (Node_ID). Also, the algorithm would identify the impact of loading at different nodes on the voltage profile of the feeder. Thus, the CMS can then negotiate with the EV charger to control the charging such that the impact on the grid is minimized without drastically hampering the user's needs. If the same CMS is then used for another feeder with different load and network characteristics, the AI-based CMS would automatically adapt to the new feeder. This reduces the need for developing custom smart charging algorithms for different feeders even if the network characteristics have changed.

3. **Blockchain technology:** Blockchain is an open, distributed, digital ledger that can record transactions between two parties. As more and more EVs are entering the market, the transactions between the different stakeholders have seen a sharp rise, with increasing quantities of data being part of these transactions. Using blockchain technology can facilitate smart charging and V2G by connecting different parties and facilitating monetary transactions between aggregators and customers through open-source standards, replacing the propriety standards of today (IRENA 2019).

10.7 Residential Smart Energy Management

EV charging is a power-intensive application and the addition of EV charging can effectively double the power drawn by the household. For a residential user with a fixed contracted demand, the maximum demand due to addition of EV charger can breach this limit. In this scenario, the user is required to upgrade their connection, which would further increase the cost of installation of chargers. However, if a smart energy controller (SEC) is utilized the user can potentially save on the electrical connection upgradation requirement. This is just one example of how a residential energy management system can benefit the user. Depending on the application, smart charging in residences can be categorized into three broader categories,

- Unidirectional Smart Charging (V1G)
- Vehicle to Home/Building (V2H/B)
- Vehicle to Grid (V2G).

10.7.1 Unidirectional Smart Charging

In unidirectional smart charging (V1G) the charging of the EV vehicle is controlled based on some external parameters. Here energy only flows from the grid to the vehicle and not the other way around. V1G can be further categorized into,

- Tariff-based V1G (Passive control)
- Coordinated V1G (Active control)

Tariff-based V1G is dependent on the availability of a time-based tariff such as time of day (ToD), time of use (ToU), real-time pricing for the customers. If the local distribution utility/energy retailer facilitates the user with such tariffs (either specific for EV charging or for the overall residential load), the user can manually control the charging of EV during periods of low price, so that the overall cost of EV charging is reduced. A smart meter is however necessary to enable logging

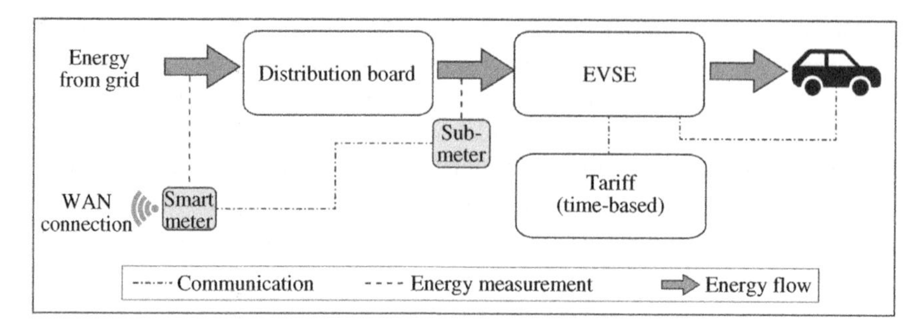

Figure 10.11 Architecture for time-based V1G (the sub-meter is optional based on the tariff structure). Source: Angshu Nath (co-author).

of energy usage with time (PG&E n.d.). In addition, a sub-meter would also be needed if the energy logging of the EV is to be done separately.[6]

The architecture for deploying time-based V1G in a residence is given in Figure 10.11. The charging of the EV is controlled by the user, based on the tariff, and the information on the energy consumption is relayed to the energy retailer through the smart meter.

Different energy retailers and distribution utilities are at different stages of provision of time-based EV tariffs to their user base. For example, Pacific Gas & Electric in California provides two different ToU-based tariffs, one specifically for EV charging and the other for the residential load along with EV charging (PG&E n.d.). Octopus Energy in the UK in addition to ToD tariffs also provides a dynamic real-time tariff with half-hourly pricing (Agile Octopus n.d.).

In *Coordinated V1G*, the charging of EV is controlled by an active load management device depending on different factors/applications such as

1. **To restrict the total demand of the residential property**: In this scenario, the SEC monitors and controls the operation of different smart devices in the property to restrict the total power drawn by the residence to its contracted demand. The SEC switches off low-priority devices, while keeping high-priority devices on. The user retains control over adjusting high and low-priority loads. The architecture to utilize this application is given in Figure 10.12. To achieve this mode 2, mode 3, or mode 4 EVSE is required.
2. **To minimize the cost of charging**: In this application, the primary goal is to minimize the cost of charging by shifting the EV charging to off-peak periods when the electricity tariff is lower. The integration of the smart meter with

6 A separate sub-meter is needed if a special EV tariff is being used.

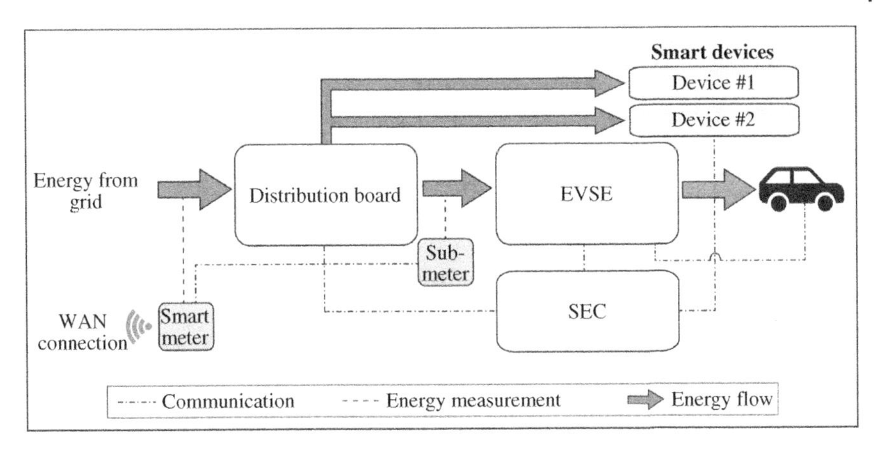

Figure 10.12 Architecture for demand limited V1G (the requirement of sub-meter and the presence of other smart devices is optional). Source: Angshu Nath (co-author).

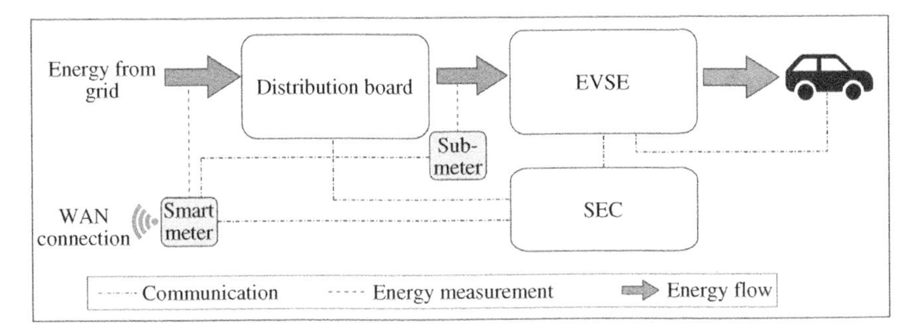

Figure 10.13 Architecture for minimization of cost of charging using V1G (the requirement of sub-meter is optional). Source: Angshu Nath (co-author).

the SEC allows the SEC to automatically respond to the time-based tariffs and schedule the EV charging accordingly. However, the user also needs to provide relevant inputs to the SEC on when the user expects to use their vehicle, as otherwise the vehicle may not be fully charged when the vehicle is plugged-out for usage. The architecture for this application is provided in Figure 10.13.

3. **To maximize the use of local renewable generation**: In this application, the user wants to maximize the utilization of local energy generation sources such as rooftop solar PV, wind generators, bio-gas plants, etc. This would reduce the user's dependency on the grid supply thereby helping in reducing the electricity bill of the residence. The architecture for this application is provided in Figure 10.14. Here, RE is prioritized to be used over grid supply and the optimum use is during peak periods, when tariffs are typically higher.

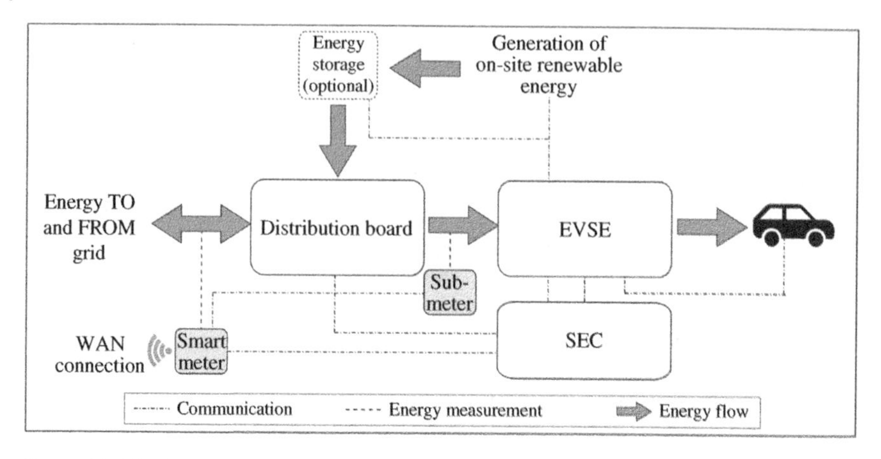

Figure 10.14 Architecture for optimization of local generation for self-use (the requirement of sub-meter and energy storage is optional). Source: Angshu Nath (co-author).

The above three applications need not be mutually exclusive of each other. The architectures can be tweaked to serve one or more of the applications simultaneously.

10.7.2 Vehicle-to-Home/Building

Vehicle-to-home (V2H) and vehicle-to-building (V2B) utilize the bidirectional charging capability of EVs to provide different advanced functionalities to the EV user. However, it is to be noted that these applications are only possible when both the EV and the EVSE are capable of bidirectional charging. Here, the EV can be used as a storage unit to feed power back to the house/building based on different use cases. The different use cases for V2H/V2B are

- To use the vehicle as storage and avoid consumption during peak periods.
- To increase the utilization of local energy generation.
- For power use beyond the contractual demand of the residence.
- To power an overall off-grid system (islanded operation).

Utilization of V2H/B can be considered as an extension of V1G as now in addition to just controlling the charging power of the vehicle, power can be also fed back from the V2H/B. This increases the amount of flexibility that can be provided by the EV.

The architecture for the utilization of V2H to avoid consumption during peak hours is provided in Figure 10.15. This architecture allows the EV to act as energy storage unit, it charges during periods of low tariff for subsequent consumption

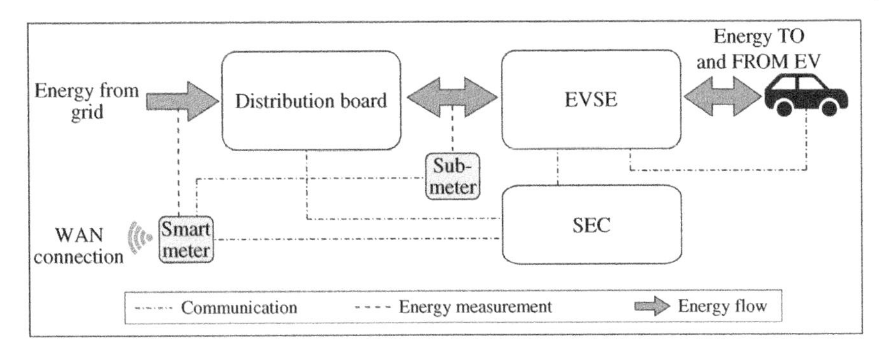

Figure 10.15 Architecture for using V2H/V2B to avoid consumption during peak periods. Source: Angshu Nath (co-author).

within the property when the tariff is much higher. Here the EV can also be used as a transporter of energy, i.e. the EV is charged at a different location at lower tariffs and then used to run the appliances at another location with higher tariffs. However, this architecture requires the integration of a sophisticated energy management tool to balance the needs of the EV and the property. The same architecture can also be extended to connect more loads than the contracted demand of the property. Here, the grid supply is utilized to power the loads up to the contracted demand, and the remaining loads are powered by the EV. Also, as the property is being powered by both the grid supply and the EV, so the EV should have the functionality to operate in parallel to the grid supply.

Figure 10.16 provides the architecture for using EVs for increased utilization of local energy generation. Similar to the architecture provided in Figure 10.14, the SEC monitors the generation of local energy and utilizes the EV battery as a storage unit to store this energy and use it for self-consumption or sell the energy at higher tariffs and earn revenue.

Using V2H/B it is also possible to provide backup power to the residence when there is no power available from the grid. Also referred to as islanded operation, this situation may arise due to,

- A power cut in the system.
- Designed as off-grid system.

The architecture and connection diagram for using the vehicle as a storage unit to power the residence during islanded operation are given in Figures 10.17 and 10.18, respectively. As shown in Figure 10.17, in the absence of supply from the grid, the SEC utilizes the energy stored in the EV and the generation from the on-site energy resource to supply power to the loads. In Figure 10.18, two separate circuits can be observed, one powered by the grid supply and the other by the

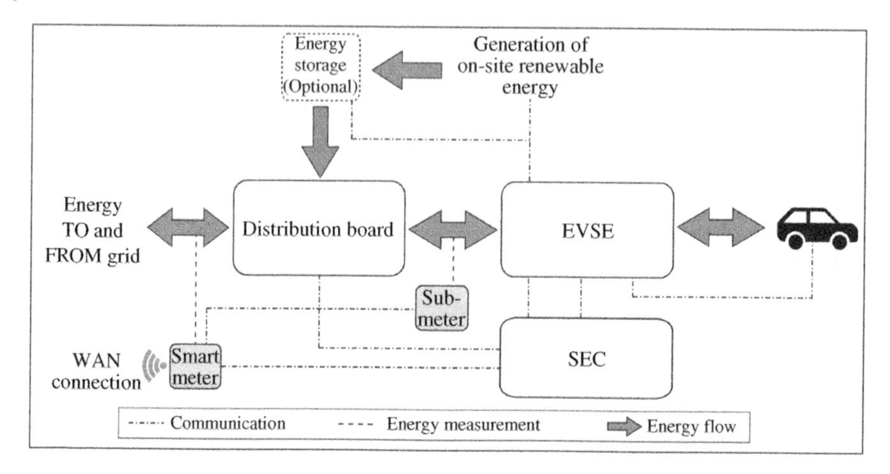

Figure 10.16 Architecture for using V2H/V2B for maximization of local energy generation usage. Source: Angshu Nath (co-author).

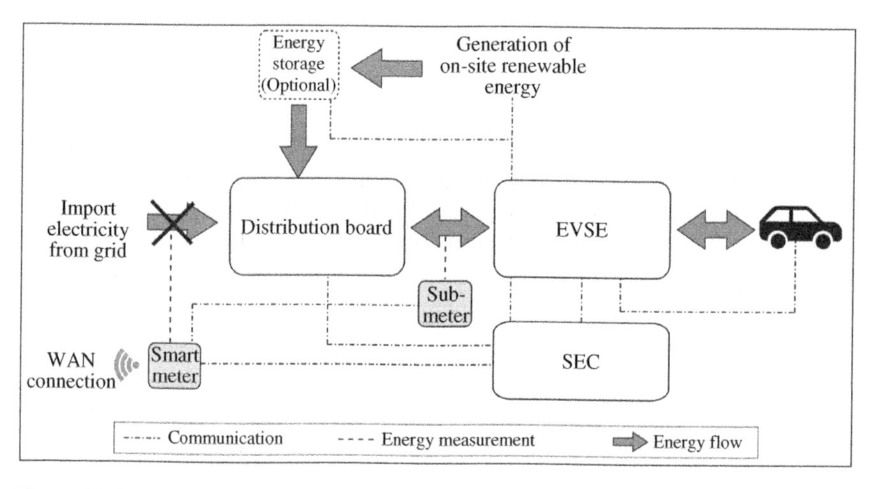

Figure 10.17 Architecture for using V2H/V2B in islanded mode. Source: Angshu Nath (co-author).

vehicle (with/without additional local stationary storage unit). If the grid is available, the grid powers the entire residence. However, if there is a power cut, i.e. the grid supply is no longer available, the backup isolator which is normally closed is automatically opened by the SEC. This isolates the circuit powered by the vehicle from the mains. This is mandatory as otherwise, the vehicle would try to power the entire energy grid.

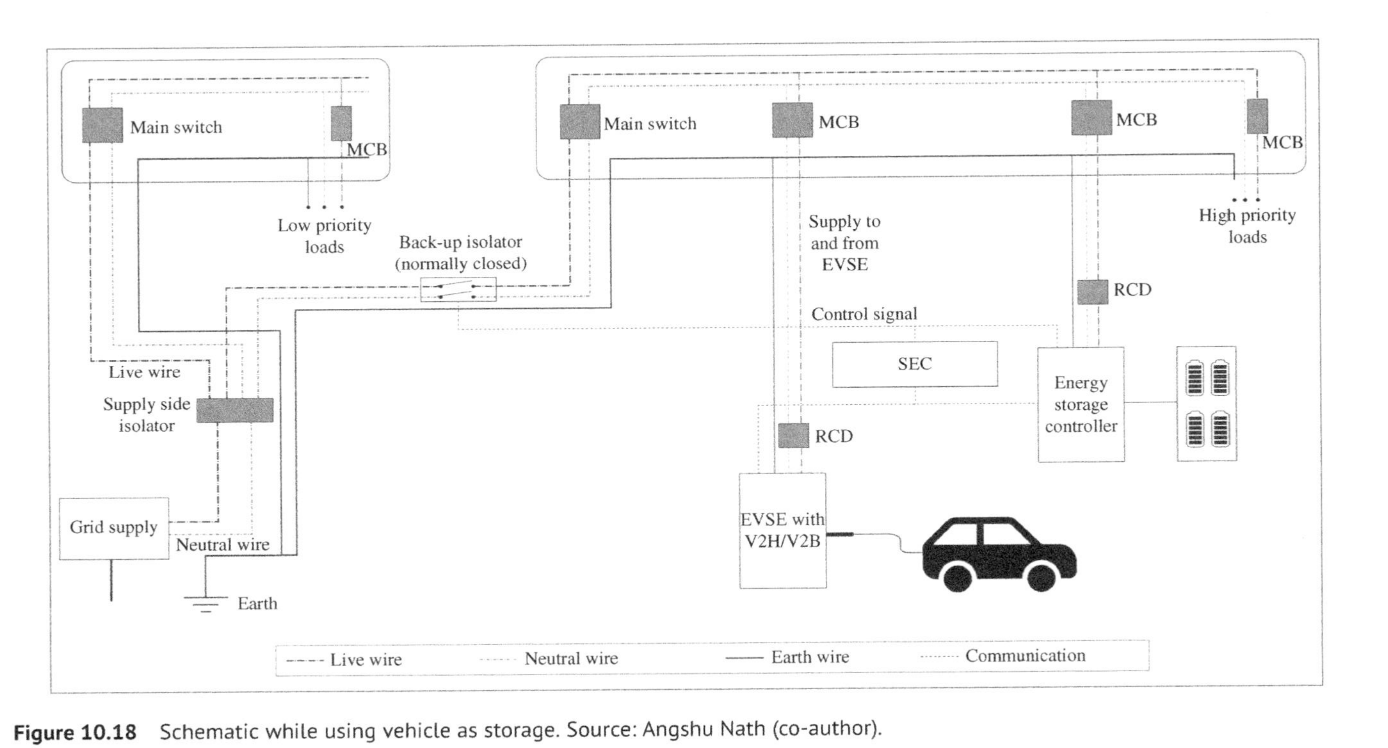

Figure 10.18 Schematic while using vehicle as storage. Source: Angshu Nath (co-author).

Although not widespread V2H/B capable EV and EVSE models are commercially available in the market today. This could change in the future as prominent vehicle OEMs have aims to provide these services to their new EV models. Some of the EV models that have V2H capability include the Nissan Leaf, Ford F-150® Lightning, and the Mitsubishi Outlander plug-in hybrid electric vehicle (PHEV).

10.7.3 Vehicle-to-Grid (V2G)

In V2G application, the vehicle is connected to the grid as a distributed energy resource is feeding energy back into the grid. V2G is one of the most advanced application of EVs and can be utilized to,

- Increase revenue earned by the EV user, by selling energy during peak periods.
- Providing grid support services such as congestion management and voltage support.
- Provision of grid ancillary services such as frequency support, ramping support, black start support, etc.

The architecture for V2G application has been provided in Figure 10.19. Here, mode 4 EV charging is utilized to export power from the vehicle via the property mains to the grid beyond the point of common coupling. This requires sophisticated smart metering and energy management integration. Further, the response needs to be type tested in accordance with the rules and regulations of the network operator. Licensing arrangements to export power would also be required from the network operator. Prior to utilization of V2G, it is also recommended to have certified protection schemes for bidirectional power flow to the grid in place.

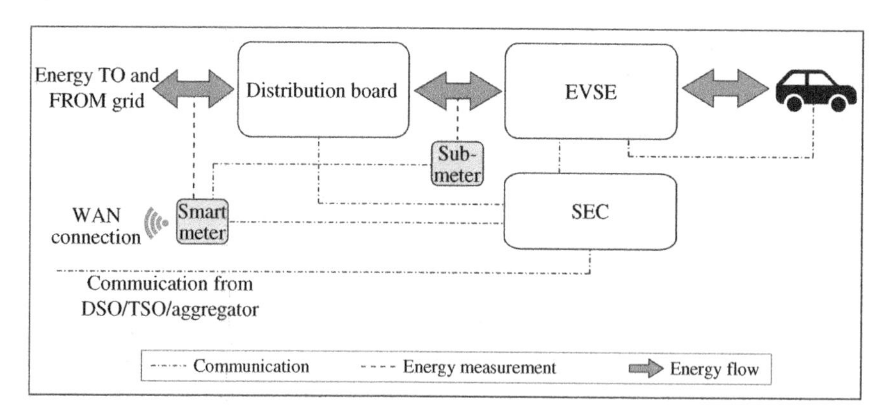

Figure 10.19 Architecture for V2G application of EV. Source: Angshu Nath (co-author).

10.8 Conclusion

Although the electrification of transportation is still in the early stages of market transformation, it has the potential to significantly reduce emissions both from the transportation and the energy sector. While public charging infrastructure is important for the mass market adoption of EVs, private residential EV charging infrastructure is equally crucial for the EV charging economy and usability. Different energy network operators around the globe have developed regulations for the smooth integration of residential chargers into the electrical network. In this chapter, different topologies and architectures for different use cases of residential smart EV charging were explored. Utilization of these architectures can provide a range of benefits to the EV user, ranging from lower cost of charging, and increased RE utilization to providing demand response services to the grid. Although not yet standardized across the globe, a coordinated approach towards making policies and regulations requiring the adoption of smart charging in EVs will act as a key enabler for seamless adoption of EVs and effective management of EV and conventional load of residential customers.

List of Abbreviations

AC	alternating current
CCS	combined charging system
CHAdeMO	CHArge de MOve
CMS	charging management system
CPO	charge point operator
DC	direct current
DER	distributed energy resources
DSO	distribution system operator
EREC	engineering recommendations
EV	electric vehicle
EVSE	electric vehicle supply equipment
G2V	grid-to-vehicle
HAN	home area network
ICCB	in cable control box
IEC	International Electrotechnical Commission
IEEE	Institute of Electrical and Electronics Engineer
IP	Internet Protocol
ISO	International Organisation for Standards
MCB	miniature circuit breaker
OCPP	open charge point protocol
OEM	original equipment manufacturer

OpenADR	open automated demand response
OSCP	open smart charging protocol
PHEV	plug-in hybrid electric vehicle
RCD	residual current device
SAE	Society of Automotive Engineers
SEC	smart energy controller
SoC	state of charge
ToD	time of day
ToU	time of use
V1G	unidirectional controlled charging
V2B	vehicle-to-building
V2G	vehicle-to-grid (bidirectional controlled charging)
V2H	vehicle-to-home
V2X	vehicle-to-everything

Glossary

Ancillary services There are a wide range of additional services/functions beyond the usual generation and transmission operations that aid the continuous flow of power from generators to consumers. These are essential in maintaining the load-generation balance, stability, and security of the grid and constitute several services that provide frequency support, voltage support, and black-start support to the grid.

Charging management system Charging Management System schedules the EV charging among the EV chargers it is connected to, based on predefined objectives. They can be either embedded in a physical device or a cloud-based platform.

Contracted demand It is the demand (in kW/kVA) agreed between the consumer and the utility engaged in providing electricity.

Electricity market The electricity market is an entity that enables the trading of electricity between energy generators and energy consumers

EVSE Electric Vehicle Supply Equipment (EVSE) also known as charging station/charging dock is a device connected to the electrical network and supplies electrical energy to the electric vehicle for charging purposes.

Grid observability A node or a substation in the grid is termed to be fully observable if the voltage and the current flows across the node are known through appropriate measurements. Consequently, a particular electricity grid is said to be completely observable if all its constituent substations are observable through direct or indirect measurements, e.g. through PMUs installed on a selected set of nodes.

MCB Miniature Circuit Breaker (MCB) is an electrical safety device that is designed to protect electrical circuits from damage caused by overcurrent or short circuits.

Off-peak demand hours The hours in a day during which the load in the network is typically the lowest are called off-peak demand hours.
Peak demand hours The hours in a day during which the load in the network is typically the highest are called peak demand hours.
Ramp limits In power system, ramp limits refer to the change in active power requirement per unit of time to maintain system stability.
Range anxiety It is the EV user's concern that the battery in the electric vehicle will drain out prior to the user reaching their destination.
RCD A residual current device (RCD) is an electrical device designed to provide protection against leakage current. The RCD detects imbalance of the electrical flow and provides protection against earth faults.
Smart meter A smart meter is an electronic device that records information on the energy usage of the property and communicates the information to the consumer as well as the local electricity retailer.
State of charge This is the amount of energy currently stored in the battery expressed in terms of percentage of the rated energy capacity of the battery.
Sub-meter A sub-meter allows a user to separately meter and bill an individual or group of loads. In electric vehicle context, if the electricity tariff for energy use for EV charging is different from the general load, a sub-meter for the EV would be required for its billing.
Temporal variability It refers to variability in time.

References

Agile Octopus (n.d.). Octopus energy. https://octopus.energy/agile/?utm_medium=website&utm_source=tarrif&utm_campaign=agile (accessed 6 January 2023).

Klettke, A., Moser, A., Bossmann, T. et al. (2018). Effect of electromobility on the power system and the integration of RES (METIS Studies) - S13 Report. B-1049 Brussels: European Commission. https://ec.europa.eu/energy/sites/ener/files/documents/metis_s13_final_report_electromobility_201806.pdf (accessed 6 January 2023).

Blonsky, M., Munankarmi, P., and Balamurugan, S. (2021). *Incorporating Residential Smart Electric Vehicle Charging in Home Energy Management Systems: Preprint*. Golden, Colorado: National Renewable Energy Laboratory.

Cazzola, P., Gorner, M., Scheffer, S. et al. (2018) Nordic EV Outlook 2018: insights from leaders in electric mobility. International Energy Agency. https://webstore.iea.org/nordic-ev-outlook-2018 (accessed 6 January 2023).

ChineseStandard.net (2016) GB/T 20234.1, connection set for conductive charging of electric vehicles - Part 1: general requirements. National Standard of the People's Republic of China. https://www.chinesestandard.net/PDF.aspx/GBT20234.1-2015.

Chuan, L. and Ukil, A. (2015). Modeling and validation of electrical load profiling in residential buildings in Singapore. *IEEE Transactions on Power Systems* 30 (5): 2800–2809. https://doi.org/10.1109/TPWRS.2014.2367509.

Deb, S., Tammi, K., Kalita, K. et al. (2018). Impact of electric vehicle charging station load on distribution network. *Energies* 11: https://doi.org/doi:10.3390/en11010178.

German National Platform for Electric Mobility (NPE) (2017). *The German Standardisation Roadman ELectric Mobility 2020*. The Federal Government's Joint Office for Electric Mobility.

IEC (2017). *IEC 61851-1: 2017: Electric Vehicle Conductive Charging System - Part 1: General Requirements*. International Electrotechnical Commission.

IRENA (2019). *Innovation Outlook: Smart Charging System for Electric Vehicles*. Abu Dhabi: International Renewable Energy Agency.

Liander (2021) A suitable connection for electric driving. Liander. https://www.liander.nl/consument/aansluitingen/elektrisch-rijden.

MacLeon, M. and Cox, C. (2018). V2G Market Study–Answering the preliminary questions for V2G: What, where and how much? CENEX. https://www.seev4-city.eu/wp-content/uploads/2018/08/V2G-Market-Study-2018.pdf.

Neaimeh, M. and Andersen, P.B. (2020). Mind the gap- open communication protocols for vehicle grid integration. *Energy Informatics* 3 (1): https://doi.org/10.1186/s42162-020-0103-1.

OpenADR Alliance (n.d.) Open ADR 2.0: demand response program implementation guide. https://www.openadr.org/assets/openadr_drprogramguide_v1.0.pdf.

PG&E (n.d.) Making sense of the rates, Pacific Gas and Electric Company. https://www.pge.com/en_US/residential/rate-plans/rate-plan-options/electric-vehicle-base-plan/electric-vehicle-base-plan.page? (accessed6 January 2023).

Rather, Z., Banerjee, R., Nath, A.P. et al. (2021a). *Fundamentals of Electric Vehicle Charging Technology and its Grid Integration*. GIZ http://www.niti.gov.in/sites/default/files/2021-09/Report1-Fundamentals-ofElectricVehicleChargingTechnology-and-its-Grid-Integration_GIZ-IITB.pdf.

Rather, Z., Nath, A.P., Banerjee, R. et al. (2021b). International review on integration of electric vehicles charging infrastructure with distribution grid. In: *GIZ*. Available at: http://www.niti.gov.in/sites/default/files/2021-09/Report2-International-Review-on-Integration-of-Electric-Vehicles-charging-infrastructure-with-distribution-grid_GIZ-IITB-compressed.pdf.

Rather, Z., Dahiwale, P. V. Lekshmi, D. et al. (2022). *Smart Charging Strategies and Technologies for Electric Vehicles*. GIZ.

Weckx, S. and Driesen, J. (2015). Load balancing with EV chargers and PV inverters in unbalanced distribution grids. *IEEE Transactions on Sustainable Energy* 6 (2): 635–643. https://doi.org/10.1109/TSTE.2015.2402834.

11

Induction Heating Appliances: Toward More Sustainable and Smart Home Appliances

Óscar Lucía, Héctor Sarnago, Jesús Acero, and José M. Burdío

University of Zaragoza, Department of Electronic Engineering and Communications. Instituto de Investigación en Ingeniería de Aragón, I3A, Zaragoza, Spain

Induction heating is nowadays the preferred heating technology in many areas due to its advantages in terms of efficiency, safety, and process quality. This is especially true in the domestic area, where it enables the design of more efficient, faster, safer, and more intelligent home appliances when compared with their counterparts based on resistive or gas technologies. In the current socio-economic context, induction heating technology provides a clear path for electrification in the residential area, leading to a reduced carbon footprint and a more sustainable development.

11.1 Introduction to Induction Heating

11.1.1 Induction Heating Fundamentals

IH is a contactless heating method based on applying an alternating magnetic field to the induction target to be heated (Figure 11.1) (Lucía et al. 2014; Davies 1990). Under these conditions, the induction target, depending on its physical properties, heats up by means of two physical phenomena: Eddy or Foucault currents and magnetic hysteresis. This induction target can be any surface to be heated and, in the case of domestic induction heating, the pan/pot is the IH target. It can be considered one of the earliest and most extended forms of wireless power transfer.

Energy Smart Appliances: Applications, Methodologies, and Challenges,
First Edition. Edited by Antonio Moreno-Munoz and Neomar Giacomini.

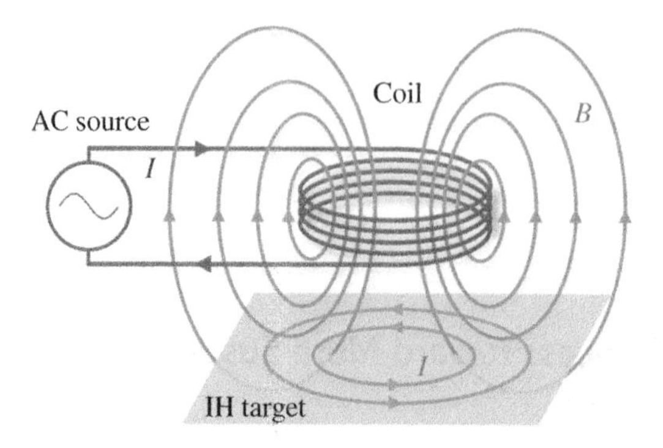

Figure 11.1 Induction heating fundamentals. Source: Guillen (2022).

Induction heating has been applied to many areas, from industrial applications, which were the seed for the success of this technology, to domestic applications later and, finally, biomedical applications. In all these areas, IH excels due to some fundamental benefits of this technology:

- **Process quality**: The contactless nature of IH makes the heating method less invasive than other technologies, avoiding cross-contamination and leading to more repeatable and higher-quality processes.
- **Safety**: The heating is directly created at the induction target, avoiding or minimizing heating in surrounding areas. Besides, no flammable fossil-fuel-based materials are used.
- **Efficiency**: IH directly heats the induction target, avoiding energy waste due to energy loss to the ambient or surrounding materials. Besides, advances in power electronics and electromagnetic design have made possible highly efficient power conversion, enabling the design of highly efficient IH systems.

These benefits make induction heating a key technology in the current path toward decarbonization through process electrification due to its benefits in terms of efficiency and the lack of direct fossil fuel usage (Clairand et al. 2019).

Although the origins of induction heating were linked to industrial applications, its benefits have allowed this technology to be introduced in millions of homes in the form of induction heating cookers (Figure 11.2). The benefits of this technology have allowed to outperform other heating technologies in some essential areas such as:

- **Fast heating**: High-power converters and inductors allow fast heating outperforming classical resistive cookers and reaching the same heating times as the gas best performers.
- **Efficiency**: Directly heating the pot minimizes the energy loss to the ambient and surrounding materials, maximizing efficiency, and reducing heating times.
- **Safety and cleanliness**: Significantly lower temperature surface avoids burns and accidents and significantly simplifies cleaning the cooking surface.

Figure 11.2 Domestic induction heating appliance: a technology enabler toward decarbonization through electrification. Source: Pérez-Tarragona et al. (2021)/from IEEE.

- **Accurate control**: Advanced control techniques allow for a more accurate cooking by precisely controlling the delivered power or even the target temperature. Optimized electromagnetic design leads to optimum heat distribution, and advanced computing capabilities enable further safety measures and improved appliance performance.

IH cookers have become one of the most energy-smart appliances due to their capabilities for high-efficiency power conversion and computing, which open the door for further developments and integrations. In terms of sustainability, IH has been pointed out as a paradigmatic example of the benefits of electrification in the domestic area. It outperforms in terms of heating efficiency to other classical heating methods such as resistive heating or gas, providing an environmentally friendly alternative and it has been recognized as a main player in an electrified future (Mai et al. 2018). As a significant achievement, the ENERGY STAR® Emerging Technology Award (ETA) has recognized residential induction cooking tops for the ETA 2021–2022 (Star 2021).

11.1.2 Induction Heating History

Since the origins of induction heating, the key enabling technologies and IH applications have evolved to map the current state of the art (Mühlbauer 2008). Figure 11.3 shows a summary of some of the important events in IH enabling technologies and domestic induction heating milestones (Lucía et al. 2018a, 2019, 2022).

The first references to domestic induction heating are patents from Mitchell (1892) and Berry (1909) where several domestic IH devices were proposed based on magnets and planar coils (Kennedy 1909). However, none of them were commercialized at that time. After that, most of the IH developments were focused on industrial applications, but enabling technologies were advancing, shaping a future for domestic applications. In 1949, induction heating was already highlighted as a promising technique in engineering for appliances (Scott 1949). It is especially important in the 1950s–1980s time frame, where several key technologies for domestic IH were developed, including power semiconductors, digital control techniques, and vitroceramic glass (Stookey 1958). It was especially remarkable in the year 1979, when insulated gate bipolar transistor (IGBT) technology (Baliga 1996, 2008) was invented, making a revolution in power electronics due to its reliability and reduced cost. This advance, together with other technologies, will make possible in subsequent years the popularization of mass-production IH appliances.

In the following years, all the required enabling technologies experienced major advances. Digital control devices expanded to many different applications, enabling the commercial expansion and the availability of inexpensive and powerful devices. Resonant power conversion techniques were also developed, enabling high-efficiency and high-power density converters to be designed. At the same time, finite-element-analysis tools were also well-established, enabling efficient design of electromagnetic systems, IH being one of them.

Last but not least, advanced user interface systems were developed, enabling smarter and simpler interaction with the user. These advances made possible the arrival during the 1990s of the first mass-production commercial units, being produced by companies in Europe and Japan.

Since then, IH popularized, and as a result of the commercial success, different new technologies and applications were developed. Digital control advancements enabled the design of smarter appliances, including not only timers but also temperature control and simple automatic cooking schemes. Power electronic advancements also enabled the use of far more efficient IGBT technology, as well as the research on the use of more efficient wide-bandgap (WBG) devices (Sarnago et al. 2015a). With these improvements, flexible cooking surfaces were targeted, developing a range of multi-coil appliances providing superior user

Pre-history

History

Future

Enabling technologies

Electrical revolution (1880s)

Mercury-arc valves (1902)

Marx generator (1924)

Vitroceramic cristal (1959)

SCR (1956)

Digital control (1970s)

Power diode (1950)

Power MOSFET (1970)

IGBT (1979)

Resonant power conversion (1980s)

FEA electromagnetic tools (1980s)

DSP (1983)

FPGA (1985)

Capacitive controls (1995)

Commercial digital controllers (2000)

IPT applications (2010)

HF power conversion (2010)

SiC SBD Diodes (2001)

SiC Transistors (2010)

HV GaN devices (2014)

- Cost-effective WBG devices
- HF magnetic materials
- Advanced Digital control
- High efficiency topologies
- Electrification incentives

1900 1950 1980 2000 202X

EM induction (Faraday 1831)

First IH furnace (Kjellin 1891)

EM theory (Maxwell 1864)

First domestic IH (Mitchell 1892)

IH furnace (Elwell 1911)

Hardening (1930s)

HF IH (spark gen.) (Northrup-Ribaud 1916)

Solid-state Generators (SCRs) (1960s)

High Efficiency High Frequency Generators (1980s)

Commercial domestic IH (1975)

CAD IH systems (1980s)

Advanced materials processing (1990s)

All-metal induction heating (1989)

Domestic IH expansion (2000)

Temperature control (2015)

Flexible cooking surfaces (2010s)

Advanced multi-domain simulation (2000)

Expansion to other appliances: oven, ironing, tools.

- Advanced flexibility
- All-metal
- Ventilation-integratd IH
- Smart Home–IoT
- Wireless Power Transfer
- Under-worktop IH
- More appliances
- Advanced UI

Domestic induction heating milestones

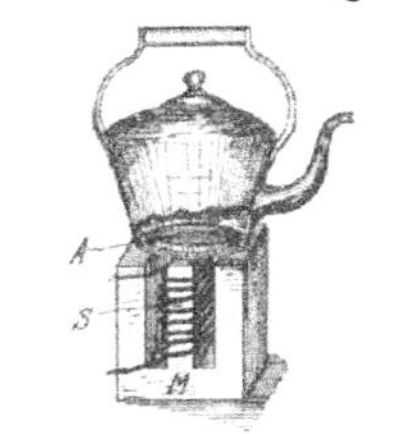

Figure 11.3 Induction heating history: main enabling technologies and technological development milestones. Source: Sarnago et al. (2019b)/from IEEE.

performance. Also, all-metal induction heating (Tanaka 1989) was popularized by Japanese brands in 2002. This also enabled the expansion of IH to other home applications, such as ovens, ironing stations, or water heaters.

Nowadays, IH technology has become a leading technology in the domestic market. According to the market research made by Grand View Research (GVR 2021), the global induction cooktops market size was valued at USD 18,667.8 million in 2020 and it is expected to grow at a compound annual growth rate (CAGR) of 8.5% from 2021 to 2028. Its penetration is especially relevant in the household segment, and this technology is dominated by the built-in product with a 62.1% market share in 2020.

This book chapter is organized as follows. Chapter 11.2 details the main domestic induction heating technology aspects, including the main power conversion blocks, electromagnetic design, and applied digital control techniques. Section 11.2 details the main advanced features and connectivity aspects included in modern induction heating appliances. Finally, Section 11.3 reviews the future challenges for domestic induction technology and summarizes the conclusions of this chapter.

11.2 Domestic Induction Heating Technology

Domestic induction heating has been popularized due to the benefits inherent to the IH technology and the special advantages for cooking applications. Its most successful implementation form is built-in appliances, although there are also free-standing appliances and countertop applications. Figure 11.4 shows a built-in induction heating appliance highlighting the main building blocks.

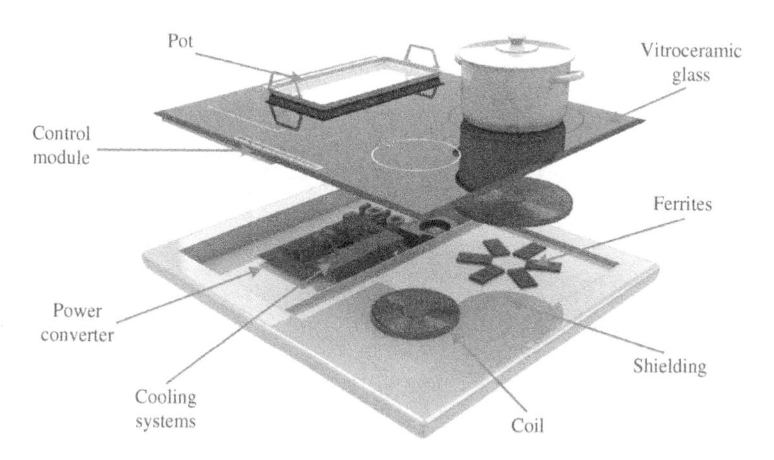

Figure 11.4 Built-in domestic induction heating appliances. Source: Authors (2022).

Power electronics are usually located at the bottom together with the cooling system. Thermal design is a critical point since the operating temperature can be above 100 °C and cooling possibilities are constrained due to geometrical factors and the presence of other nearby elements such as ovens and cabinets. The induction coil is placed above the electronics using typically an aluminum shielding plus ferrite to improve the coupling with the pot and to avoid any interference with the electronics. Finally, a vitroceramic glass is placed between the inductor and the pot to provide mechanical support and esthetics. Depending on the implementation, a thermal and/or electrical isolator is also placed to improve safety and performance. The interaction with the user is made through the user interface. Nowadays, advancements in user interface technologies have made it possible to implement a wide range of human-machine interfaces (HMIs) ranging from simple capacitive buttons to advanced thin film transistor (TFT) capacitive touchscreens. As it will be later discussed, these HMIs also include the possibility of adding connectivity for remote use and programming.

IH appliances rely on the use of many electronic technologies as key enabling technologies to build up what is considered one of the most advanced home appliances. Among the involved technologies, three of them are considered the key enablers and will be discussed in this chapter: power electronics, electromagnetic design, and digital control.

Figure 11.5 shows the block diagram of the main elements of an induction heating cooker. Input power is taken from the mains, typically up to 3.6 kW per

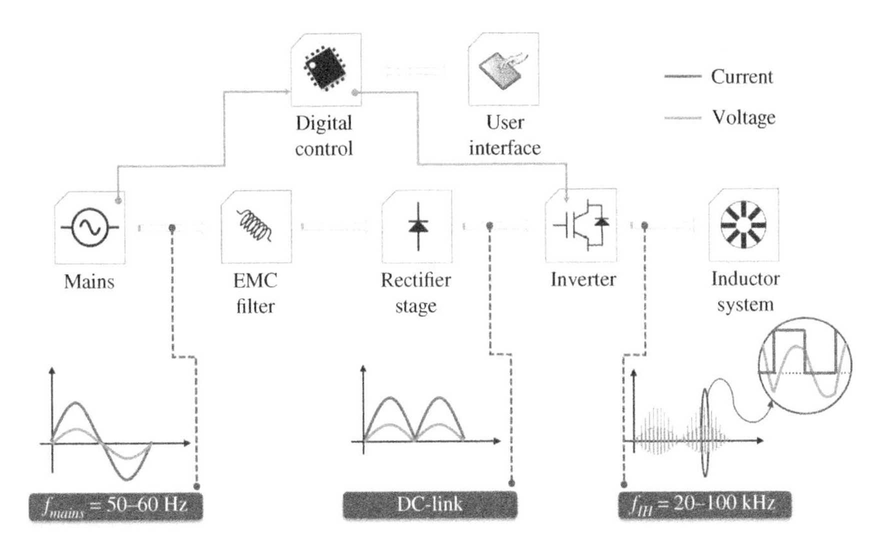

Figure 11.5 Induction heating appliance power conversion block diagram. Source: Lucía et al. (2022).

phase. After that, an Electromagnetic Compatibility (EMC) filter is used to comply with the EMC standards. The main power conversion structure is composed of an indirect alternating current (AC)–AC converter. As a first step, a high-ripple direct current (DC)-link bus voltage is generated using a non-controlled full-bridge diode rectifier and a low-capacitance capacitive filter. This provides a cost-effective way to ensure unity power factor without the need for an additional power factor correction (PFC) stage. After that, an amplitude-modulated DC-link is used by the inverter to generate the medium-frequency voltage to supply the induction coil. Typically, the operating frequency ranges from 20 kHz, to avoid audible noise, up to 100 kHz, to minimize switching losses. All these converters are controlled by a digital control scheme which ensures efficient and proper operation of the IH appliances.

The aim of the previously described power conversion architecture is to achieve an efficient power conversion that ensures fast and safe heating with the highest efficiency. Figure 11.6 summarizes the typical power flow in an induction heating appliance (Lucía et al. 2022), from the input power P_{in} to the power delivered to the actual load, P_h. Power losses can be classified as internal power losses, due to the induction heating appliance itself, and external, due to the cookware. Internal power losses include power loss in the rectifier, P_{Rect}, in the inverter, P_{Inv}, and in the induction coil, P_{Ind}. At this point, it is important to note that the coil losses are not only due to its design but also to the cookware because its materials and geometry may affect the system's performance. Not all the power transmitted to the pot, P_{RF},

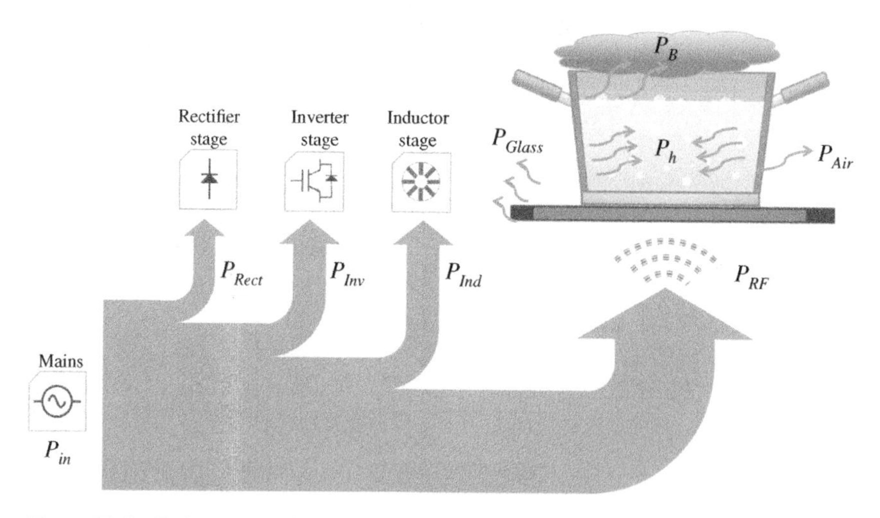

Figure 11.6 Typical power flow in an induction heating appliance. Source: Lucía et al. (2022).

is effectively used to heat, P_h. Part of it is lost to the ambient through the cooking surface, P_{Glass}, through the pot walls, P_{air}, and through evaporation, P_B.

In order to optimize the overall appliance efficiency, the internal power losses must be optimized. Sections 11.2.1–11.2.4 will cover the main enabling technologies used to optimize the converter operation: power electronics, electromagnetic design, and digital control.

11.2.1 Power Electronics

As previously mentioned, the inverter is the main building block of the power conversion architecture inside induction heating appliances. It must provide the medium frequency current to the coil to generate the required magnetic field. From an electrical point of view, the induction load is usually modeled as a series connection of an inductor and a resistor, R_l-L_R (Acero et al. 2013). Assuming this equivalent load, a useful parameter is the load quality factor, defined as $Q = \omega L_R / R_l$, where ω is the excitation angular frequency. This model is usually valid for a given operation point, and it will be used in this chapter to detail the power electronic converter operation. However, it is important to keep in mind that these parameters vary with the excitation frequency, the temperature, the geometry of both inductor and cookware, and, in general, the coupling between the material to be heated and the coil. From the point of view of the inverters, two main technologies are used: single-switch quasi-resonant inverters and half/full-bridge resonant inverters. The former is used in the low-power range, i.e. typically below 2 kW, whereas the latter is used for higher-power appliances up to 4.4 kW per coil.

Single-switch quasi-resonant inverters are cost-effective topologies derived from the class-E inverter that enable efficient and cost-effective implementations of induction heating appliances (Wang et al. 1998; Sarnago et al. 2014a, 2018; Omori and Nakaoka 1989). Figure 11.7 shows implementation examples of the

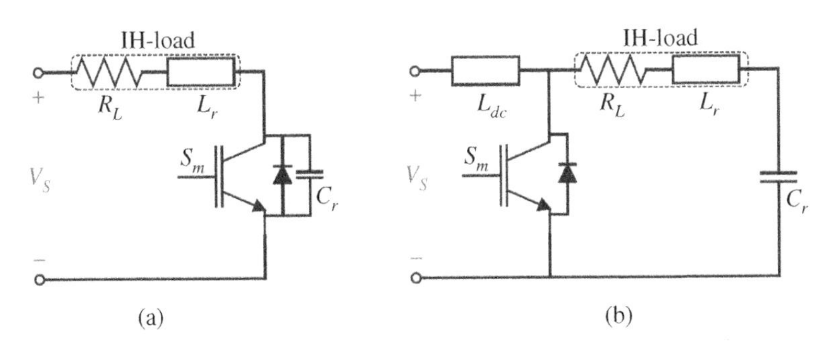

Figure 11.7 Single-switch quasi-resonant inverters for induction heating applications: (a) ZVS implementation and (b) ZCS implementation. Source: Authors (2022).

Zero Voltage Switching (ZVS) and Zero Current Switching (ZCS) configurations, both using an added resonant capacitor C_r. Usually, the ZVS configuration is preferred because a filtering inductor L_{dc} is not needed.

Both single-switch topologies achieve a quasi-resonant operation, i.e. depending on the status of the switching transistor, a resonant tank is formed by the inductor-load system, R_l, L_r, and the resonant capacitor C_r. This operation ensures high efficiency and cost-effective implementation. However, its main limitations are derived from the difficulty to achieve soft-switching conditions and the high voltage in the switching device. The first limitation severely constraints the implementation of multi-load systems due to highly variable switching frequencies, control constraints, and audible noise. The second limitation, switching device voltage, limits the operation range because the voltage in the power device, which is the sum of the bus voltage, V_s, plus the inductor voltage, can easily exceed the device ratings. Figure 11.8 shows a summary of the main waveforms of the single-switch quasi-resonant ZVS inverter and its normalized switching device

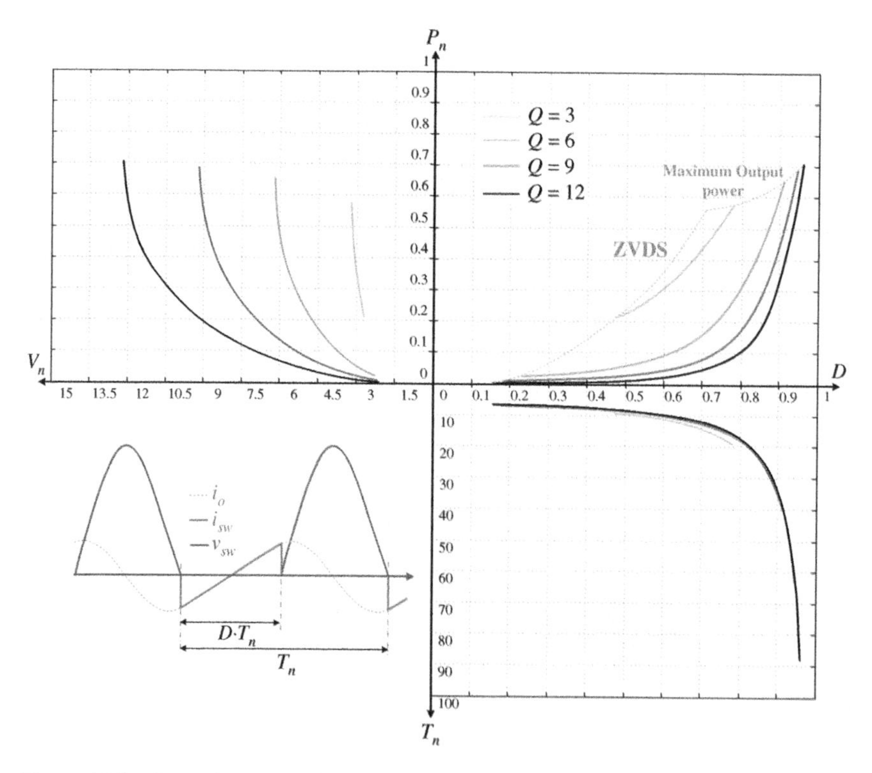

Figure 11.8 Operation areas of a ZVS single-switch quasi-resonant inverter for domestic induction heating. Source: Sarnago et al. (2014a).

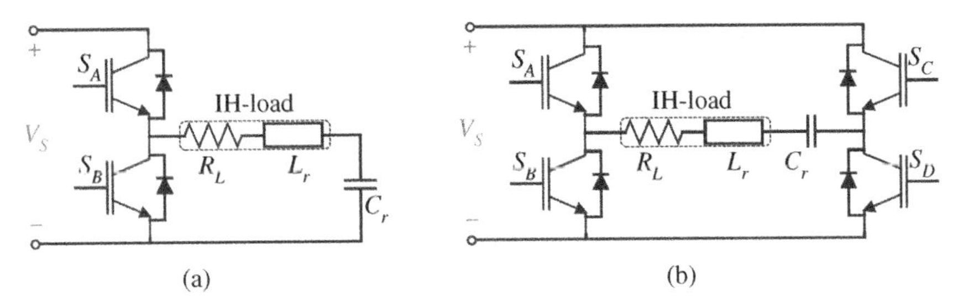

Figure 11.9 Half-bridge (a) and full-bridge (b) series resonant inverters. Source: Authors (2022).

voltage and power as a function of the control parameters, duty cycle (D), and period (T_n), for different quality factor (Q) loads (Sarnago et al. 2014a). From this figure, higher output power rapidly leads to high voltage in the switching power device. For this reason, these topologies are often limited to single-inductor structures with output power below 2.2 kW using 1200 V IGBT technology.

When higher output power is required, the full/half-bridge topologies are often preferred (see Figure 11.9) (Sarnago et al. 2015b). These topologies are formed of two or four devices and are used to deliver up to 4.4 kW using IGBT technology (Fernández et al. 2020). The preferred implementation is the series resonant inverter, where a resonant capacitor C_r is placed in series with the inductor-load system, R_l, L_r. Under these conditions, the resonant frequency is defined as $f_0 = 1/2\pi\sqrt{L_r C_r}$.

One of the main advantages of these topologies is easy control and high efficiency (Lucía et al. 2010b). As it is shown in Figure 11.10, the output power curve vs. the switching frequency is continuous and predictable over a wide range. Typically, operation above the resonant frequency f_0, is preferred because ZVS soft switching is achieved during the turn-on transition. Consequently, a monotonous power curve is obtained and power control in variable-load systems becomes easier. Nowadays, the half-bridge series resonant inverter is considered

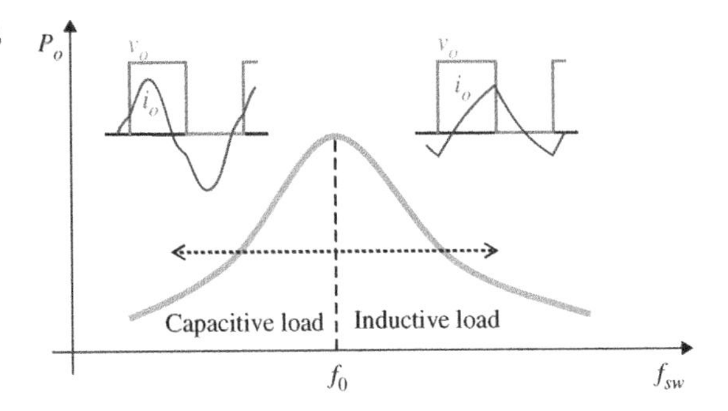

Figure 11.10 Output power, P_0, vs. switching frequency, f_{sw}, in a series resonant inverter. In frequencies below the resonant frequency, f_0, ZCS is achieved, whereas in frequencies higher than the resonant frequency ZVS is achieved. Source: Authors (2022).

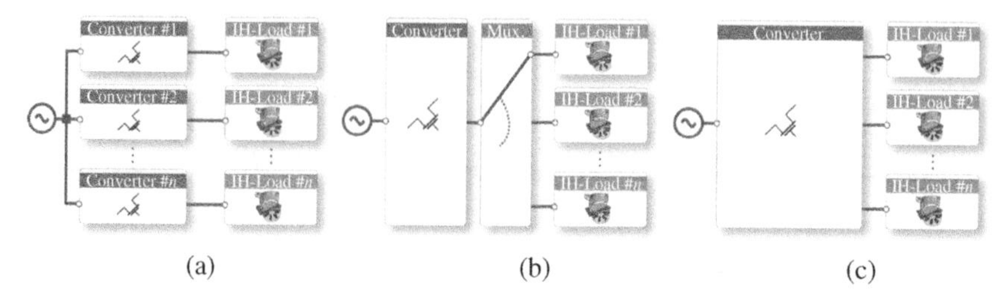

Figure 11.11 Multiple-output induction heating architectures: (a) single-inverter single-coil, (b) relay-multiplexed, and (c) multiple-output full-solid-state configurations. Source: Sarnago et al. (2019b)/from IEEE.

to have the best balance between cost and performance, and it is the selected power conversion topology by most manufacturers.

Finally, there are certain applications that require specific power electronic converters to meet the required performance. Among these, there are two remarkable cases: multi-load inverters for flexible cooking zones and high-power/current conversion stages with PFC.

The first group of specific applications includes multiple-output inverters for flexible cooking zones (Lucía et al. 2013). These appliances, as it will be later discussed, feature large cooking areas where the user can place any pot, with any shape, anywhere. To power such structures, multi-output power converters are needed. Nowadays, different strategies are used (Figure 11.11), including using expensive single-inverter single-coil configurations (i), relay-multiplexed structures (ii), and multiple-output full-solid-state converters (iii).

Nowadays, the most popular configurations are the single-inverter single-coil structures for appliances below 12 induction coils, and the relay-multiplexed structures when more than 20 coils are present. However, recent research lines have also proposed the use of multi-output inverters (Lucía et al. 2010a, 2011) that enable power in an efficient and cost-effective way with a set of inductors with a single inverter. Figure 11.12 shows an example of a series-resonant ZVS matrix inverter (Sarnago et al. 2019a, 2019b) that enables to power a flexible IH cooking appliance with a significant reduction in the number of required switching power devices.

The second group of application-specific power converters is related to high-performance applications where high power or high-performance heating is required. Nowadays, all-metal heating is a design trend where heating cookware made from highly conductive materials such as aluminum or copper is desired (Li et al. 2021; Huang et al. 2021). To achieve this, several strategies have been proposed from higher voltage topologies (Figure 11.13) to new modulation strategies.

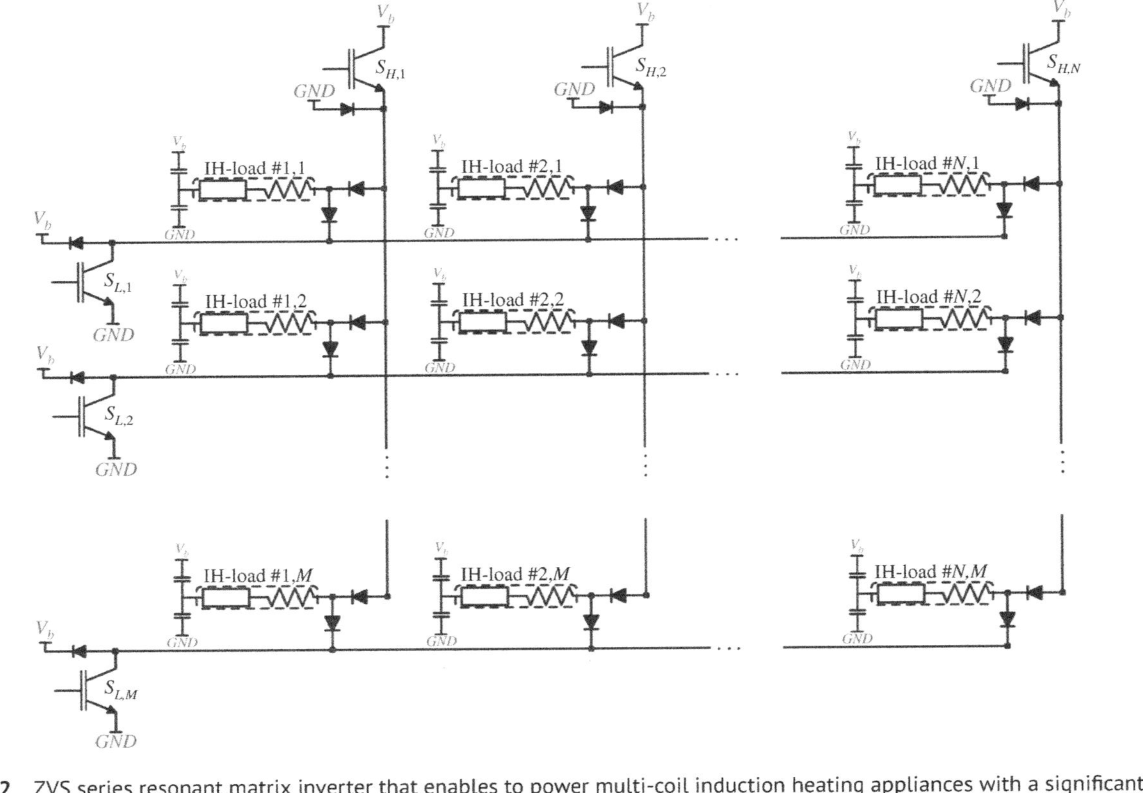

Figure 11.12 ZVS series resonant matrix inverter that enables to power multi-coil induction heating appliances with a significant reduction in the number of required power devices. Source: Sarnago et al. (2019b)/IEEE/CC BY 4.0.

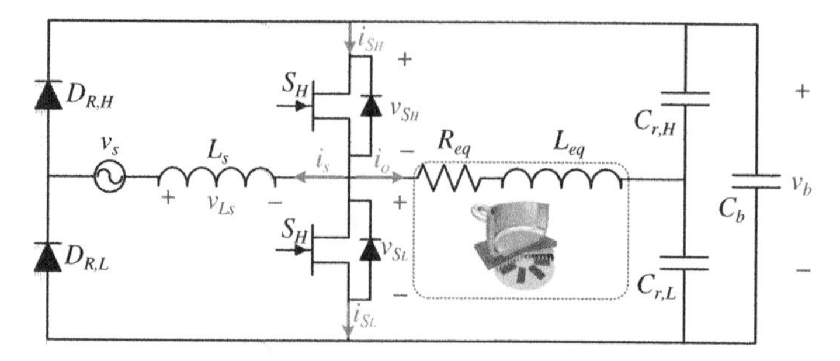

Figure 11.13 Direct AC–AC boost series resonant inverter that achieves higher efficiency and performance due to the increased bus voltage. Source: Sarnago et al. (2014c)/from IEEE.

Besides, when higher power or higher performance is required, PFC is considered, following the power flow shown in Figure 11.14. This implementation achieves high power factor while enabling the design of industrial induction heating cookers with output powers higher than 11 kW.

11.2.2 Electromagnetic Design

The second key technology powering modern IH appliances is electromagnetic design applied to the induction coil and auxiliary elements. Figure 11.15 shows a representative example of the structure of an induction heating coil system, including aluminum shielding, flux concentrators made of ferrite, the coil itself, and the vitroceramic glass that provides both electrical isolation for the user and a structural surface.

IH appliances can be seen as wireless power transfer applications where the induction coil acts as the primary side or emitter and the pot is the secondary side or receiver. Usually, frequencies between 20 and 100 kHz are applied. At the beginning of IH technology, solid strand magnet wire was used due to the reduced cost and the limited performance requirements in terms of output power or performance. However, with the advance of technology, higher output powers and efficiency levels were required. Since single-strand wires were limited by skin effect that led to high power loss, foil-based inductors were used to improve thermal performance. However, increasing currents and frequencies required higher performance, leading to the use of Litz wire, i.e. multi-stranded wire optimized for high frequency (HF) operation (Lucía et al. 2022).

Nowadays, Litz wire has become an industry standard to achieve high-performance and high-efficiency appliances (Acero et al. 2006). These designs are optimized to minimize both conduction losses and proximity losses, due to induction heating in the coil itself, to ensure an efficient operation. Design using finite element analysis (Kim et al. 2016; Aoyama et al. 2021; Carretero et al. 2011;

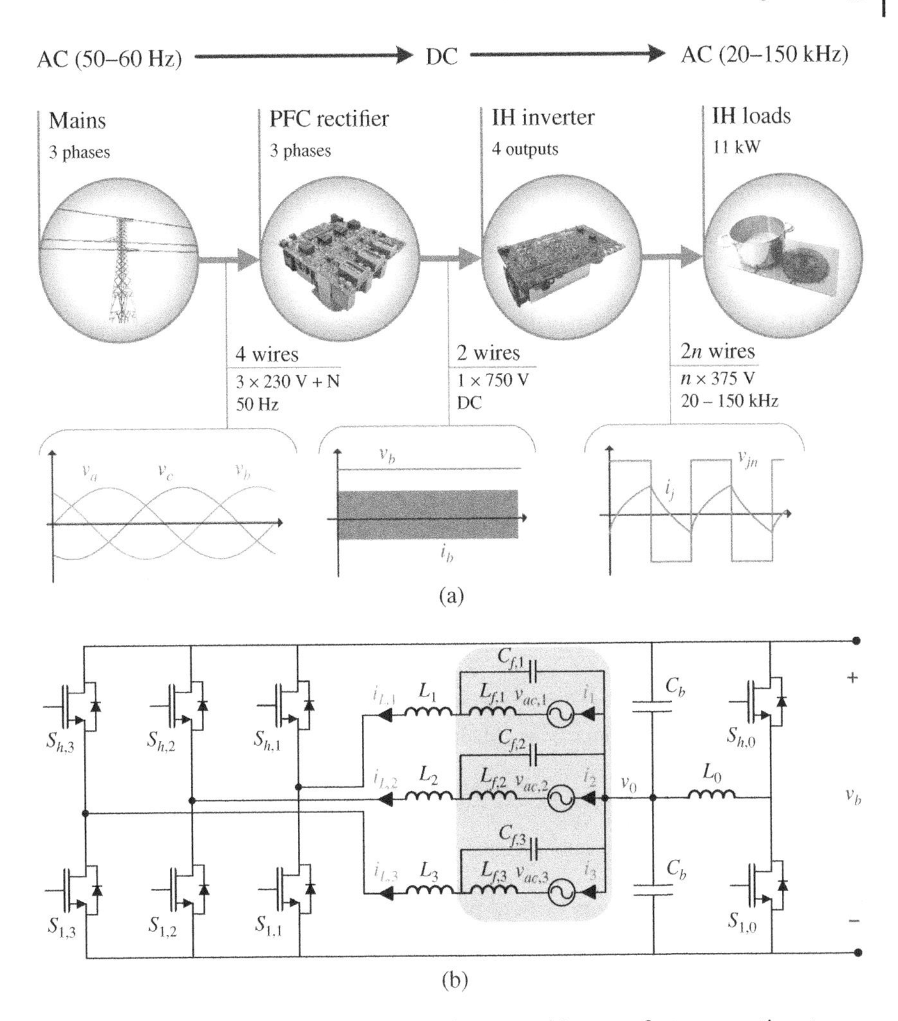

Figure 11.14 Three-phase series resonant inverter with power factor correction stage that enables the design of high-power induction heating cookers for industrial and professional applications for multi-load systems (waveforms for j-load shown). Source: Pérez-Tarragona et al. (2022)/from John Wiley & Sons.

Acero et al. 2011) enables to optimize its performance from three different points of view: electromagnetic operation, electrical operation of the induction coil and inverter, and thermal distribution in the pot to be heated (Figure 11.16).

11.2.3 Digital Control

Finally, digital control is the third key technology that enables the design of safe, accurate, and high-performance appliances. Modern control architectures (Lucía

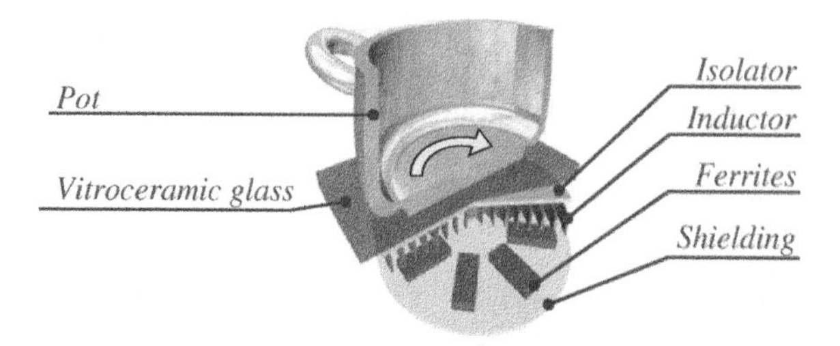

Figure 11.15 Structure of an induction heating coil system. Source: Lucía et al. (2014)/from IEEE.

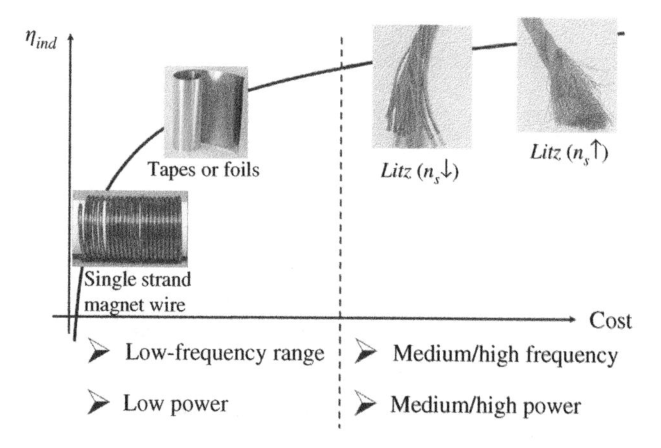

Figure 11.16 Inductor wire evolution: efficiency vs. cost. Source: Authors (2022).

et al. 2018b) include the use of field-programmable gate array (FPGA) devices for research and prototype purposes (Navarro et al. 2012), and microcontrollers, digital signal processors (DSPs) and/or application-specific integrated circuits (ASICs).

The digital control architecture of an induction heating appliance (Figure 11.17) performs three key functions classified according to the application and frequency ranges: monitors and controls the modulation of the power converter (kHz range); applies control strategies to achieve the desired performance in terms of output power, temperature, etc. (Hz) range; and serves as user interface (Jiménez et al. 2013, 2014a, 2014b).

Given these highly differentiated tasks, a wide range of control architectures can be used. Figure 11.17 shows an example where a fast-response ASIC is used for measuring and control the power converter modulation (Jimenez et al. 2014, 2014c; Pérez-Tarragona 2020), whereas the high-level control algorithms for output power or advanced features are performed in a low-cost microcontroller.

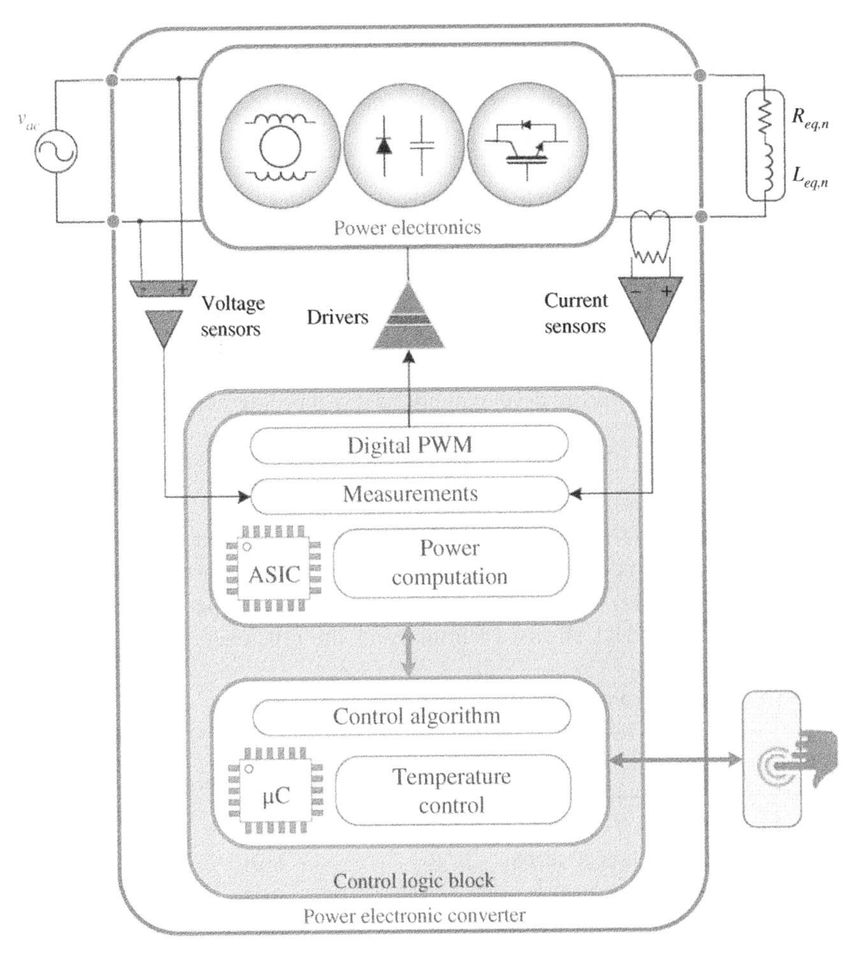

Figure 11.17 Digital control architecture of an induction heating appliance. Source: Pérez-Tarragona (2020)/Universidad de Zaragoza Servicio de Publicaciones.

Additionally, an external user interface is used to interact using different tactile and visual technologies.

In order to achieve the required performance and safe operation, many different elements must be implemented in the control architecture, from induction load monitoring and identification techniques, to new output power measurement for high-frequency resonant converters. Besides, new modulation strategies enable higher performance, higher efficiency, and/or reduced electromagnetic emissions (Villa et al. 2022; Lucía et al. 2009; Guillén et al. 2022); and new control algorithms enable faster and better performance (Lucia et al. 2020, 2021). As it can be seen in the later efficiency analysis, these modulation strategies and control algorithms can have a great impact on the overall efficiency of the induction heating appliance. Typically, square wave (SW) or asymmetrical duty cycle (ADC) modulations are applied. However, certain applications required applying pulse density modulation (PDM) (Guillén et al. 2021) or discontinuous modulations (DMs) in the low power range to increase efficiency. Finally, these modulations are also used to improve power factor and harmonic content at the appliance input.

Last but not least, given the high sensing and computation capabilities of IH appliance electronics, a wide range of advanced features are possible to implement. These include, but are not limited, the inverter control optimization to maximize output power, temperature control (Paesa et al. 2009), or audible noise reduction (Guillen et al. 2021). Also, they provide enhanced means for connectivity, opening the design window for future developments as it will be later discussed.

11.2.4 Efficiency

The main aim of the previously discussed power conversion architectures is to achieve a high-efficiency power conversion when transforming the mains input power (50/60 Hz) to the medium frequency voltage applied to the inductor (20–100 kHz). Table 11.1 shows a summary of the efficiency achieved by the

Table 11.1 Efficiency comparison.

Topology/modulation	$\hat{\eta}$ (%)
Half-bridge/SW (Lucía et al. 2009)	95.0
Half-bridge/ADC (Lucía et al. 2010b)	96.0
Half-bridge/PWM + PDM (Ahmed 2011)	94.5
Class D-DE HB/SW (Sarnago et al. 2013)	95.3
Multi-MOSFET HB/SW (Sarnago et al. 2014d)	98.3
Direct AC–AC boost/ADC (Sarnago et al. 2014c)	99.0
AC–AC full ZCS/HF-PDM (Sarnago et al. 2014b)	97.2
Direct AC–AC class-E/P1N0 (Sarnago et al. 2014a)	95.0
AC–AC class-E/P1N0+P1N1 (Sarnago et al. 2014a)	96.1
SR multi-inverter/SW (Lucía et al. 2010a)	95.8
SR multi-inverter/HF-PDM (Lucía et al. 2013)	97.8
Twin half-bridge/HF-R (Mishima et al. 2014)	96.5
Full-bridge/AVC (Burdío et al. 2004)	96.0
Full-bridge/AVCo (Barragán et al. 2005)	96.0
PFC + HB/CCM + DCM (Kawaguchi et al. 2010)	94.5

Source: Lucía et al. (2022).

main combinations of topologies and control strategies (Lucía et al. 2022). From this table, it can be seen that conventional topologies and control strategies, i.e. half-bridge with SW or ADC control strategies, achieve efficiencies well above 95%. Only when an additional PFC stage is added, the efficiency drops slightly below 95%. In some cases, when either more complex topologies with boosting capabilities or wide bandgap devices are used, efficiencies are significantly increased and can reach levels close to 99%. It is important to note that these values reflect only the efficiency of the inverter, not the coil due to the difficult measurement and that the rectifier stage is not included in all topologies.

As a conclusion, it can be seen that induction heating technologies are sustained by key enabling technologies in the power electronics, electromagnetic design, and digital control areas that enable to obtain high-efficiency systems. This provides a clear path toward decarbonization through the electrification of household appliances.

11.3 Advanced Features and Connectivity

Section 11.2 has covered the key enabling technologies of induction heating and their state-of-the-art in the market and scientific sphere nowadays. This section covers advanced topics enabling high-performing appliances, some of them already in development or in early market deployment.

11.3.1 High-Performance Power Electronics

Power electronics determine in a great manner the final performance and efficiency of IH appliances. In this context, IH designers are always in pursue of high-performance power electronic converters. In the last decade, WBG devices have arisen as a powerful tool to implement high-performance converters (She et al. 2017). Their advantages in terms of blocking voltage and high switching speeds make them perfect candidates for implementing IH appliances. Both silicon carbide (SiC) and gallium nitride (GaN) exhibit excellent properties that make them suitable for high-voltage high-power implementations and high-frequency designs, respectively. Figure 11.18 shows an example of the on-state voltage (a) and efficiency of several WBG technologies operating in induction heating appliances (Sarnago et al. 2015a). From this figure, it can be seen that there is great room for improvement with the introduction of WBG technology.

Whereas the cost and some performance issues on WBG have been a great barrier to their introduction in domestic induction appliances, in the last years there have been profound technical and industrial changes that have opened new opportunities for their introduction. On the one hand, GaN technology has

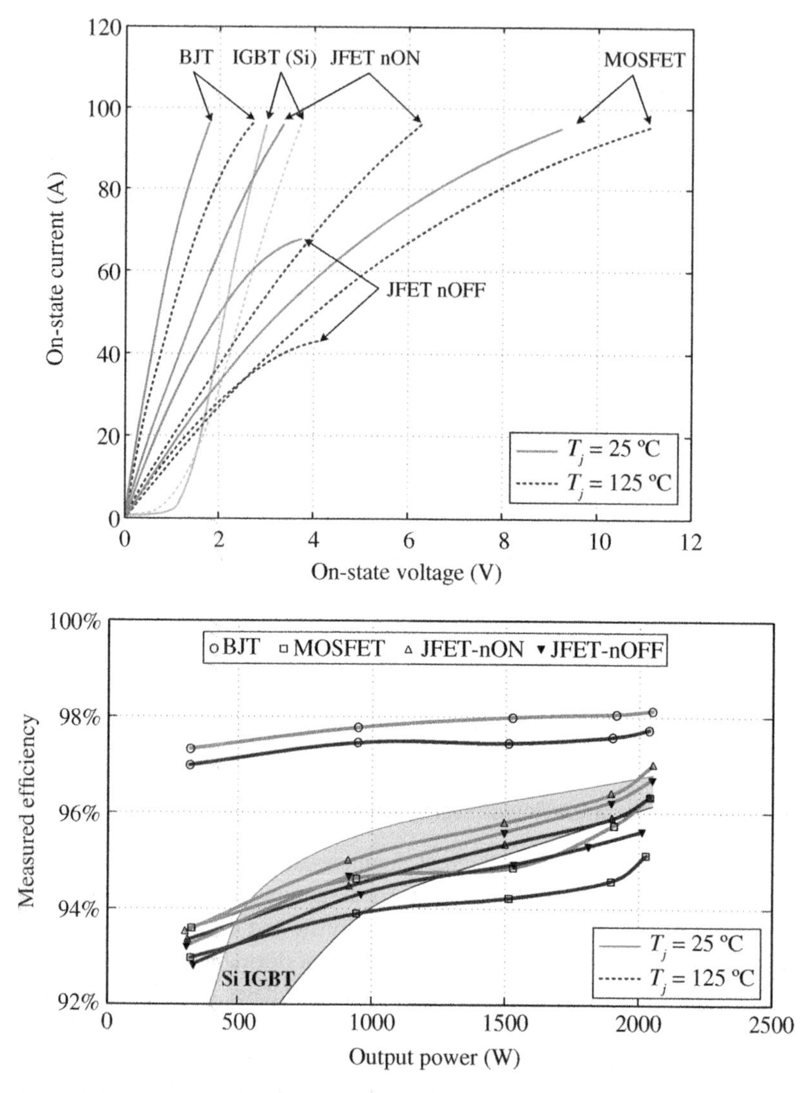

Figure 11.18 Some silicon carbide power devices applied to domestic induction heating: (a) measured on-state voltage and (b) measured efficiency vs. output power. Source: Sarnago et al. (2015a).

evolved toward higher voltage devices. These devices, which were once limited to the sub 200 V range, are now designed in the 650 V range, being suitable for mains-connected applications. SiC has greatly evolved by introducing trench technology and, more importantly, production costs have dramatically decreased due to both the improvement of industrial processes with bigger wafers and volume increase due to the fast development of renewable energy and electric vehicle markets.

11.3.2 Advanced Control

IH appliances have one of the most advanced control architectures that can be found in our homes. Whereas basic control strategies deal with inverter operation using classical linear control techniques and power control, there are advanced features that can be included.

Regarding modulation strategies, modern IH appliances feature advanced modulation techniques that allow obtaining improved efficiency, electromagnetic emissions, or the capability to heat highly conductive materials in so-called all-metal heating (Han et al. 2018; Millán et al. 2011). Besides, classical linear control techniques are evolving toward the use of more advanced non-linear control. Among these techniques, model predictive control has risen as a powerful tool that can include advanced constraints in the control design to optimize both electrical and user performance (Lucia et al. 2017, 2019, 2021). One of the main challenges of these controllers are their implementation in the appliance-embedded control architecture. Recently, artificial intelligence (AI) implementations based on using deep neural networks (DNNs) have been proposed to obtain a high-performance and cost-effective control implementation (Figure 11.19).

Finally, higher-level control algorithms provide advanced features for the user compared with classical power-control algorithms. One of the most promising control techniques is temperature control. Considering the target of these appliances is cooking performance, temperature control makes more sense than power controls to achieve the desired culinary targets. For this reason, modern IH appliances feature temperature control systems that may rely either on in-product or off-product temperature sensors (Franco et al. 2012).

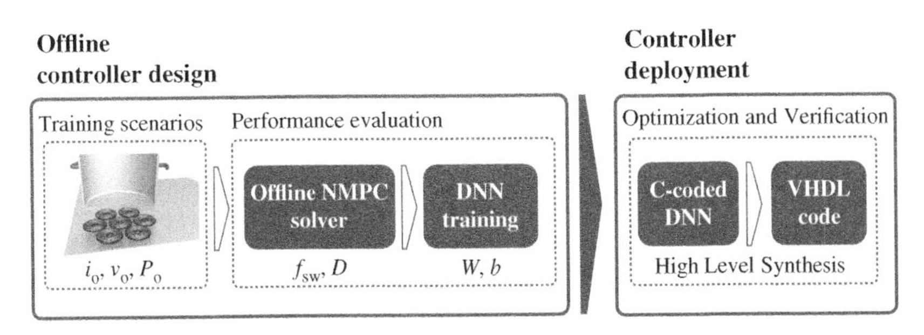

Figure 11.19 Deep neural network implementation workflow for a model predictive controller for IH. Source: Lucia et al. (2021)/IEEE.

11.3.3 Flexible Cooking Surfaces

Flexible cooking surfaces intend to provide the user with a much more adaptable cooking area with additional freedom degrees (Millán et al. 2010). These appliances can be implemented either using concentric windings that adapt to different pot sizes (Figure 11.20) or using multiple coil systems (Figure 11.21). Nowadays, these appliances have gained great attention from consumers and the so-called flexible heating zones have become a priority.

In order to achieve this advanced feature, significant technological challenges must be addressed in the areas of inductor design and partial coupling (Figure 11.21b), multiple-output converter design (Forest et al. 2000, 2007; Lucía et al. 2010a, 2011, 2012), and load-pot identification (Sarnago et al. 2016) and the associated control algorithms. This represents one of the most promising research areas for high-end IH appliances.

11.3.4 Connectivity

One of the main benefits of the advanced control architecture of modern IH appliances is the possibility of connecting with internal and external systems to share information and provide advanced user performance (Lawton 1997; Nakakita et al. 2003). Connectivity in appliances can be classified into internal connectivity, external connectivity, and power connectivity.

Internal connectivity refers to communications inside the appliance ecosystem. These communications have evolved from simple diagnostics systems for monitoring and repairing purposes, to more advanced communications to interact with other elements in the kitchen. The most common examples are communications with sensors for temperature control, which may be embedded in smart pots or attached inside or outside as an external sensor, or communication with the hood

Figure 11.20 Flexible induction heating appliance using concentric windings. Source: Pérez-Tarragona et al. (2018)/from IEEE.

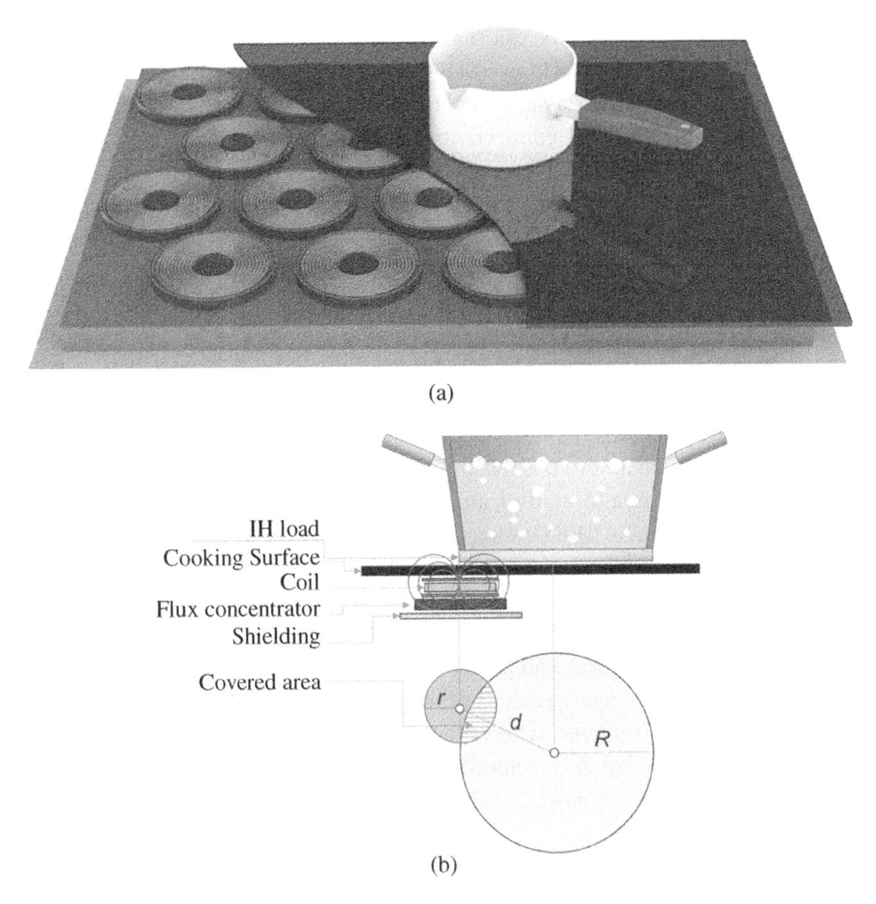

Figure 11.21 Flexible induction heating appliance using multiple coils: (a) general structure and (b) partial coupling challenge. Source: (a) Sarnago et al. (2019b)/from IEEE.

to synchronize air extraction with the cooking process. These elements may use proprietary communication protocols or, more recently, standard protocols such as Bluetooth low energy (BLE).

External connectivity refers to the possibility to include communications with elements external to the appliance ecosystem. In the current context of the Internet of Things (IoTs) era, this includes remote communications, usually via WiFi™, with external servers for more complex monitoring and control tasks. This includes controlling the appliance remotely, monitoring its main parameters, and transferring cooking recipes, among others. These tasks can be done using user-friendly apps for mobile phones or computers. In addition to this, more advanced integrations include integrating IH appliances in popular home voice-controlled platforms such as Amazon Alexa, Google Assistant, and others.

Another important aspect of external connectivity is the added capability to interact with Smart Energy Home (SHE) Smart Energy Controllers (SECs) as introduced in Chapter 6 for advanced energy use cases. To reduce peak demand in the household this external connectivity enables the IH appliance to publish to the SEC that a cooking process is about to start, therefore enabling the SEC to consider temporarily turning off other major loads not critical at that moment. One example would be to stop the water heating process and re-enabling it after cooking has finished. This would of course depend on the current temperature of the water in the water heater, but it is a possibility to consider. Other actions could be stopping air conditioner in case the temperature, even not yet at target temperature, is already at acceptable levels, or publishing a signal to avoid the dishwasher or dryer to start for a given period, unless absolutely critical to the consumer.

An important aspect to enable external connectivity in an efficient and broadly accepted manner is to assure the IH appliance is compliant with a common language in terms of command, control, and data publishing to the SEC. As presented in Chapter 7, the ETSI SAREF Ontology is an interesting solution for that.

Finally, a new recent form of connectivity is referred to as power connectivity. This takes advantage of the inherent wireless power transfer capability of IH appliances to serve as a general wireless power transfer platform (Kim et al. 2020). In this context, IH appliances will power wirelessly virtually any element capable of receiving energy wirelessly such as other appliances or consumer electronics. This is an area of great interest where a new standard, Ki standard (Figure 11.22; Consortium 2019), is being developed following the previous steps of the Qi standard oriented to lower power levels.

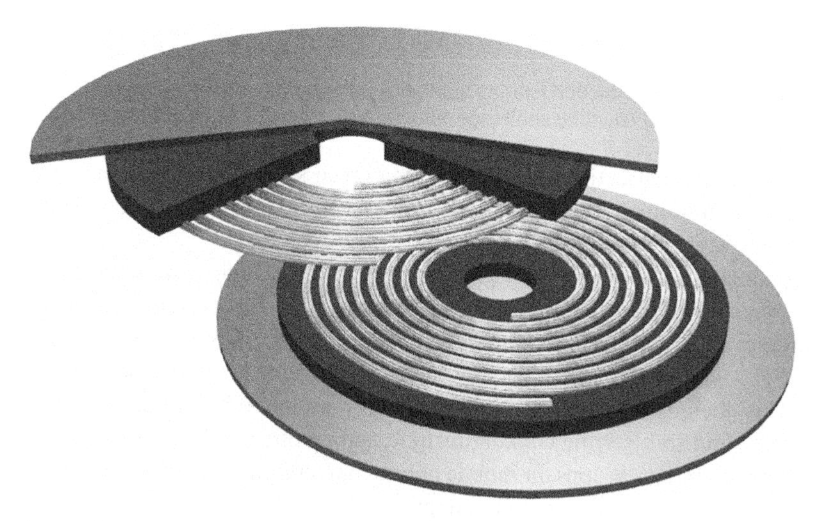

Figure 11.22 Wireless power transfer using IH appliances. Source: Authors (2022).

11.4 Conclusion and Future Challenges

This chapter has presented the main benefits and technologies behind induction heating appliances. These devices are among the most technologically advanced appliances and provide a clear path toward decarbonization through the electrification of one of the most energy-consuming processes, i.e. heating in cooking processes.

Domestic IH relies on some key enabling technologies to sustain its high performance and efficiency: power electronics, electromagnetic design, and digital control. These provide useful tools to implement high-performance systems and, at the same time, create significant technological challenges to implement future higher efficiency and higher performance systems, and cost-effective implementations.

Among the future features and challenges discussed, connectivity arises as one of the most relevant ones due to its key relevance for the user and the current household ecosystem. Connectivity will play a key role to provide additional features and performance for the user, and to optimize electrical energy usage in residential areas. Features such as integration in smart ecosystems, advanced sensors, and smart cooking will be possible in the future with smarter connected appliances. Moreover, the development of new wireless power transfer standards will enable a completely new set of applications with still to unveil new features for the user.

Symbols and Abbreviations

AC	alternating current
ADC	asymmetrical duty cycle modulation
AI	artificial intelligence
ASIC	application specific integrated circuit
BLE	bluetooth low energy
DC	direct current
DNN	deep neural networks
DSP	digital signal processor
EMC	electromagnetic compatibility
FPGA	field programmable gate array
f_{sw}	switching frequency
GaN	gallium nitride
HF	high frequency
IGBT	insulated gate bipolar transistor
IH	induction heating

i_o output current
IoT internet of things
P_o mean output power
SiC silicon carbide
SW square wave modulation
v_o output voltage
WBG wide bandgap
ZCS zero current switching
ZVS zero voltage switching

References

Acero, J., Alonso, R., Burdio, J.M. et al. (2006). Frequency-dependent resistance in Litz-wire planar windings for domestic induction heating appliances. *IEEE Transactions on Power Electronics* 21: 856–866.

Acero, J., Carretero, C., Millán, I. et al. (2011). Analysis and modeling of planar concentric windings forming adaptable-diameter burners for induction heating appliances. *IEEE Transactions on Power Electronics* 26: 1546–1558.

Acero, J., Carretero, C., Alonso, R. et al. (2013). Mutual impedance of small ring-type coils for multi-winding induction heating appliances. *IEEE Transactions on Power Electronics* 28: 1025–1035.

Ahmed, N.A. (2011). High-frequency soft-switching AC conversion circuit with dual-mode PWM/PDM control strategy for high-power IH applications. *IEEE Transactions on Industrial Electronics* 58: 1440–1448.

Aoyama, M., Thimm, W., Knoch, M., and Ose, L. (2021). Proposal and challenge of Halbach array type induction coil for cooktop applications. *IEEE Open Journal of Industry Applications* 2: 168–177.

Baliga, B.J. (1996). Trends in power semiconductor devices. *IEEE Transactions on Electron Devices* 43: 1717–1731.

Baliga, B.J. (2008). *Fundamentals of Power Semiconductor Devices*. New York: Springer.

Barragán, L.A., Burdío, J.M., Artigas, J.I. et al. (2005). Efficiency optimization in ZVS series resonant inverters with asymmetrical voltage-cancellation control. *IEEE Transactions on Power Electronics* 20: 1036–1044.

Berry, A.F. (1909). Improvements in electrically heated apparatus. Gb190920639 Patent Application, filed 09 September 1909 and issued 08 September 1910. https://worldwide.espacenet.com/publicationDetails/biblio?II=0&ND=3&adjacent=true&locale=en_EP&FT=D&date=19100908&CC=GB&NR=190920639A&KC=A.

Burdío, J.M., Barragán, L.A., Monterde, F. et al. (2004). Asymmetrical voltage-cancelation control for full-bridge series resonant inverters. *IEEE Transactions on Power Electronics* 19: 461–469.

Carretero, C., Lucía, O., Acero, J. et al. (2011). An application of the impedance boundary condition for the design of coils used in domestic induction heating systems. *COMPEL-The International Journal for Computation and Mathematics in Electrical Engineering* 30: 1616–1625.

Clairand, J.M., Arriaga, M., Cañizares, C.A., and Álvarez-Bel, C. (2019). Power generation planning of Galapagos' microgrid considering electric vehicles and induction stoves. *IEEE Transactions on Sustainable Energy* 10: 1916–1926.

Consortium, W. P. (2019). *Ki Cordless Kitchen: From Concept To Industry Standard.* https://www.wirelesspowerconsortium.com/data/downloadables/2/3/7/5/ki-cordless-kitchen-white-paper-september-2019.pdf.

Davies, J. (1990). *Conduction and Induction Heating*. London: Peter Peregrinus Ltd.

Fernández, M., Perpiñà, X., Vellvehi, M. et al. (2020). Power losses and current distribution studies by infrared thermal imaging in soft- and hard-switched IGBTs under resonant load. *IEEE Transactions on Power Electronics* 35: 5221–5237.

Forest, F., Labouré, E., Costa, F., and Gaspard, J.-Y. (2000). Principle of a multi-load/single converter system for low power induction heating. *IEEE Transactions on Industrial Electronics* 15: 223–230.

Forest, F., Faucher, S., Gaspard, J.-Y. et al. (2007). Frequency-synchronized resonant converters for the supply of multiwindings coils in induction cooking appliances. *IEEE Transactions on Industrial Electronics* 54: 441–452.

Franco, C., Acero, J., Alonso, R. et al. (2012). Inductive sensor for temperature measurement in induction heating applications. *IEEE Sensors Journal* 12: 996–1003.

Guillen, P. (2022). Multi-output matrix resonant converters for domestic induction heating. Tesis Doctoral, Universidad De Zaragoza.

Guillen, P., Sarnago, H., Lucia, O., and Burdio, J.M. (2021). Asymmetrical noncomplementary modulation strategies for independent power control in multioutput resonant inverters. *IEEE Journal of Emerging and Selected Topics in Power Electronics* 9: 629–637.

Guillén, P., Sarnago, H., Lucia, O., and Burdio, J.M. (2021). Mains-synchronized pulse density modulation strategy applied to a ZVS resonant matrix inverter. *IEEE Transactions on Industrial Electronics* 68: 10835–10844.

Guillén, P., Sarnago, H., Lucia, O., and Burdio, J.M. (2022). Series-resonant matrix inverter with asymmetrical modulation for improved power factor correction in flexible induction heating appliances. *IEEE Transactions on Industrial Electronics* https://doi.org/10.1109/TIE.2022.3161789.

GVR (2021). *Induction Cooktops Market Size, Share & Trends Analysis Report By Product (Built-In, Free-Standing), By Application (Household, Commercial),*

By Distribution Channel (Online, Specialty Stores), And Segment Forecasts, 2021–2028. Grand View Research. https://www.grandviewresearch.com/industry-analysis/induction-cooktops-market.

Han, W., Chau, K.T., Jiang, C., and Liu, W. (2018). All-metal domestic induction heating using single-frequency double-layer coils. *IEEE Transactions on Magnetics* 54: 1–5.

Huang, M.S., Liao, C.C., Li, Z.F. et al. (2021). Quantitative design and implementation of an induction cooker for a copper pan. *IEEE Access* 9: 5105–5118.

Jiménez, O., Lucía, O., Barragán, L.A. et al. (2013). FPGA-based test-bench for resonant inverter load characterization. *IEEE Transactions on Industrial Informatics* 9: 1645–1654.

Jimenez, O., Lucia, O., Urriza Parroque, I. et al. (2014). Design and evaluation of a low-cost high-performance sigma-delta ADC for embedded control systems in induction heating appliances. *IEEE Transactions on Industrial Informatics* 61: 2601–2611.

Jiménez, O., Lucia, O., Urriza, I. et al. (2014a). An FPGA-based gain-scheduled controller for resonant converters applied to induction cooktops. *IEEE Transactions on Power Electronics* 29: 2143–2152.

Jiménez, O., Lucía, O., Urriza, I. et al. (2014b). Analysis and implementation of FPGA-based online parametric identification algorithms for resonant power converters. *IEEE Transactions on Industrial Informatics* 10: 1144–1153.

Jiménez, O., Lucía, O., Urriza, I. et al. (2014c). Power measurement for resonant power converters applied to induction heating applications. *IEEE Transactions on Power Electronics* 29: 6779–6788.

Kawaguchi, Y., Hiraki, E., Tanaka, T. et al. (2010). A comparative evaluation of DCM control and CCM control for soft-switching PFC converter. In: *36th Annual Conference of the IEEE Industrial Electronics Society* (7–10 November 2010). USA: IEEE, 250–255.

Kennedy, R. (1909). *Electrical Installations*. London: Caxton.

Kim, D., So, J., and Kim, D. (2016). Study on heating performance improvement of practical induction heating rice cooker with magnetic flux concentrator. *IEEE Transactions on Applied Superconductivity* 26: 1–4.

Kim, M., Park, H.P., and Jung, J.H. (2020). Practical design methodology of IH and IPT dual-functional apparatus. *IEEE Transactions on Power Electronics* 35: 8897–8901.

Lawton, G. (1997). Dawn of the internet appliance. *Computer* 30: 16.

Li, Z., Chen, Q., Zhang, S. et al. (2021). A novel domestic SE-IH with high induction efficiency and compatibility of nonferromagnetic vessels. *IEEE Transactions on Industrial Electronics* 68: 8006–8016.

Lucía, O., Burdío, J.M., Millán, I. et al. (2009). Load-adaptive control algorithm of half-bridge series resonant inverter for domestic induction heating. *IEEE Transactions on Industrial Electronics* 56: 3106–3116.

Lucía, O., Burdío, J.M., Barragán, L.A. et al. (2010a). Series-resonant multiinverter for multiple induction heaters. *IEEE Transactions on Power Electronics* 24: 2860–2868.

Lucía, O., Burdío, J.M., Millán, I. et al. (2010b). Efficiency oriented design of ZVS half-bridge series resonant inverter with variable frequency duty cycle control. *IEEE Transactions on Power Electronics* 25: 1671–1674.

Lucía, O., Burdío, J.M., Barragán, L.A. et al. (2011). Series resonant multi-inverter with discontinuous-mode control for improved light-load operation. *IEEE Transactions on Industrial Electronics* 58: 5163–5171.

Lucía, O., Carretero, C., Burdío, J.M. et al. (2012). Multiple-output resonant matrix converter for multiple induction heaters. *IEEE Transactions on Industry Applications* 48: 1387–1396.

Lucía, O., Acero, J., Carretero, C., and Burdío, J.M. (2013). Induction heating appliances: towards more flexible cooking surfaces. *IEEE Industrial Electronics Magazine* 7: 35–47.

Lucía, O., Maussion, P., Dede, E., and Burdío, J.M. (2014). Induction heating technology and its applications: past developments, current technology, and future challenges. *IEEE Transactions on Industrial Electronics* 61: 2509–2520.

Lucia, S., Navarro, D., Lucia, O. et al. (2017). Optimized FPGA implementation of model predictive control for embedded systems using high level synthesis tool. *IEEE Transactions on Industrial Informatics* 14: 137–145.

Lucía, O., Domínguez, A., Sarnago, S., and Burdío, J.M. (2018a). Induction heating. In: *Control of Power Electronic Converters and Systems*, 1e (ed. F. Blaabjerg). USA: Elsevier.

Lucía, O., Monmasson, E., Navarro, D. et al. (2018b). Modern control architectures and implementation. In: *Control of Power Electronic Converters and Systems*, 1e (ed. F. Blaabjerg), 477–500. USA: Elsevier.

Lucia, O., Navarro, D., Guillén, P. et al. (2019). Deep learning-based magnetic coupling detection for advanced induction heating appliances. *IEEE Access* 7: 181668–181677.

Lucía, O., Sarnago, H., Acero, J. et al. (2019). Evolution and future challenges of domestic induction heating. In: *Heating by Electromagnetic Sources HES19*, 291–296. Italy: S.G.E. Servizy Grafici Editoriali.

Lucía, S., Navarro, D., Sarnago, H., and Lucía, O. (2020). Development of new high-performance induction heating systems using model predictive control. *International Journal of Applied Electromagnetics and Mechanics* 63: 101–108.

Lucia, S., Navarro, D., Karg, B. et al. (2021). Deep learning-based model predictive control for resonant power converters. *IEEE Transactions on Industrial Informatics* 17: 409–420.

Lucía, O., Sarnago, H., Acero, J. et al. (2022). Induction heating cookers: a path towards decarbonization using energy saving cookers. In: *International Power Electronics Conference 2022 IPEC22*, 1435–1439. Italy: S.G.E. Servizy Grafici Editoriali.

Mai, T., Steinberg, D., Logan, J. et al. (2018). An electrified future: initial scenarios and future research for U.S. energy and electricity systems. *IEEE Power and Energy Magazine* 16: 34–47.

Millán, I., Burdío, J.M., Acero, J. et al. (2010). Resonant inverter topologies for three concentric planar windings applied to domestic induction heating. *Electronics Letters* 46: 1225–1226.

Millán, I., Burdío, J.M., Acero, J. et al. (2011). Series resonant inverter with selective harmonic operation applied to all-metal domestic induction heating. *IET Power Electronics* 4: 587–592.

Mishima, T., Takami, C., and Nakaoka, M. (2014). A new current phasor-controlled ZVS twin half-bridge high-frequency resonant inverter for induction heating. *IEEE Transactions on Industrial Electronics* 61: 2531–2545.

Mitchell, W. (1892). Electrical heater. US487285 Patent Application.

Mühlbauer, A. (2008). *History of Induction Heating and Melting*. Essen: Vulkan-Verlag GmbH.

Nakakita, H., Yamaguchi, K., Hashimoto, M. et al. (2003). A study on secure wireless networks consisting of home appliances. *IEEE Transactions on Consumer Electronics* 49: 375–381.

Navarro, D., Lucía, O., Barragán, L.A. et al. (2012). Synchronous FPGA-based implementations of digital pulse width modulators. *IEEE Transactions on Power Electronics* 27: 2515–2525.

Omori, H. and Nakaoka, M. (1989). New single-ended resonant inverter circuit and system for induction-heating apparatus. *International Journal of Electronics* 67: 277–296.

Paesa, D., Llorente, S., Sagues, C., and Aldana, O. (2009). Adaptive observers applied to pan temperature control of induction hobs. *IEEE Transactions on Industry Applications* 45: 1116–1125.

Pérez-Tarragona, M. (2020). Multi-phase power factor correction for domestic induction heating. Tesis Doctoral, Universidad De Zaragoza.

Pérez-Tarragona, M., Sarnago, H., Lucía, O., and Burdío, J.M. (2018). Design and experimental analysis of pfc rectifiers for domestic induction heating applications. *IEEE Transactions on Power Electronics* 33: 6582–6594.

Pérez-Tarragona, M., Sarnago, H., Burdio, J.M., and Lucia, O. (2021). Multi-phase PFC rectifier and modulation strategies for domestic induction heating applications. *IEEE Transactions on Industrial Electronics* 68: 6424–6433.

Pérez-Tarragona, M., Sarnago, H., Lucía, O., and Burdío, J.M. (2022). Power factor correction stage and matrix ZVS resonant inverter for domestic induction heating appliances. *IET Power Electronics* https://doi.org/10.1049/pel2.12297.

Sarnago, H., Lucía, O., Mediano, A., and Burdío, J.M. (2013). Class-D/DE dual-mode-operation resonant converter for improved-efficiency domestic induction heating system. *IEEE Transactions on Power Electronics* 28: 1274–1285.

Sarnago, H., Lucia, O., Mediano, A., and Burdio, J. (2014a). A class-E direct AC–AC converter with multi-cycle modulation for induction heating systems. *IEEE Transactions on Industrial Electronics* 61: 2521–2530.

Sarnago, H., Lucia, O., Mediano, A., and Burdio, J. (2014b). Efficient and cost-effective ZCS direct AC–AC resonant converter for induction heating. *IEEE Transactions on Industrial Electronics* 61: 2546–2555.

Sarnago, H., Lucía, O., Mediano, A., and Burdío, J.M. (2014c). Direct AC–AC resonant boost converter for efficient domestic induction heating applications. *IEEE Transactions on Power Electronics* 29: 1128–1139.

Sarnago, H., Lucía, O., Mediano, A., and Burdío, J.M. (2014d). Multi-MOSFET-based series resonant inverter for improved efficiency and power density induction heating applications. *IEEE Transactions on Power Electronics* 29: 4301–4312.

Sarnago, H., Lucía, O., and Burdío, J.M. (2015a). A comparative evaluation of SiC power devices for high performance domestic induction heating. *IEEE Transactions on Industrial Electronics* 62: 4795–4804.

Sarnago, H., Lucía, O., Mediano, A., and Burdío, J.M. (2015b). Analytical model of the half-bridge series resonant inverter for improved power conversion efficiency and performance. *IEEE Transactions on Power Electronics* 30: 4128–4143.

Sarnago, H., Lucía, O., Navarro, D., and Burdío, J.M. (2016). Operating conditions monitoring for high power density and cost-effective resonant power converters. *IEEE Transactions on Power Electronics* 31: 488–496.

Sarnago, H., Lucía, O., and Burdío, J.M. (2018). High-performance and cost-effective single-ended induction heating appliance using new MOS-controlled thyristors. In: *IEEE Applied Power Electronics Conference And Exposition*, 3505–3509. USA: IEEE.

Sarnago, H., Burdio, J.M., and Lucia, O. (2019a). High performance and cost effective ZCS matrix resonant inverter for total active surface induction heating appliances. *IEEE Transactions on Power Electronics* 34: 117–125.

Sarnago, H., Guillén, P., Burdío, J.M., and Lucía, O. (2019b). Multiple-output soft-switching resonant inverter architecture for flexible induction heating appliances. *IEEE Access* 7: 157046–157056.

Scott, C.F. (1949). Engineering for appliances. *Electrical Engineering* 68: 205–211.

She, X., Huang, A.Q., Lucia, O., and Ozpineci, B. (2017). Review of silicon carbide power devices and their applications. *IEEE Transactions on Industrial Electronics* 64: 8193–8205.

Star, E. (2021). 2021-2022 Residential Induction Cooking Tops. https://www.energystar.gov/about/2021_residential_induction_cooking_tops (accessed 6 January 2023).

Stookey, S.D. (1958). Low expansion glass-ceramic and method of making it. US 3157522 Patent Application, filed 03 March 1958 and issued 17 November 1964.

Tanaka, T. (1989). A new induction cooking range for heating any kind of metal vessels. *IEEE Transactions on Consumer Electronics* 35: 635–641.

Villa, J., Domínguez, A., Barragán, L.A. et al. (2022). Conductance control for electromagnetic-compatible induction heating appliances. *IEEE Transactions on Power Electronics* 37: 2909–2920.

Wang, S., Izaki, K., Hirota, I. et al. (1998). Induction-heated cooking appliance using new quasi-resonant ZVS-PWM inverter with power factor correction. *IEEE Transactions on Industry Applications* 34: 705–712.

Index

Energy Smart Appliances: Applications, Methodologies, and Challenges, First Edition. Edited by Antonio Moreno-Munoz and Neomar Giacomini.

Z

Printed and bound by CPI Group (UK) Ltd, Croydon, CR0 4YY

19/06/2023

03228420-0001

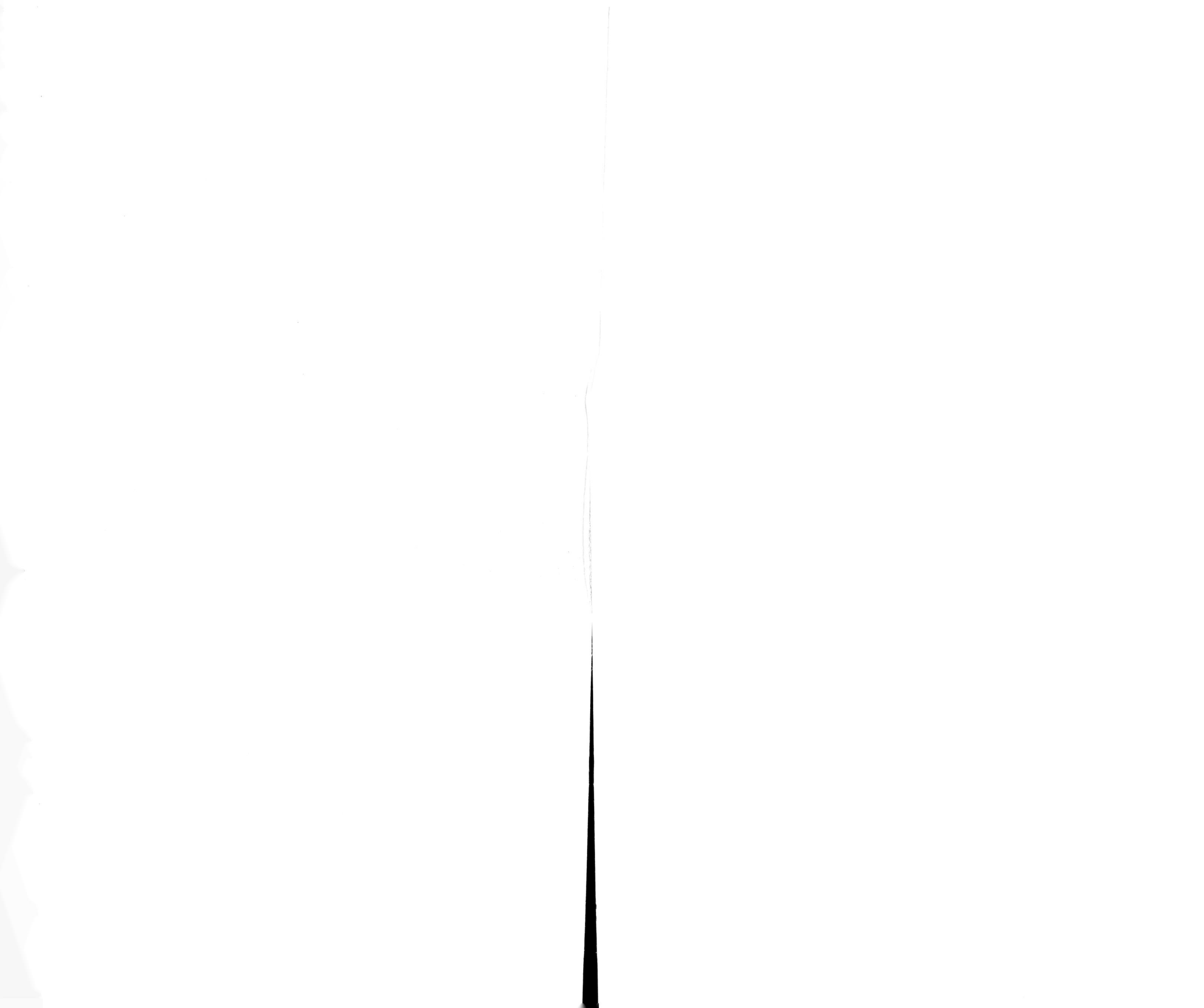